Communications and Control Engineering

Series editors

Alberto Isidori, Roma, Italy
Jan H. van Schuppen, Amsterdam, The Netherlands
Eduardo D. Sontag, Piscataway, USA
Manfred Thoma, Hannover, Germany
Miroslav Krstic, La Jolla, USA

More information about this series at http://www.springer.com/series/61

Yoshio Ebihara · Dimitri Peaucelle
Denis Arzelier

S-Variable Approach to LMI-Based Robust Control

Yoshio Ebihara
Department of Electrical Engineering
Kyoto University
Kyoto
Japan

Dimitri Peaucelle
Denis Arzelier
Laboratory for Analysis and Architecture
   of Systems Science
National Centre for Scientific Research
Toulouse
France

ISSN 0178-5354
ISBN 978-1-4471-6605-4
DOI 10.1007/978-1-4471-6606-1

ISSN 2197-7119 (electronic)
ISBN 978-1-4471-6606-1 (eBook)

Library of Congress Control Number: 2014948741

Springer London Heidelberg New York Dordrecht

Printed on acid-free paper

Springer is part of Springer Science+Business Media (www.springer.com)

Preface

At the end of 1990s, we discovered with enthusiasm the publication by Mauricio C. de Oliveira, Jacques Bernussou, and José C. Geromel concerning a "New Discrete-Time Stability Condition." Since then, we have dedicated most of our attention to understanding and applying the underlying powerful methodology. We have extended the results in many directions and witnessed major interest of our colleagues for these extensions. The technique is now widely used but never was completely formalized. Because of that, it has been sometimes missused or missinterpreted, some central issues have not been clarified, and there is not even a unique denomination for it. In this book, we decide to call that methodology the *S*-variable approach, following the denomination *S*-procedure that is a powerful tool in dealing with constrained inequality conditions. We explain this choice, expose in details our understanding of it, and provide important robust control results related to it.

The *S*-variable approach, or SV approach for short, enters the well developed field of convex-optimization-based control theory. Since the pioneering results on linear matrix inequality (LMI) and semidefinite programming (SDP) in the 1980s, convex optimization has become a standard strategy for system analysis and synthesis. Not only existing closed-form analytic solutions can be losslessly reformulated as the feasibility of SDPs, efficient polynomial-time powerful softwares are available, but the convex optimization point of view also allows the derivation of new major results, especially for robust control.

Even though the convex optimization approach to robust control problems has provided answers to many practical problems, these answers happen to be, except in few simple cases, "conservative." The control problems can in general be formulated as finding some control (defined as a gain or a dynamic equation) that guarantees one or several performances for the controlled system (some physical processes most often conceived by engineers, but it could as well be an economic or an ecologic system). To perform the design one assumes a mathematical model for the system. This model is imperfect and robust control aims at guaranteeing the performances whatever bounded model uncertainties. The design thus formulates as a minimax problem: finding the best controller against the worst-case uncertainty.

The exact LMI characterization of most such robust control problems happens to be out of range, but LMIs do provide upper-bounds. The gap with the exact value characterizes the conservatism of the result. LMI results issued from the SV approach prove theoretically to reduce this conservatism, and do reduce largely the gap in practice. The conservatism reduction provided by the SV approach is the central property discussed in the book.

The basics of the SV approach can be seen immediately when considering a simple example. Let a linear time-invariant system be described in state-space by the differential equation $\dot{x}(t) = Ax(t)$. Its asymptotic stability (i.e., the convergence of the state $x(t)$ to zero as times goes to infinity) is ensured by the existence of a positive quadratic function $V(x) = x^T Px$, called a *Lyapunov function*, such that its derivatives are negative along all trajectories of the system:

$$\dot{V}(x) = \dot{x}^T Px + x^T P\dot{x} < 0 \quad \forall (x, \dot{x}) \text{ such that } \dot{x} = Ax, \ x \neq 0.$$

LMI formulas to that problem are derived by including the equality constraint into the inequality constraint. The first classical way of doing so is to replace $\dot{x}$ by its value Ax. This leads to the following matrix valued inequalities

$$P \succ 0, \quad A^T P + PA \prec 0$$

where the $\succ 0$ ($\prec 0$) stands for "positive (negative) definiteness." These matrix inequalities are linear with respect to the decision variable P, it is an LMI feasibility problem. On the other hand, the second way is rewrite the condition explicitly as a constrained inequality condition as in

$$\begin{bmatrix} \dot{x} \\ x \end{bmatrix}^T \begin{bmatrix} 0 & P \\ P & 0 \end{bmatrix} \begin{bmatrix} \dot{x} \\ x \end{bmatrix}^T < 0 \quad \forall \begin{bmatrix} \dot{x} \\ x \end{bmatrix} \text{ such that } \begin{bmatrix} I & -A \end{bmatrix} \begin{bmatrix} \dot{x} \\ x \end{bmatrix} = 0, \ x \neq 0.$$

By closely following the idea of *S*-procedure, we obtain

$$P \succ 0, \quad \begin{bmatrix} 0 & P \\ P & 0 \end{bmatrix} + F[1 - A] + \begin{bmatrix} I \\ A^T \end{bmatrix} F^T \prec 0$$

where the decision variables are now P and F. The second formulation corresponds to the SV approach and the LMIs will be called SV-LMIs. The two LMI results are of course mathematically equivalent in the sense that the former is feasible if and only if the latter is feasible. However, in the SV-LMIs the matrix F introduces additional degrees of freedom and somehow relaxes or dilates the inequalities. This matrix F is called the *S*-variable, following the denomination *S*-procedure. The book is dedicated to exploring the benefits (and the drawbacks) of this approach, shortened by SV-approach.

The main drawback to be mentioned all at once is in terms of numerical computation. The size of the SV-LMI problem roughly doubles (both in terms of size of the inequalities and of the number of variables). Therefore, one can expect increased computation time when solving SV-LMIs with available SDP solvers. But, can conservatism reduction be obtained for free?

The main advantage is the decoupling SV produces between the decision variable and the data (P and A, respectively, in the above example). This decoupling is the key for conservatism reduction and new powerful robust analysis results follow from it. Chapters 2, 3 and 7 of the book are dedicated to these robust analysis issues (computing upper bounds on worst case performances for a given control). But as soon as the control design is considered the decoupling is no more absolute. In the above example, the design problem amounts to replacing A by a controller dependent matrix $A(K)$, thus creating couplings between the controller and the S-variable (K and F, respectively). Handling this issue while remaining in an LMI context and keeping the conservatism reduction is the central question of SV-LMI-based control design. Chapters 4–6 and 8 of the book are dedicated to this question.

The more precise scope of the monograph is now presented by outlining each chapter contents.

Chapter 1 is dedicated to the origins of the SV approach. We trace the contributions of independent authors that participated to establish the fundamentals and justify our choice for the denomination "S-variable approach". This attentive detailed study is the occasion to show several interpretations of the S-variables with respect to technical results such as Finsler's lemma and elimination lemma. The chapter concludes with the brief exposure of all the problems to be tackled in the remaining part of the book, justifying the importance of these selected problems.

The primary goal of Chap. 2 is founding the basic idea of the SV approach. Before generalizing the technique (which is done in the following chapters) we show its effectiveness on simple essential control problems. We mainly consider the robust performance analysis problems of linear time-invariant systems affected by parametric uncertainties, and clarify why the SV-LMIs do perform well on these intractable infinite-dimensional semi-infinite problems. We also highlight the improvements in terms of conservatism both theoretically and on examples.

Chapter 3 is an extension of the SV-LMIs of the preceding chapter both in terms of generalization of the results and for further conservatism reduction. First the SV results are generalized to descriptor systems. Not only this result is valuable in itself, but also combined to a model manipulation technique, where an infinite sequence of SV-LMIs can be built. This sequence of SV-LMIs is shown to be easy to construct and proved to be of decreasing conservatism. Tests are provided for checking if the conservatism gap vanishes. With examples it is shown that the conservatism gap indeed vanishes, and this is obtained by early elements of the sequence (i.e., the convergence is rather fast). The chapter concludes with the mathematical description of SV approach for a class of robust LMI problems.

In Chap. 4 the SV-LMIs of Chap. 2 are reconsidered for robust state-feedback design. The results rely on the structuring of the S-variables. An interpretation in terms of virtual stable model is given to this structure. Moreover, we show the effect of this structuring on conservatism reduction. It happens to be of different nature in the discrete-time and continuous-time cases.

Chapter 5 focuses on multi-objective control design. The goal here is to design a controller that satisfy multiple design specifications. In the state-feedback case, the advantage of the SV-LMIs readily follows from the results in Chap. 4. To handle output-feedback case, we first extend the results of Chap. 4 and provide SV-LMI-based formulas for dynamic output feedback controller synthesis. Then, the effectiveness of the SV-LMIs in conservatism reduction can be shown almost the same way as in the state-feedback case.

While the two previous chapters address control design cases, which have LMI solutions, Chap. 6 tackles the hard problem of static-output feedback design. No polynomial-time optimization method exists for that problem, and if keeping in the matrix inequality framework, one has to resort to iterative LMI algorithms. For that important and hard problem the SV approach produces an sophisticated version of such iterative LMI algorithm. In this case, the structuring of the S-variables reveals a virtual stabilizing state-feedback. This original design procedure is detailed and tested on examples.

Finally, Chaps. 7 and 8 are dedicated to the case of discrete-time periodic systems. For that special case the SV-LMIs have interesting noncausal system interpretations in analysis (Chap. 7). Similarly to the LTI case, SV-LMIs are effective in reducing the conservatism of the analysis and synthesis results when dealing with discrete-time periodic systems affected by polytopic uncertainties.

Acknowledgments

We owe a debt of gratitude to many of our colleagues and friends. Among them we particularly thank Olivier Bachelier, Jacques Bernussou, Guilherme Chevarria, Christophe Farges, Alexander Fradkov, Frédéric Gouaisbaut, Elena Gryazina, Tomomichi Hagiwara, Didier Henrion, Christelle Pittet, Boris Polyak, Masayuki Sato, and Jean-Franois Tregouët with whom we coauthored publications dealing with the S-variable approach. This book is largely theirs.

The theory of S-variable approach has developed by the efforts of many researchers. Among them, we acknowledge the contributions of Mauricio C. de Oliveira, José C. Geromel, Alexandro Trofino, Emilia Fridman, and Uri Shaked that inspired us in many ways.

We are grateful to Kyoto University, Japan and CNRS, France, that kindly supported this joint research.

Above all, we are grateful to Charlotte Cross from Springer for her understanding and patience for our slow progress in writing the book.

Kyoto, Japan   Yoshio Ebihara
Toulouse, France   Dimitri Peaucelle
                                                                         Denis Arzelier

Contents

Notations

Some Specific Sets

$\mathbb{R}$	Real numbers
$\mathbb{R}^n$	Real n-vectors ($n \times 1$ matrices)
$\mathbb{R}^{m \times n}$	Real $m \times n$ matrices
$\mathbb{C}$	Complex numbers
$\mathbb{C}^n$	Complex n-vectors
$\mathbb{C}^{m \times n}$	Complex $m \times n$ matrices
$\mathbb{C}_-$	Complex numbers with strictly negative real part
$\mathbb{D}$	Open unit disc on the complex plane
$\partial \mathbb{D}$	Unit circle on the complex plane
$\mathbb{S}^n$	Real symmetric $n \times n$ matrices
$\mathbb{S}^n_+, \mathbb{S}^n_{++}$	Real symmetric positive semidefinite, positive definite, $n \times n$ matrices
$\mathbb{H}^n$	Complex Hermitian $n \times n$ matrices
$\mathbb{H}^n_+, \mathbb{H}^n_{++}$	Complex Hermitian positive semidefinite, positive definite, $n \times n$ matrices
$\lambda(A)$	The set of the eigenvalues of $A \in \mathbb{C}^{n \times n}$
$\mathscr{I}_{n_1,n_2}, \mathscr{I}_n$	Positive integers from n_1 to n_2 and 1 to n, i.e., $\mathscr{I}_{n_1,n_2} := \{n_1, \cdots, n_2\}$ and $\mathscr{I}_n := \{1, \cdots, n\}$.

Vectors and Matrices

$\mathbf{1}(\mathbf{1}_n)$	Vector with all components one (of size n)
x_i	i-th element of vector x
e_i	i-th standard basis vector
$I\,(I_n)$	Identity matrix (of size n)
X^T	Transpose of matrix X
X^*	Complex conjugate transpose of matrix X

trace (X) Trace of matrix X
rank (X) Rank of matrix X
               The following two notations assume $X \in \mathbb{R}^{n \times m}$ and rank $(X) = r$
$X^{\perp}$ Full rank matrix such that $X^{\perp} \in \mathbb{R}^{(n-r) \times n}$, $X^{\perp}X = 0$ (the image of X is
               the kernel of $X^{\perp}$, (null (X'))' in Matlab©)
X° Full rank matrix such that $X^{\circ} \in \mathbb{R}^{m \times r}$, XX° is full rank (the image of X°
               is orthogonal to the kernel of X, orth(X') in Matlab©).

Norms for Vectors, Matrices and Signals

$\|x\|$ Euclidean norm of vector x
$\|X\|$ Maximum singular value of matrix X
$\|z\|_2$ L_2 norm of the signal $z(t)$, i.e., $\| z \|_2 := \sqrt{\int_0^{\infty} \| z(t) \|^2 \, dt}$

Inequalities

$x \geq y$ Elementwise inequality between real vectors x and y
$x > y$ Strict elementwise inequality between real vectors x and y
$X \succeq Y$ Matrix inequality between real symmetric matrices X and Y
$X \succ Y$ Strict matrix inequality between real symmetric matrices X and Y

Norms for Stable LTI Systems

$\|G\|_2$ H_2 norm of stable LTI system G
$\|G\|_{\infty}$ H_{∞} norm of stable LTI system G

Functions

$\mathrm{He} : \mathbb{R}^{n \times n} \to \mathbb{R}^{n \times n}$ $\mathrm{He}(X) := X + X^T$
$\mathrm{Sq} : \mathbb{R}^{m \times m} \to \mathbb{R}^{m \times m}$ $\mathrm{Sq}(X) := XX^T$

Simplex

Let us define

$$\mathbb{E}^n := \{\theta \in \mathbb{R} : \theta \geq 0, \mathbf{1}^T \theta = 1\}$$

Then, $\mathbb{E}^n$ is said to be a standard simplex on $\mathbb{R}^n$.

Polytopes

For given matrices $A^{[j]} \in \mathbb{R}^{n \times n}$ $(j \in \mathscr{I}_L)$, let us define $A := \{A^{[j]}, \cdots, A^{[L]}\}$. Then, the convex hull of the set A, denoted by $\mathrm{conv}(A)$, is defined by

$$\mathrm{conv}(A) := \left\{ \sum_{j=1}^{L} \theta_j A^{[j]} : \theta \in \mathbb{E}^L \right\}$$

In the control literature, $\mathrm{conv}(A)$ is also called as a polytope with L vertex matrices $A^{[j]}$ $(j \in \mathscr{I}_L)$.

Chapter 1
Introduction

1.1 On the Origin and History of S-Variable Approach

To the best of the author's knowledge, the origin of S-variable approach (in our denomination) dates back to a sequence of papers written by Geromel and his colleagues in the late 1990s [1–3]. In the paper [1] entitled "LMI Characterization of Structural and Robust Stability," the authors consider stability analysis of continuous-time linear systems affected by structural (and possibly nonlinear) uncertainties. Still, the main result can be simplified in the way that a matrix $A \in \mathbb{R}^{n \times n}$ is Hurwitz stable (i.e., all the eigenvalues of A have strictly negative real parts) if and only if there exists $P \in \mathbb{S}^n_{++}$ and $F_1, F_2 \in \mathbb{R}^{n \times n}$ such that

$$\begin{bmatrix} 0 & P \\ P & 0 \end{bmatrix} + \mathrm{He}\left\{ \begin{bmatrix} F_1 \\ F_2 \end{bmatrix} \begin{bmatrix} A & -I \end{bmatrix} \right\} \prec 0 \tag{1.1}$$

where $\mathrm{He}\{Y\} := Y + Y^{\mathrm{T}}$. Here, the variables F_1 and F_1 are introduced as coefficient matrices of a multiplier in the frequency domain (s-domain). More precisely, it is shown that A is Hurwitz stable if and only if there exists F_1 and F_2 such that

$$\left(F_2^T s + F_1^T \right) (sI - A)^{-1} = \left(F_2^{\mathrm{T}} A + F_1^T \right) (sI - A)^{-1} + F_2^{\mathrm{T}}$$

becomes extended strictly positive real (ESPR). By applying ESPR-lemma [4], the above system is ESPR if and only if

$$\begin{bmatrix} PA + A^{\mathrm{T}} P & P - \left(F_2^{\mathrm{T}} A + F_1^T \right)^T \\ P - \left(F_2^{\mathrm{T}} A + F_1^T \right) & -F_2^{\mathrm{T}} - F_2 \end{bmatrix} \prec 0.$$

Applying a congruence transformation with

© Springer-Verlag London 2015
Y. Ebihara et al., *S-Variable Approach to LMI-Based Robust Control*,
Communications and Control Engineering, DOI 10.1007/978-1-4471-6606-1_1

$$\begin{bmatrix} I & -A^{\mathrm{T}} \\ 0 & I \end{bmatrix},$$

we can readily obtain (1.1). From this brief review, we notice that S-variables F_1 and F_2 have the meaning of multipliers in the frequency-domain at the first advent.

On the other hand, in the succeeding papers [2, 3] dealing with discrete-time systems, the situation is rather different. The main results in these papers can be simplified in the way that a matrix $A \in \mathbb{R}^{n \times n}$ is Schur stable (i.e., all the eigenvalues of A have absolute values less tha one) if and only if there exists $P \in \mathbb{S}_{++}^n$ and $F_1, F_2 \in \mathbb{R}^{n \times n}$ such that

$$\begin{bmatrix} -P & 0 \\ 0 & P \end{bmatrix} + \mathrm{He} \left\{ \begin{bmatrix} F_1 \\ F_2 \end{bmatrix} \begin{bmatrix} A & -I \end{bmatrix} \right\} \prec 0. \tag{1.2}$$

Moreover, we can let $F_1 = 0$ without loss of generality. Here the proof in [2, 3] has been done by showing the equivalence of (1.2) with the Lyapunov inequality given by

$$- P + A^{\mathrm{T}} P A \prec 0 \tag{1.3}$$

or its equivalent condition

$$\begin{bmatrix} -P & A^{\mathrm{T}} P \\ PA & -P \end{bmatrix} \prec 0. \tag{1.4}$$

Indeed, if (1.2) holds, then we readily obtain (1.3) by multiplying

$$\begin{bmatrix} I \\ A \end{bmatrix}$$

from the right-hand side of (1.2) and its transpose from the left-hand side. On the other hand, if (1.4) holds then it is easy to see that (1.2) holds with $F_1 = 0$ and $F_2 = P$. In this way, contrary to the continuous-time system case, S-variables F_1 and F_2 are introduced without any system-theoretic interpretation, and they behave as if they are auxiliary variables that decouples the product between P and A existing in the standard Lyapunov inequality (1.3) or (1.4). After the pioneering works [1–3], plenty of results have been obtained along the same line and for almost all cases new variables are introduced through matrix operations without any system-theoretic interpretation. Due to this reason, the denomination of the approach conceived by Geromel has been a difficult issue. In this book, we decided to use "S-variable" and "S-variable LMIs" for the corresponding LMIs, whose reason and possible validation will be given in the next subsection.

We next review the history of S-variable after the publication of and [1–3] and show that S-variable approach has been one of the core issues in control commu-nity. Just after [1–3], Peaucelle et al. succeeded in extending the stability results to regional stability conditions (*D*-stability conditions) [5]. It was gradually recognized that the *S*-variable approach is effective in dealing with robustness analysis of linear systems, and many papers are published along this line by, for example, Oliveira and Skelton [6], Henrion et al. [7], Leite and Peres [8], and Ebihara and Hagiwara [9]. Shaked et al. also provided effective methods for the analysis and synthesis of time-delay systems using SV-LMIs [10, 11]. On the other hand, Geromel et al. made continuing effort to establish SV-LMI-based discrete-time controller synthesis, and published excellent papers for robust filtering [12] and multiobjective control [13]. In particular, the paper [13] clearly shows that SV-LMIs for discrete-time H_2 and H_∞ controller synthesis can be derived almost the same way as stabilizing controller synthesis. After the publication of [13], intensive research effort has been made to establish counterpart results for continuous-time systems, and partial results are obtained by Apkarian et al. [14] and Ebihara and Hagiwara [15]. Still, the SV-LMI for continuous-time H_∞ controller synthesis is not available to this date. As noted previously, it is commonly recognized that SV-LMI-based controller synthesis has different nature in the discrete-time and continuous-time cases. Finally, we note that *S*-variable approach is successfully extended for the analysis and synthesis of discrete-time periodic systems recently. Early publication by Farges et al. [16] introduces *S*-variables for the LMIs simply derived from periodic-Lyapunov-lemma. On the other hand, recent papers by Ebihara et al. [17, 18] and Trégouët [19] focus on memory structure of periodic systems and provide useful SV-LMIs for analysis and synthesis.

1.2 On the Denomination *S*-Variable LMI

Our denomination, "*S*-Variable Approach" and "*S*-Variable LMI" are motivated by the *S*-procedure [4] given in the following.

Theorem 1.1 *For given* $\Theta, \Pi \in \mathbb{S}_n$, *the following two conditions are equivalent.*

(i) $\xi^{\mathrm{T}} \Theta \xi < 0$ $\forall \xi \neq 0$ *such that* $\xi^{\mathrm{T}} \Pi \xi \leq 0$.
(ii) $\exists \tau > 0$ *such that* $\Theta - \tau \Pi \prec 0$.

Note that the constrained inequality condition in (i) is hard to handle directly. On the other hand, the condition (ii) is an LMI with respect to $\tau > 0$ and hence easy to solve at least numerically. It is known that (generalized) *S*-procedure is the core for deriving LMI conditions that characterizes frequency domain properties of LTI systems [20–22].

To reveal how the *S*-variable approach is closely related to the *S*-procedure, let us recall the stability analysis of the LTI system $\dot{x} = Ax$ where $A \in \mathbb{R}^{n \times n}$. The corresponding SV-LMI is given by (1.1). We emphasize that if (1.1) is obtained in some way, it is easy to see that (1.1) is a sufficient condition for the stability (i.e., Hurwitz stability of A) since (1.1) implies

$$\begin{bmatrix} I \\ A \end{bmatrix}^{\mathrm{T}} \left\{ \begin{bmatrix} 0 & P \\ P & 0 \end{bmatrix} + \mathrm{He} \left\{ \begin{bmatrix} F_1 \\ F_2 \end{bmatrix} [\, A \ \ -I \,] \right\} \right\} \begin{bmatrix} I \\ A \end{bmatrix}^{\mathrm{T}} \prec 0 \qquad (1.5)$$
$$\Longleftrightarrow PA + A^{\mathrm{T}} P \prec 0.$$

Therefore, the key issue is how we can derive (1.1) from the stability condition. To show that this can be done by means of the S-procedure, recall that the stability is ensured by the existence of a Lyapunov function $V(x) = x^{\mathrm{T}} P x$ with $P \in \mathbb{S}_{++}^{n}$ such that its derivatives are negative along all trajectories of the system:

$$\dot{V}(x) = \dot{x}^{\mathrm{T}} P x + x^{\mathrm{T}} P \dot{x} < 0 \quad \forall (x, \dot{x}) \text{ such that } \dot{x} = Ax, \quad x \neq 0.$$

This condition can be restated equivalently as

$$\begin{bmatrix} \dot{x} \\ x \end{bmatrix}^{\mathrm{T}} \begin{bmatrix} 0 & P \\ P & 0 \end{bmatrix} \begin{bmatrix} \dot{x} \\ x \end{bmatrix}^{\mathrm{T}} < 0 \ \ \forall \begin{bmatrix} \dot{x} \\ x \end{bmatrix} \text{ such that } [\, I \ \ -A \,] \begin{bmatrix} \dot{x} \\ x \end{bmatrix} = 0, \quad x \neq 0. \tag{1.6}$$

Here, it is elementary to see that we can replace $x \neq 0$ with $[x^{\mathrm{T}} \ \dot{x}^{\mathrm{T}}] \neq 0$ since the following two sets are the same:

$$\Xi_1 := \left\{ \begin{bmatrix} \dot{x} \\ x \end{bmatrix} : x \neq 0 \text{ and } [\, I \ \ -A \,] \begin{bmatrix} \dot{x} \\ x \end{bmatrix} = 0 \right\},$$
$$\Xi_2 := \left\{ \begin{bmatrix} \dot{x} \\ x \end{bmatrix} : \begin{bmatrix} \dot{x} \\ x \end{bmatrix} \neq 0 \text{ and } [\, I \ \ -A \,] \begin{bmatrix} \dot{x} \\ x \end{bmatrix} = 0 \right\}.$$

Indeed, if $\xi \in \Xi_1$ then it is clear that $\xi \in \Xi_2$. On the other hand, if $\xi = [\, \dot{x}^{\mathrm{T}} \ x^{\mathrm{T}} \,]^{\mathrm{T}} \in \Xi_2$ then $x \neq 0$ since if $x = 0$ then $[\, I \ - \ A \,][\, \dot{x}^{\mathrm{T}} \ x^{\mathrm{T}} \,]^{\mathrm{T}} = 0$ implies $\dot{x} = 0$, which contradicts to $[\, \dot{x}^{\mathrm{T}} \ x^{\mathrm{T}} \,]^{\mathrm{T}} \neq 0$. It follows that (1.6) can be replaced by

$$\begin{bmatrix} \dot{x} \\ x \end{bmatrix}^{\mathrm{T}} \begin{bmatrix} 0 & P \\ P & 0 \end{bmatrix} \begin{bmatrix} \dot{x} \\ x \end{bmatrix}^{\mathrm{T}} < 0 \ \ \forall \begin{bmatrix} \dot{x} \\ x \end{bmatrix} \neq 0 \text{ such that } [\, I \ \ -A \,] \begin{bmatrix} \dot{x} \\ x \end{bmatrix} = 0 \quad (1.7)$$

or equivalently,

$$\xi^{\mathrm{T}} \begin{bmatrix} 0 & P \\ P & 0 \end{bmatrix} \xi < 0 \ \ \forall \xi \neq 0 \text{ such that } \xi^{\mathrm{T}} \begin{bmatrix} I \\ -A^{\mathrm{T}} \end{bmatrix} [\, I \ \ -A \,] \xi \leq 0.$$

By the S-procedure, this constrained inequality condition holds if and only if there exists $\tau > 0$ such that

$$\begin{bmatrix} 0 & P \\ P & 0 \end{bmatrix} - \tau \begin{bmatrix} I \\ -A^{\mathrm{T}} \end{bmatrix} [\, I \ \ -A \,] \prec 0. \tag{1.8}$$

This is essentially nothing but the SV-LMI (1.1) where $F_1 = -\frac{1}{2}\tau I$ and $F_2 = \frac{1}{2}\tau A^{\mathrm{T}}$. In this way, the SV-LMI (1.1) can be derived by the S-procedure, and the variables

F_1 and F_2 can be seen as a generalization of the variable τ originally existing in the S-procedure. Due to these reasons, we decided to use the denomination S-variable approach and SV-LMI in this book.

Once we have shown (1.5) and (1.8), we can reveal the relationship between SV-LMIs and Finsler's lemma or Elimination lemma, both of which are closely related to the S-procedure as well. To this end, we summarize the equivalent conditions for the Hurwitz stability of A obtained so far and recall the two lemmas.

Theorem 1.2 *For matrix $A \in \mathbb{R}^{n \times n}$, the following conditions are equivalent:*

(i) *The matrix A is Hurwitz stable.*
(ii) *There exists $P \in \mathbb{S}_{++}^n$ such that $PA + A^{\mathrm{T}}P \prec 0$.*
(iii) *There exist $P \in \mathbb{S}_{++}^n$ and $\tau > 0$ such that (1.8) holds.*
(iv) *There exist $P \in \mathbb{S}_{++}^n$ and $F_1, F_2 \in \mathbb{R}^{n \times n}$ such that (1.1) holds.*

Lemma 1.1 (Finsler's Lemma) *For given $B \in \mathbb{R}^{n \times m}$ and $Q \in \mathbb{S}^n$, the following conditions are equivalent:*

(i) *There exists $\tau > 0$ such that $Q - \tau BB^{\mathrm{T}} \prec 0$.*
(ii) *$B^{\perp}QB^{\perp \mathrm{T}} \prec 0$.*

Lemma 1.2 (Elimination Lemma) *For given $B \in \mathbb{R}^{n \times m}$, $C \in \mathbb{R}^{l \times n}$ and $Q \in \mathbb{S}^n$, the following conditions are equivalent:*

(i) *There exists $K \in \mathbb{R}^{m \times l}$ such that $Q + \mathrm{He}\{BKC\} \prec 0$.*
(ii) *$B^{\perp}QB^{\perp \mathrm{T}} \prec 0$ and $C^{\mathrm{T}\perp}QC^{\mathrm{T}\perp \mathrm{T}} \prec 0$.*

The equivalence of conditions in Theorem 1.2, and in particular the soundness of SV-LMI given in (iv), can be verified directly by these lemmas. First the equivalence of (ii) and (iii) can be seen by Lemma 1.1 if we identify

$$\begin{bmatrix} 0 & P \\ P & 0 \end{bmatrix}, \quad \begin{bmatrix} I \\ -A^{\mathrm{T}} \end{bmatrix}$$

with Q and B, respectively. Similarly, the equivalence of (ii) and (iv) can be seen by Lemma 1.2 if we identify

$$\begin{bmatrix} 0 & P \\ P & 0 \end{bmatrix}, \quad \begin{bmatrix} I & -A \end{bmatrix}$$

with Q and C, respectively and let $B = I_n$ (i.e., $B^{\perp}$ is null). The variable K corresponds to $[F_1^{\mathrm{T}} \ F_2^{\mathrm{T}}]^T$ in (1.1). Usually, the elimination lemma moves from (i) to (ii) in Lemma 1.2 to "eliminate" the variable K. In this sense, the SV-LMI in (iv) of Theorem 1.2 has been obtained by *reciprocally* applying the elimination lemma and *generating S-variables* F_1 and F_2.

1.3 Other Denominations Used in Existing Literature

Even though we have stated our standpoint to use the denomination SV-LMI in this book, there are other denominations already used in existing literature. In the paper [13] that is an advanced version of [1–3], de Oliveira et al. use the terminology "extended" and this is used in subsequent papers, for example, in [23, 24]. This terminology is naturally introduced since analysis and synthesis are performed in extended space by incorporating derivative of the state as in (1.6), and the resulting LMI conditions are extended in size. See [6] for relevant discussions. Following similar considerations, Ebihara, the first author of this book, has used the terminology "dilated LMIs" in [9, 15]. In these papers, he studied possible extension of the discrete-time case results in [1–3] to continuous-time system cases and showed that LMI conditions are particularly dilated in continuous-time system cases. This terminology is used by many authors, for example, [25–29]. On the other hand, Peaucelle, the second author of this book, used terminology "slack-variable LMIs" in [30–33], identifying the role of S-variables that "normalize" wide variety of LMI conditions with slack variables in linear programming. From other point of view, Apkarian used "enhanced LMIs" in [14], emphasizing the superior ability of SV-LMIs over the standard LMIs in handling "difficult" problems for which definitive approaches are not available.

It is obvious that each denomination has its own advantages and disadvantages. We decided to use S-variable approach and SV-LMI in this book, believing that this terminology achieves better compromise between compactness and underlying mathematical foundation (i.e., the S-procedure).

1.4 Overview of Selected Topics

In the following, we give a list of selected topics in this book.

1 **Robust performance analysis of LTI systems affected by polytopic-type uncertainties** for giving concise understanding of S-variable approach and illustrate its advantages/disadvantages (Chap. 2),
2 **Mathematical formulas for deriving SV-LMIs** for unifying the methodology to derive SV-LMIs from standard LMIs, the latter of which can be naturally obtained from control system theory (Chap. 2),
3 **Derivation procedure of standard LMIs for linear system analysis**: we try to give elementary proofs on how standard LMIs for linear system analysis can be derived. Once those standard LMIs can be obtained, corresponding SV-LMIs readily follow from the mathematical formulas stated above (Chap. 2),
4 **Descriptor representation of systems and system augmentation** for reducing the conservatism of S-variable approach by means of descriptor representation and system augmentation (Chap. 3),

5 **Robust state-feedback controller synthesis using SV-LMIs** to show the effectiveness of SV-LMIs for synthesis problems in a concise fashion (Chap. 4),

6 **Multiobjective controller synthesis using SV-LMIs** to illustrate the effectiveness of SV-LMIs even in the case of output-feedback controller synthesis cases, and offer a list of standard- and SV-LMI conditions for output-feedback controller synthesis and associated linearizing change-of-variables techniques (Chap. 5),

7 **Static output-feedback controller (SOF) synthesis using SV-LMIs** to explicate the potentials of SV-LMIs in conceiving effective iterative methods for nonconvex SOF synthesis problems (Chap. 6),

8 **Analysis and synthesis of discrete-time periodic systems using SV-LMIs**: to clarify that the effectiveness of SV-LMIs, widely known for the analysis and synthesis of discrete-time LTI systems, can be directly inherited to periodic systems, and to show that SV-LMIs have its own advantage by means of noncausal system interpretation of S-variables (Chaps. 7 and 8).

For each selected topic, we include plenty of numerical examples that validate the effectiveness (and show possible shortcomings) of S-variable approach.

References

1. Geromel JC, de Oliveira MC, Hsu L (1998) LMI characterization of structural and robust stability. Linear Algebra Appl 285:68–80
2. de Oliveira MC, Geromel JC, Hsu L (1999) LMI characterization of structural and robust stability: The discrete-time case. Linear Algebra Appl 296(1–3):27–38
3. de Oliveira MC, Bernussou J, Geromel JC (1999) A new discrete-time stability condition. Syst Control Lett 37(4):261–265
4. Boyd S, El Ghaoui L, Feron E, Balakrishnan V (1994) Linear matrix inequalities in system and control theory. SIAM Studies in Applied Mathematics, Philadelphia
5. Peaucelle D, Arzelier D, Bachelier O, Bernussou J (2000) A new robust D-stability condition for real convex polytopic uncertainty. Syst Control Lett 40(1):21–30
6. de Oliveira MC, Skelton RE (2001) Stability tests for constrained linear systems. In: Reza Moheimani S.O. (ed) Lecture notes in control and information sciences, pp 241–257. Springer-Verlag, New York
7. Henrion D, Arzelier D, Peaucelle D (2003) Positive polynomial matrices and improved LMI robustness conditions. Automatica 39(8):1479–1485
8. Leite V, Peres P (2003) An improved LMI condition for robust D-stability of uncertain polytopic systems. IEEE Trans Autom Control 48(3):500–504
9. Ebihara Y, Hagiwara T (2005) A dilated LMI approach to robust performance analysis of linear time-invariant uncertain systems. Automatica 41(11):1933–1941
10. Fridman E, Shaked U (2002) An improved stabilization method for linear time-delay systems. IEEE Trans Autom Control 47(11):1931–1937
11. Fridman E, Shaked U (2002) A descriptor system approach to H_∞ control of time-delay systems. IEEE Trans Autom Control 47:253–270
12. Geromel JC, de Oliveira MC, Bernussou J (2002) Robust filtering of discrete-time linear systems with parameter dependent lyapunov functions. SIAM J Control Optim 41:700–711
13. de Oliveira MC, Geromel JC, Bernussou J (2002) Extended H_2 and H_∞ norm characterizations and controller parametrizations for discrete-time systems. Int J Control 75:666–679
14. Apkarian P, Tuan HD, Bernussou J (2001) Continuous-time analysis and H_2 multi-channel synthesis with enhanced LMI characterizations. IEEE Trans Autom Control 46(12):1941–1946

15. Ebihara Y, Hagiwara T (2004) New dilated LMI characterizations for continuous-time multi-objective controller synthesis. Automatica 40(11):2003–2009
16. Farges C, Peaucelle D, Arzelier D, Daafouz J (2007) Robust H_2 performance analysis and synthesis of linear polytopic discrete-time periodic systems via LMIs. Syst Control Lett 56:159–166
17. Ebihara Y, Peaucelle D, Arzelier D (2011) Periodically time-varying memory state-feedback controller synthesis for discrete-time linear systems. Automatica 47(1):14–25
18. Ebihara Y (2013) Periodically time-varying memory state-feedback for robust H_2 control of uncertain discrete-time linear systems. Asian J Control 15(2):409–419
19. Trégouët JF, Peaucelle D, Arzelier D, Ebihara Y (2013) Periodic memory state-feedback controller: New formulation, analysis and design results. IEEE Trans Autom Control 58(8):1986–2000
20. Rantzer A (1996) On the Kalman-Yakubovitch-Popov lemma. Syst Control Lett 28:7–10
21. Iwasaki T, Hara S (2005) Generalized KYP lemma: Unified frequency domain inequalities with design applications. IEEE Trans Autom Control 50(1):41–59
22. Ebihara Y, Maeda K, Hagiwara T (2008) Generalized S-procedure for inequality conditions on one-vector-lossless sets and linear system analysis. SIAM J Control Optim 47(3):1547–1555
23. Shimomura T, Takahashi M, Fujii T (2001) Extended-space control design with parameter-dependent lyapunov functions. In: Proceedings of the conference on decision and control, pp 2157–2162
24. Pipeleers Goele, Demeulenaere Bram, Swevers Jan, Vandenberghe Lieven (2009) Extended LMI characterizations for stability and performance of linear systems. Syst Control Lett 58(7):510–518
25. Sebe N (2007) A new dilated LMI characterization and iterative control system synthesis. In: Proceedings of the 11th IFAC symposium on large scale complex systems theory and applications, pp 250–255, 2007
26. Fujisaki Y, Befekadu GK (2009) Reliable decentralised stabilisation of multi-channel systems: A design method via dilated LMIs and unknown disturbance observers. Int J Control 82(11):2040–2050
27. Feng Y, Yagoubiab M, Chevrel P (2010) Dilated LMI characterisations for linear time-invariant singular systems. Int J Control 83(11):2276–2284
28. Bara GL (2011) Dilated LMI conditions for time-varying polytopic descriptor systems: The discrete-time case. Int J Control 84(6):1010–1023
29. Sajjadi-Kia S, Jabbari F (2012) On bounded real matrix inequality dilation. Int J Control 85(10):1593–1601
30. Sato M, Peaucelle D (2007) Comparison between SOS approach and slack variable approach for non-negativity check of polynomial functions: Single variable case. In: Proceedings of the american control conference, New-York, July 2007
31. Sato M, Peaucelle D (2007) Comparison between SOS approach and slack variable approach for non-negativity check of polynomial functions: Multiple variable case. In: Proceedings of the european control conference, Kos
32. Sato M, Peaucelle D (2007) Robust stability/performance analysis for uncertain linear systems via multiple slack variable approach: Polynomial LTIPD systems. In: Proceedings of the IEEE conference on decision and control, New-Orleans
33. Peaucelle D, Sato M (2009) LMI tests for positive definite polynomials: Slack variable approach. IEEE Trans Autom Control 54(4):886–891

Chapter 2
Robust Performance Analysis of LTI Systems

2.1 Introduction

In this chapter, we study simple control system analysis problems that embrace many, if not most, engineering issues with respect to performances of dynamical systems close to an equilibrium point. Systems are modeled in state-space and assumed linear time-invariant. Inevitable modeling errors are considered in the simplest affine polytopic form. Having chosen this rather well-established and highly studied framework, the aim of the chapter is to give a comprehensive, step by step understanding of key characteristics of the S-variable approach. Meanwhile, the chapter recalls some fundamental proofs and reasonings that structure the robust control field. One fundamental understanding is that results should be formalized in terms of convex optimization (linear matrix inequalities) for which polynomial-time solvers exist. It is shown that the S-variable approach fits perfectly this specification and has a numerical burden that remains reasonable compared to prior existing results. The other fundamental understanding is conservatism, that is the ability of an approach to merge the sometimes inevitable gap between necessary and sufficient conditions. With respect to conservatism it is shown that the S-variable approach is much less conservative than the prior existing methods as long as robustness issues are considered. This feature, combined with the very easy understanding of mathematical manipulations, explains its rapid success in the control literature.

2.2 Robust Stability Analysis of Uncertain LTI Systems

Let us consider the linear time-invariant (LTI) system described by

$$\dot{x}(t) = Ax(t). \tag{2.1}$$

© Springer-Verlag London 2015

Y. Ebihara et al., *S-Variable Approach to LMI-Based Robust Control*,
Communications and Control Engineering, DOI 10.1007/978-1-4471-6606-1_2

Here, $x \in \mathbb{R}^n$ is the state variable and $A \in \mathbb{R}^{n \times n}$ is the coefficient matrix. This system is said to be *stable* if

$$\lim_{t \to \infty} x(t) \to 0 \quad (\forall x(0) \in \mathbb{R}^n). \tag{2.2}$$

Stability is of course the premier requirement in any engineering system. Therefore, analyzing stability of a given system, and moreover, designing a controller ensuring stability of closed-loop systems, appear to be important issues in control theory.

If a practical system of interest can be modeled accurately as (2.1), it is straightforward to determine its stability. It is well known that the system (2.1) is stable if and only if the matrix A is Hurwitz stable, i.e., every eigenvalue of A is located in $\mathbb{C}_-$. Here, $\mathbb{C}_-$ stands for the open left-half plane in $\mathbb{C}$. Therefore, we can easily check the stability of (2.1) by computing the eigenvalues of A.

However, due to inevitable uncertainties, it is impossible to model a given practical system accurately as (2.1) even if it is LTI. In particular, if the system of interest includes physical parameters whose exact values are hardly available, it is often natural to employ a "model set" described by

$$\dot{x}(t) = A(\theta)x(t), \quad \theta \in \mathbb{E}^L \subset \mathbb{R}^L. \tag{2.3}$$

Here, the matrix-valued function $A(\cdot) : \mathbb{R}^L \to \mathbb{R}^{n \times n}$ is linear and given by

$$A(\theta) := \sum_{j=1}^{L} \theta_j A_j \tag{2.4}$$

where $A_j \in \mathbb{R}^{n \times n}$ $(j \in \mathscr{I}_L)$ are known matrices. On the other hand, the set $\mathbb{E}^L$ is a standard simplex in $\mathbb{R}^L$ defined by

$$\mathbb{E}^L := \left\{ \theta \in \mathbb{R}^L : \theta \geq 0,\ \mathbf{1}^T \theta = 1 \right\}.$$

We emphasize that the parameter θ is time-invariant, but whose exact value is unknown. That is, θ denotes *uncertain parameters* in the system (2.3), and the only available information is $\theta \in \mathbb{E}^L$. The model of the form (2.3) is often called *polytopic uncertain model*, and A_j $(j \in \mathscr{I}_L)$ in (2.4) are *vertex matrices* of $A(\theta)$.

For a concrete example, let us consider a quarter-car suspension model depicted in Fig. 2.1. This model is frequently used for designing active suspensions.

The model is composed of a chassis of weight M and a wheel of weight m, connected by a spring of constant k_1 and a damper of constant c. The spring of constant k_2 stands for the wheel stiffness. The control input is the force to chassis and wheel and denoted by f. The vertical positions of chassis and wheel (from their equilibrium points) are represented by z_c and z_w, respectively.

In this model, the parameters M and k_2 should be treated uncertain, since in practice they depend on the number of passengers and wheel conditions, etc. Therefore,

Fig. 2.1 Quarter-car suspension model

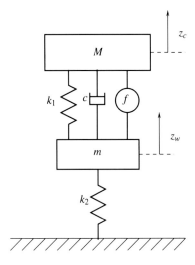

we assume that these two parameters are unknown but bounded with known (or say, reasonably estimated) minimum and maximal values as follows:

$$M \in [M_{\min}, M_{\max}], \quad k_2 \in [k_{2,\min}, k_{2,\max}].$$ (2.5)

On the other hand, we assume that other parameters are exactly known.

The equations of motion for the quarter-car suspension model are given by

$$\begin{aligned} M\ddot{z}_c &= -k_1(z_c - z_w) - c(\dot{z}_c - \dot{z}_w) + f, \\ m\ddot{z}_w &= k_1(z_c - z_w) + c(\dot{z}_c - \dot{z}_w) - k_2 z_w - f. \end{aligned}$$ (2.6)

If we define $x := [z_c \; \dot{z}_c \; z_w \; \dot{z}_w]^T$ and $u := f$, the above equations can be converted to the state-space form as follows:

$$\dot{x}(t) = A(M, k_2)x(t) + B_2(M)u(t), \quad M \in [M_{\min}, M_{\max}], \quad k_2 \in [k_{2,\min}, k_{2,\max}],$$

$$A(M, k_2) := \begin{bmatrix} 0 & 1 & 0 & 0 \\ -\frac{k_1}{M} & -\frac{c}{M} & \frac{k_1}{M} & \frac{c}{M} \\ 0 & 0 & 0 & 1 \\ \frac{k_1}{m} & \frac{c}{m} & -\frac{k_1+k_2}{m} & -\frac{c}{m} \end{bmatrix}, \quad B_2(M) := \begin{bmatrix} 0 \\ \frac{1}{M} \\ 0 \\ -\frac{1}{m} \end{bmatrix}.$$ (2.7)

For this system, suppose a static state-feedback control $u = Kx$ is designed to achieve typical requirements for active suspensions such as satisfactory ride comfort and road-holding ability, etc. If we apply this state-feedback, then the closed-loop system can be written as

$$\dot{x}(t) = A_{cl}(M, k_2)x(t), \quad M \in [M_{\min}, M_{\max}], \quad k_2 \in [k_{2,\min}, k_{2,\max}],$$
$$A_{cl}(M, k_2) := A(M, k_2) + B_2(M)K.$$

It is now elementary to see that this can be rewritten, equivalently, to a polytopic model of the form (2.3) and (2.4) as follows:

$$\dot{x}(t) = A_{cl}(\theta)x(t) \quad \theta \in \mathbb{E}^4, \quad A_{cl}(\theta) = \sum_{j=1}^{4} \theta_j A_{cl,j},$$

$$A_{cl,1} = A_{cl}(M_{\max}, k_{2,\max}), \quad A_{cl,2} = A_{cl}(M_{\max}, k_{2,\min}),$$
$$A_{cl,3} = A_{cl}(M_{\min}, k_{2,\max}), \quad A_{cl,4} = A_{cl}(M_{\min}, k_{2,\min}).$$

In this example, we emphasize that the open-loop system (without feedback $u = Kx$) is, of course, stable irrespective of the parameters $M(>0)$ and $k_2(>0)$. However, such physically ensured stability property cannot retain in general if we apply feedback control to achieve desired control objectives. Therefore, analyzing robust stability of designed controllers, or moreover, designing feedback controllers ensuring robust stability against parameter variations, are definitely important issues in control engineering.

We now go back to the discussion on the general uncertain system description (2.3). Since the parameter θ is assumed to be time-invariant, it is clear that the uncertain system (2.3) is *robustly stable* (i.e., it is stable for all $\theta \in \mathbb{E}^L$) if and only if all the eigenvalues of $A(\theta)$ remain to be $\mathbb{C}_-$. However, it is hard to check the variations of the eigenvalues of $A(\theta)$ over $\theta \in \mathbb{E}^L$ rigorously. One of the effective ways to get around this difficulty is relying on the Lyapunov's stability theorem described below.

Theorem 2.1 *The system (2.1) is stable, or equivalently, the coefficient matrix in (2.1) is Hurwitz stable if and only if there exists $P \in \mathbb{S}^n$ such that*

$$P \succ 0, \quad PA + A^T P \prec 0. \tag{2.8}$$

The inequality above is known as *Lyapunov inequality*, which is an LMI for the matrix variable P to be determined. The matrix P is often called *Lyapunov (matrix) variable*. In the following, we give brief remarks on the Lyapunov inequality (2.8) since it forms an important basis for stability analysis of (uncertain) LTI systems. First, the inequality is strongly related to the existence of quadratic-in-the-state Lyapunov function $V_P(x) := x^T Px$ that certificates the stability of (2.1). Indeed, if we define $V_P(x)$ by means of a feasible solution P for (2.8), we see that the time-derivative of $V_P(x)$ along the trajectory of (2.1) satisfies

$$\frac{dV_P(x)}{dt} = x^T(PA + A^T P)x < 0 \quad (\forall x \neq 0).$$

On the other hand, since $P \succ 0$, it is obvious that $V_P(x) > 0$ ($\forall x \neq 0$). With this fact and the above inequality, we have $V_P(x) \to 0$ ($t \to \infty$) for any $x(0) \in \mathbb{R}^n$. Again, since $P \succ 0$, this also ensures $x \to 0$ ($t \to \infty$), that is, the system (2.1) is stable.

It is also possible to prove the sufficiency of (2.8) in a purely algebraic manner (without invoking the dynamical behavior of (2.1)). Indeed, if (2.8) holds, then we have

$$\xi^*(PA + A^T P)\xi < 0 \quad \forall \xi \in \mathbb{C}^n \setminus \{0\}.$$

In particular, if we let ξ as an eigenvector of A corresponding to an eigenvalue $\lambda \in \mathbb{C}$ (i.e., $A\xi = \lambda\xi$), the above inequality implies

$$(\lambda + \lambda^*)\xi^* P \xi < 0.$$

Since $P \succ 0$, we have $\xi^* P \xi > 0$ and hence we can conclude

$$\lambda + \lambda^* < 0.$$

This clearly shows that every eigenvalue of A is located in $\mathbb{C}_-$. The exposition above proves the sufficiency of (2.8) for stability. On the other hand, its necessity also follows from elementary arguments on Lyapunov equality. See [1] for details.

We next apply Theorem 2.1 to the robust stability analysis of the uncertain system (2.3). Since $A(\theta)$ in (2.4) is linear on θ, the following theorem readily follows:

Theorem 2.2 *The uncertain LTI system (2.3) is stable for all $\theta \in \mathbb{E}^L$ if there exists $P \in \mathbb{S}^n$ such that*

$$P \succ 0, \quad PA_j + A_j^T P \prec 0 \ (j \in \mathscr{I}_L). \tag{2.9}$$

Proof Suppose (2.9) holds. Then, for any $\theta \in \mathbb{E}^L$, we have

$$P \succ 0, \quad \sum_{j=1}^{L} \theta_j(PA_j + A_j^T P) \prec 0$$

or equivalently,

$$P \succ 0, \quad PA(\theta) + A(\theta)^T P \prec 0. \tag{2.10}$$

From this inequality and Theorem 2.1, we can conclude that the uncertain LTI system (2.3) is stable for all $\theta \in \mathbb{E}^L$. □

The condition (2.9) is numerically tractable since it is finite-dimensional finite-number of LMI feasibility problems. This is the main advantage in working with the

LMI condition (2.8) instead of directly examining the variations of the eigenvalues of $A(\theta)$. The condition (2.9) is known as *quadratic stability condition* for the uncertain system (2.3), since, as clearly shown in (2.10), the condition (2.9) is necessary and sufficient for the existence of a single quadratic-in-the-state Lyapunov function $V_P(x)$ that certificates the stability of (2.3) for all $\theta \in \mathbb{E}^L$. Finally, we emphasize that the condition (2.9) is *conservative*, i.e., it is only sufficient for the robust stability of (2.3) and far from necessary in general. The conservatism stems from the fact we seek for a single, or say, common Lyapunov function that uniformly ensures the stability of the uncertain system (2.3). Therefore, we need to derive a condition that allows us to narrow the "gap" toward the exact analysis, but at the same time we need to keep the condition still being computationally tractable. The *S*-variable approach was conceived to achieve these objectives.

2.3 Robust Stability Analysis Using *S*-Variable Approach

We now state how the stability of the LTI system (2.1) is characterized by an LMI with *S*-variables.

Theorem 2.3 *The system* (2.1) *is stable, or equivalently, the coefficient matrix in* (2.1) *is Hurwitz stable if and only if there exist* $P \in \mathbb{S}^n$ *and* $F_i \in \mathbb{R}^{n \times n}$ ($i = 1, 2$) *such that*

$$P > 0, \quad \begin{bmatrix} 0 & P \\ P & 0 \end{bmatrix} + \text{He} \left\{ \begin{bmatrix} F_1 \\ F_2 \end{bmatrix} \begin{bmatrix} I & -A \end{bmatrix} \right\} < 0. \tag{2.11}$$

Proof We will establish the validity by showing the equivalence of (2.8) and (2.11). To this end, let us note that the following choice is valid:

$$\begin{bmatrix} -I \\ A^T \end{bmatrix}^{\perp} = \begin{bmatrix} A^T & I \end{bmatrix}.$$

It also reads as

$$\begin{bmatrix} I & -A \end{bmatrix} \begin{bmatrix} -I \\ A^T \end{bmatrix}^{\perp T} = \begin{bmatrix} I & -A \end{bmatrix} \begin{bmatrix} A \\ I \end{bmatrix} = 0$$

and hence

$$\begin{bmatrix} -I \\ A^T \end{bmatrix}^{\perp} \left(\begin{bmatrix} 0 & P \\ P & 0 \end{bmatrix} + \text{He} \left\{ \begin{bmatrix} F_1 \\ F_2 \end{bmatrix} \begin{bmatrix} I & -A \end{bmatrix} \right\} \right) \begin{bmatrix} -I \\ A^T \end{bmatrix}^{\perp T}$$
$$= \begin{bmatrix} A^T & I \end{bmatrix} \begin{bmatrix} 0 & P \\ P & 0 \end{bmatrix} \begin{bmatrix} A \\ I \end{bmatrix}$$
$$= PA + A^T P.$$

Therefore, the equivalence of (2.8) and (2.11) readily follows from Lemma 1.2, i.e., Elimination Lemma [2, 3].   □

As in the proof, the equivalence of (2.8) and (2.11) can be proved by eliminating the variables F_1 and F_2 from (2.11) via Elimination lemma. Since (2.8) is originally given, this elimination procedure can be restated conversely that the condition (2.11) is derived from (2.8) by introducing (or creating) S-variables F_1 and F_2.

The main advantage in working with (2.11) instead of (2.8) lies in the fact that in (2.11) the Lyapunov variable P is free from multiplication with A, whereas F_1 and F_2 are. This *decoupling* property is the main feature of SV-LMIs in general. In control system analysis and synthesis, we often encounter matrix inequality formulations that include undesirable multiplications among data matrices, Lyapunov variables, scalings, and controller variables, etc. Such multiplications can be decoupled by introducing SVs, and this decoupling property indeed brings great advantage when we deal with wide range of "difficult" problems for which definitive approaches are hardly available. On the other hand, a drawback of the SV-LMI (2.11) is of course obvious; it is computationally more demanding than the original one (2.8) due to the increase of number of variables and size of LMIs. This is, again, a typical drawback of SV-LMIs in general.

As noted, the SV-LMI (2.11) can be obtained by dilating (2.8) with S-variables F_1 and F_2. However, we see that the proof based on Elimination lemma does not provide any concrete ways to create F_1 and F_2 satisfying (2.11) from the original LMI (2.8). Indeed, deriving explicit expressions of SVs satisfying SV-LMIs is an important issue in the study of SV-LMIs and therefore this issue will be discussed repeatedly in this book. In the following, we derive such an explicit expression for (2.8) and (2.11). To this end, suppose (2.8) holds with $P = \Pi (\succ 0)$. Then, there exists $\bar{\varepsilon} > 0$, which depends on Π, such that for any $\varepsilon \in (0, \bar{\varepsilon})$ the following inequality holds:

$$\Pi A + A^T \Pi + \frac{1}{2} \varepsilon A^T \Pi A \prec 0. \tag{2.12}$$

By Schur complement, we see that this is equivalent to

$$\begin{bmatrix} -2\varepsilon \Pi & \varepsilon \Pi A \\ \varepsilon A^T \Pi & \Pi A + A^T \Pi \end{bmatrix} \prec 0.$$

Furthermore, we can rewrite this matrix inequality as

$$\begin{bmatrix} 0 & \Pi \\ \Pi & 0 \end{bmatrix} + \mathrm{He} \left\{ \begin{bmatrix} -\varepsilon \Pi \\ -\Pi \end{bmatrix} \begin{bmatrix} I & -A \end{bmatrix} \right\} \prec 0.$$

This clearly shows that (2.11) holds with $P = \Pi$, $F_1 = -\varepsilon \Pi$ and $F_2 = -\Pi$. This result is summarized in the next lemma.

Lemma 2.1 *Suppose* (2.8) *holds with* $P = \Pi$. *Then, there exists* $\bar{\varepsilon} > 0$ *such that the LMI* (2.11) *holds with* $(P, F_1, F_2) = (\Pi, -\varepsilon \Pi, -\Pi)$ *for all* $\varepsilon \in (0, \bar{\varepsilon})$.

We are now ready to apply the SV-LMI (2.11) to the robust stability analysis of the uncertain system (2.3). The usefulness of the the SV-LMI will be clarified immediately.

Theorem 2.4 *The uncertain LTI system* (2.3) *is stable for all* $\theta \in \mathbb{E}^L$ *if there exist* $P_j \in \mathbb{S}^n$ $(j \in \mathscr{I}_L)$, $F_1, F_2 \in \mathbb{R}^{n \times n}$ *such that*

$$P_j \succ 0, \quad \begin{bmatrix} 0 & P_j \\ P_j & 0 \end{bmatrix} + \mathrm{He} \left\{ \begin{bmatrix} F_1 \\ F_2 \end{bmatrix} \begin{bmatrix} I & -A_j \end{bmatrix} \right\} \prec 0 \; (j \in \mathscr{I}_L). \qquad (2.13)$$

Proof Suppose (2.13) holds. Then, for any $\theta \in \mathbb{E}^L$, we have

$$\sum_{j=1}^{L} \theta_j P_j \succ 0, \quad \sum_{j=1}^{L} \theta_j \left(\begin{bmatrix} 0 & P_j \\ P_j & 0 \end{bmatrix} + \mathrm{He} \left\{ \begin{bmatrix} F_1 \\ F_2 \end{bmatrix} \begin{bmatrix} I & -A_j \end{bmatrix} \right\} \right) \prec 0$$

or equivalently,

$$P(\theta) \succ 0, \quad \begin{bmatrix} 0 & P(\theta) \\ P(\theta) & 0 \end{bmatrix} + \mathrm{He} \left\{ \begin{bmatrix} F_1 \\ F_2 \end{bmatrix} \begin{bmatrix} I & -A(\theta) \end{bmatrix} \right\} \prec 0,$$

$$P(\theta) := \sum_{j=1}^{L} \theta_j P_j. \qquad (2.14)$$

From the first two inequalities above and Theorem 2.3, we can conclude that the uncertain LTI system (2.3) is stable for all $\theta \in \mathbb{E}^L$. □

This theorem provides an SV-LMI-based condition for the robust stability of (2.3). Unfortunately, the condition (2.13) is still sufficient to conclude robust stability in general. Since (2.9) is also sufficient, we are naturally lead to the following question: Is the condition (2.13) better (or more precisely, no worse) than (2.9)? The answer is "yes," as formally stated in the following theorem.

Theorem 2.5 *Suppose* (2.9) *holds with* $P = \Pi$. *Then, there exists* $\varepsilon > 0$ *such that* (2.13) *holds with* $P = \Pi$, $F_1 = -\varepsilon \Pi$ *and* $F_2 = -\Pi$.

Proof Suppose (2.9) holds with $P = \Pi$. Then, from Lemma 2.1, there exist $\overline{\varepsilon}_j$ $(j \in \mathscr{I}_L)$ such that the jth LMI in (2.13) holds with $P_j = \Pi$, $F_1 = -\varepsilon_j \Pi$ and $F_2 = -\Pi$ for any $\varepsilon_j \in (0, \overline{\varepsilon}_j)$. Therefore, by letting $\varepsilon \in (0, \min_j \overline{\varepsilon}_j)$, the assertion is verified. □

Theorem 2.5 says that if we can conclude robust stability of (2.3) by the quadratic stability condition (2.9), then we can always conclude the same by the SV-LMI condition (2.13) as well. The converse is not true in general. As illustrated by numerical examples in Sect. 2.6, the effectiveness of (2.13) is remarkable, and we frequently experience that (2.13) works fine to conclude robust stability even for those problem instances where (2.9) fails.

As we have seen in the proof of Theorem 2.5, Lemma 2.1 plays a key role in ensuring an explicit advantage of (2.13) over (2.9). However, qualitatively speaking, the effectiveness of (2.13) can be explained from (2.14), which implies that the condition (2.13) ensures the robust stability of (2.3) via *parameter-dependent* quadratic-in-the state Lyapunov function $V_P(\theta, x) := x^T P(\theta)x$. Namely, the restriction to a common Lyapunov function in the quadratic stability approach (2.9) is relaxed to a Lyapunov function that can depend linearly on the parameter θ. In the literature, there are several ways to introduce parameter-dependent Lyapunov functions so that less-conservative analysis/synthesis results can be achieved. Among them, the SV approach can be regarded as a powerful tool, and the most attractive feature is that it allows us to derive LMI conditions for (robust) controller synthesis as well. See Chaps. 4, 7, 8 for detailed discussions on this point.

Anticipating slightly the discussions of these chapters, we can state some issues for deriving the control design results. The first one is related to the fact that state-feedback design method have classically nice formulations when LMI conditions are derived not for the original system $\dot{x} = Ax$ but for the *dual system* $\dot{x}_d = A^T x_d$. Related to Theorems 2.1 and 2.3, the following corollaries readily follow for the dual system.

Corollary 2.1 *The system* (2.1) *is stable, or equivalently, the coefficient matrix in* (2.1) *is Hurwitz stable if and only if there exists* $X \in \mathbb{S}^n$ *such that*

$$X \succ 0, \quad AX + XA^T \prec 0. \tag{2.15}$$

Corollary 2.2 *The system* (2.1) *is stable, or equivalently, the coefficient in* (2.1) *is Hurwitz stable if and only if there exist* $X \in \mathbb{S}^n$ *and* $F_i \in \mathbb{R}^{n \times n}$ $(i = 1, 2)$ *such that*

$$X \succ 0, \quad \begin{bmatrix} 0 & X \\ X & 0 \end{bmatrix} + \mathrm{He}\left\{ \begin{bmatrix} I \\ -A \end{bmatrix} \begin{bmatrix} F_1 & F_2 \end{bmatrix} \right\} \prec 0. \tag{2.16}$$

Corollary 2.3 *Suppose* (2.15) *holds with* $X = \Xi$. *Then, there exists* $\bar{\varepsilon} > 0$ *such that the LMI* (2.16) *holds with* $(X, F_1, F_2) = (\Xi, -\varepsilon\Xi, -\Xi)$ *for all* $\varepsilon \in (0, \bar{\varepsilon})$.

Corollary 2.4 *The uncertain LTI system* (2.3) *is stable for all* $\theta \in \mathbb{E}^L$ *if there exist* $X_j \in \mathbb{S}^n$ $(j \in \mathscr{I}_L)$ *and* $F_i \in \mathbb{R}^{n \times n}$ $(i = 1, 2)$ *such that*

$$X_j \succ 0, \quad \begin{bmatrix} 0 & X_j \\ X_j & 0 \end{bmatrix} + \mathrm{He}\left\{ \begin{bmatrix} I \\ -A_j \end{bmatrix} \begin{bmatrix} F_1 & F_2 \end{bmatrix} \right\} \prec 0. \tag{2.17}$$

In the LMIs (2.15), (2.16) and (2.17), the matrix A has multiplication with variables from the right-hand side and this is different from (2.8), (2.11) and (2.13). Due to this reason, the LMIs (2.15) and (2.16) are effective for feedback controller synthesis as well (see related discussions in Chaps. 4 and 8).

Theorem 2.4 and Corollary 2.4 are two versions of a same result but applied to either the original system $\dot{x} = A(\theta)x$ or to its dual $\dot{x}_d = A^T(\theta)x_d$. Both conclude on the robust stability of $\dot{x} = A(\theta)x$ and one could therefore expect the two results to be equivalent. This is not the case.

Theorem 2.6 *The LMI condition* (2.13) *may be feasible and* (2.17) *not, and the converse is true as well.*

The proof of this result needs only to give one such example. This is done later on in the Sect. 2.6 devoted to numerical examples. To understand the reasons for this result recall that if $x_d^T X x_d$ is a Lyapunov function proving the stability of the dual system, then $x^T X^{-1} x$ is a Lyapunov function for the original system. Now, from the proof of Theorem 2.4 it can be noticed that the LMI conditions (2.13) imply that $x^T P(\theta) x$ is a Lyapunov function for the system, with $P(\theta) = \sum_{j=1}^{L} \theta_j P_j$ linear with respect to θ. Similarly, the solution of (2.17) implies that $x_d^T X(\theta) x_d$ is a Lyapunov function for the dual system, with $X(\theta) = \sum_{j=1}^{L} \theta_j X_j$ linear with respect to θ. Hence, Theorem 2.4 is related to the existence of a linear parameter-dependent Lyapunov matrix $P(\theta)$ while Corollary 2.4 is related to the existence of a Lyapunov matrix $X^{-1}(\theta)$ which is not linear. These two properties are trivially not equivalent.

The second issue for deriving design conditions that will be discussed in the next chapters is that F_1 and F_2 give too many degrees of freedom. To make the control design conditions convex it is needed to structure the S-variables. To this end, Theorem 2.5 can give the impression that the choice $F_1 = -\varepsilon \Pi$ and $F_2 = -\Pi$ for some sufficiently small ε is appropriate when solving (2.13). Unfortunately, it is not quite the case for the following reason.

Theorem 2.7 *For the choice $F_1 = -\varepsilon \Pi$ and $F_2 = -\Pi$ with $\varepsilon > 0$ and $\Pi \in \mathbb{R}^{n \times n}$, if ε tends to zero then* (2.13) *converges to the quadratic stability conditions of* (2.9).

Proof Assume a solution to (2.13) with S-variables structured as $F_1 = -\varepsilon \Pi$ and $F_2 = -\Pi$ where $\varepsilon > 0$:

$$
\begin{bmatrix} 0 & P_j \\ P_j & 0 \end{bmatrix} + \mathrm{He} \left\{ \begin{bmatrix} -\varepsilon \Pi \\ -\Pi \end{bmatrix} \begin{bmatrix} I & -A_j \end{bmatrix} \right\}
$$
$$
= \begin{bmatrix} -\varepsilon \Pi - \varepsilon \Pi^T & P_j - \Pi^T + \varepsilon \Pi A_j \\ P_j - \Pi + \varepsilon A_j^T \Pi^T & \Pi A_j + A_j^T \Pi^T \end{bmatrix} \prec 0.
$$

After a Schur complement argument applied to the (1,1) block, this LMI also reads as

$$
\mathrm{He} \left\{ \left((P_j - \Pi)(\Pi + \Pi^T)^{-1} + I \right) \Pi A_j \right\}
$$
$$
\prec -\varepsilon A_j^T \Pi^T (\Pi + \Pi^T)^{-1} \Pi A_j - \varepsilon^{-1} (P_j - \Pi)(\Pi + \Pi^T)^{-1} (P_j - \Pi^T).
$$

Assume now that $\varepsilon > 0$ is indeed small, i.e., assume $\varepsilon \to 0$. Then the right-hand side of the inequality is a matrix with eigenvalues going to $-\infty$ except if $\Pi = P_j$ ($j \in \mathscr{I}_L$). Since the left-hand side is independent of ε, the only possible finite solution is indeed that $(P_j - \Pi) \to 0$ as ε goes to zero. Hence we conclude that for small values of ε the SV-LMI with structured S-variables $F_1 = -\varepsilon \Pi$ and $F_2 = -\Pi$

implies asymptotically

$$\Pi \succ 0, \quad \Pi A_j + A_j^T \Pi \prec 0 \ (j \in \mathscr{I}_L)$$

which is the condition of the quadratic stability Theorem 2.2. □

Theorem 2.7 is a negative result in the sense that the structure $F_1 = -\varepsilon \Pi$ and $F_2 = -\Pi$ with ε small brings no advantage compared to quadratic stability conditions. The question of structuring the S-variables is left open at this point of the book. The discussion is continued in the next chapters.

2.4 Lemmas for SV-LMI Derivation

In the previous section, SV results are presented for the case of stability analysis of continuous-time LTI systems. Some properties of the result are discussed for this specific criterion. These properties happen to be valid for many more cases some of which are discussed in the following section. Before that, we expose the technical lemmas that formalize the properties.

Lemma 2.2 *For given* $\mathsf{M}_{11} \in \mathbb{H}_+^n$, $\mathsf{M}_{12} \in \mathbb{C}^{n \times m}$, $\mathsf{M}_{22} \in \mathbb{H}^m$, $\mathsf{A} \in \mathbb{C}^{n \times m}$, $\mathsf{E} \in \mathbb{C}^{n \times n}$ *the following two conditions are equivalent.*

(i) The matrix E *is invertible and*

$$\begin{bmatrix} \mathsf{E}^{-1}\mathsf{A} \\ I \end{bmatrix}^* \begin{bmatrix} \mathsf{M}_{11} & \mathsf{M}_{12} \\ \mathsf{M}_{12}^* & \mathsf{M}_{22} \end{bmatrix} \begin{bmatrix} \mathsf{E}^{-1}\mathsf{A} \\ I \end{bmatrix} \prec 0. \tag{2.18}$$

(ii) There exist matrices $\mathsf{F}_1 \in \mathbb{C}^{n \times n}$ *and* $\mathsf{F}_2 \in \mathbb{C}^{m \times n}$ *such that*

$$\begin{bmatrix} \mathsf{M}_{11} & \mathsf{M}_{12} \\ \mathsf{M}_{12}^* & \mathsf{M}_{22} \end{bmatrix} + \mathrm{He}\left\{ \begin{bmatrix} \mathsf{F}_1 \\ \mathsf{F}_2 \end{bmatrix} \begin{bmatrix} \mathsf{E} & -\mathsf{A} \end{bmatrix} \right\} \prec 0. \tag{2.19}$$

Moreover,

(a) If A *and* E *are real valued, then* F_1 *and* F_2 *can be chosen real valued.*
(b) If (2.18) holds, then whatever $\mathsf{W} \in \mathbb{H}_{++}^n$, *(2.19) holds as well with* $\mathsf{F}_1 = -(\mathsf{M}_{11} + \varepsilon \mathsf{W})\mathsf{E}^{-1}$ *and* $\mathsf{F}_2 = -\mathsf{M}_{12}^* \mathsf{E}^{-1}$ *and a sufficiently small* $\varepsilon > 0$.
(c) If (2.18) holds and $\mathsf{M}_{11} \succ 0$, *then (2.19) holds as well with* $\mathsf{F}_1 = -\mathsf{M}_{11}\mathsf{E}^{-1}$ *and* $\mathsf{F}_2 = -\mathsf{M}_{12}^* \mathsf{E}^{-1}$.

Proof The equivalence of (i) and (ii) is the direct application of the elimination lemma [2, 3] due to the fact that for invertible E:

$$\begin{bmatrix} \mathsf{E}^{-1}\mathsf{A} \\ I \end{bmatrix}^* = \begin{bmatrix} \mathsf{E}^* \\ -\mathsf{A}^* \end{bmatrix}^{\perp}.$$

Note that $M_{11} \in \mathbb{H}_+^n$ ensures the invertibility of E in (2.19). Property (a) comes from Finsler's lemma [3]. That lemma states that if (2.18) holds then there exists some positive scalar $\tau \geq 0$ such that

$$\begin{bmatrix} M_{11} & M_{12} \\ M_{12}^* & M_{22} \end{bmatrix} \prec \tau \begin{bmatrix} E^T \\ -A^T \end{bmatrix} \begin{bmatrix} E & -A \end{bmatrix}.$$

This inequality provides a possible choice of S-variables is $F_1 = \frac{\tau}{2} E^T$ and $F_2 = -\frac{\tau}{2} A^T$ which are real valued.

We shall now prove property (b). Assume (2.18) holds, then for any positive definite matrix $W \succ 0$ and a sufficiently small $\varepsilon > 0$ the following inequality holds as well:

$$\begin{bmatrix} E^{-1}A \\ I \end{bmatrix}^* \begin{bmatrix} M_{11} + \varepsilon W & M_{12} \\ M_{12}^* & M_{22} \end{bmatrix} \begin{bmatrix} E^{-1}A \\ I \end{bmatrix}$$
$$= A^*E^{-*}(M_{11} + \varepsilon W)E^{-1}A + A^*E^{-*}M_{12} + M_{12}^*E^{-1}A + M_{22} \prec 0.$$

The matrix $M_{11} + \varepsilon W$ is positive definite, one can therefore apply the Schur complement result and get

$$\begin{bmatrix} -M_{11} - \varepsilon W & (M_{11} + \varepsilon W)E^{-1}A \\ A^*E^{-*}(M_{11} + \varepsilon W) & A^*E^{-*}M_{12} + M_{12}^*E^{-1}A + M_{22} \end{bmatrix} \prec 0,$$

which implies

$$\begin{bmatrix} -M_{11} - 2\varepsilon W & (M_{11} + \varepsilon W)E^{-1}A \\ A^*E^{-*}(M_{11} + \varepsilon W) & A^*E^{-*}M_{12} + M_{12}^*E^{-1}A + M_{22} \end{bmatrix} \prec 0,$$

This inequality is exactly (2.19) for the choice $F_1 = -(M_{11} + \varepsilon W)E^{-1}$ and $F_2 = -M_{12}^*E^{-1}$.

Property (c) is a special case of (b) for the choice $\varepsilon = 0$. This choice can be done only if M_{11} is positive definite.   □

To illustrate how Lemma 2.2 works fine, let us derive (2.11) from (2.8) along the line of the lemma. We first note that the second inequality (2.8) can be identified with (2.18) by the correspondences

$$M_{11} = 0, \quad M_{12} = P, \quad M_{22} = 0, \quad E = I, \quad A = A. \tag{2.20}$$

Then, we obtain directly (2.11) from (2.19) by letting $F_i = F_i$ ($i = 1, 2$). The results in Lemma 2.1 follow from property (b) in Lemma 2.2 for the choice $W = P$.

As illustrated above, we can use Lemma 2.2 as follows for the derivation of SV-LMIs. First, we identify a known LMI condition with (2.18) by finding out appropriate correspondences as in (2.20). Once this step is done, it is straightforward to derive a corresponding SV-LMI of the form (2.19).

In comparison with (2.18), we see that the SV-LMI (2.19) has the following preferable properties:

(i) *Decoupling*: In (2.19), variables that may be contained in M_{11}, M_{12}, and M_{22} have no multiplication relation with E and A, while F_1 and F_2 do.
(ii) *Eliminating inverse*: In (2.19), the inverse of the matrix E is eliminated.

As partly illustrated in the preceding section, the property (i) is effective when we deal with robust performance analysis problems for polytopic uncertain systems. Indeed, it enables us to employ parameter-dependent Lyapunov functions so that less-conservative analysis/synthesis conditions can be achieved.

Before formulating the central technical results of this chapter, we first give an intermediate technical lemma. It states in which case testing an LMI over a polytope or over its vertices is equivalent.

Lemma 2.3 *Assume G and M are given where $G \succeq 0$. Then the following two conditions are equivalent*

$$H_j + \text{He}\{K_j^* M\} + J_j^* G J_j \prec 0, \quad \forall j \in \mathscr{I}_L \tag{2.21}$$

$$H(\theta) + \text{He}\{K^*(\theta)M\} + J^*(\theta)G J(\theta) \prec 0, \quad \forall \theta \in \mathbb{E}_L \tag{2.22}$$

where $H(\theta) = \sum_{j=1}^{L} \theta_j H_j$, $K(\theta) = \sum_{j=1}^{L} \theta_j K_j$, $J(\theta) = \sum_{j=1}^{L} \theta_j J_j$.

Proof The implication (2.22)⇒(2.21) is obtained by taking the choices $\theta_j = 1$, $\theta_{i \neq j} = 0$. The implication (2.21)⇒(2.22) is obtained by convexity. Since G is positive semi-definite it has a factorization $G = L^*L$. Applying a Schur complement argument (2.21) implies that

$$\begin{bmatrix} H_j + \text{He}\{K_j^* M\} & J_j^* L^* \\ L J_j & -I \end{bmatrix} \prec 0, \quad \forall j \in \mathscr{I}_L$$

Summing these conditions with weights θ_j gives

$$\sum_{j=1}^{L} \theta_j \begin{bmatrix} H_j + \text{He}\{K_j^* M\} & J_j^* L^* \\ L J_j & -I \end{bmatrix}$$
$$= \begin{bmatrix} H(\theta) + \text{He}\{K^*(\theta)M\} & J^*(\theta)L^* \\ L J(\theta) & -I \end{bmatrix} \prec 0, \quad \forall \theta \in \mathbb{E}^L.$$

A converse Schur complement argument allows to conclude with (2.22).   □

We now give the two central results which explicate the advantage of the SV-LMIs in case of polytopic systems. For the greatest generality of the result, the matrix M_{22} from Lemma 2.2 is decomposed as $M_{22} = N + C^* V C$ where $V \succeq 0$ is assumed to be positive semidefinite.

Lemma 2.4 *For given* $M_{11j} \in \mathbb{H}^n$, $M_{12j} \in \mathbb{C}^{n \times m}$, $M_{22j} \in \mathbb{H}^m$ *with* $M_{22j} = N_j + C_j^* V C_j$ $(j \in \mathscr{I}_L)$, $A_j \in \mathbb{C}^{n \times m}$, *and* $E_j \in \mathbb{C}^{n \times n}$, *suppose* $V \succeq 0$, $E(\theta) = \sum_{j=1}^L \theta_j E_j$ *is invertible for all* $\theta \in \mathbb{E}^L$ *and there exist* $F_1 \in \mathbb{C}^{n \times n}$ *and* $F_2 \in \mathbb{C}^{m \times n}$ *such that*

$$\begin{bmatrix} M_{11j} & M_{12j} \\ M_{12j}^* & N_j + C_j^* V C_j \end{bmatrix} + \mathrm{He}\left\{ \begin{bmatrix} F_1 \\ F_2 \end{bmatrix} \begin{bmatrix} E_j & -A_j \end{bmatrix} \right\} \prec 0 \quad (\forall j \in \mathscr{I}_L). \quad (2.23)$$

Then, the following condition hold for all $\theta \in \mathbb{E}^L$:

$$\begin{bmatrix} E^{-1}(\theta) A(\theta) \\ I \end{bmatrix}^* \begin{bmatrix} M_{11}(\theta) & M_{12}(\theta) \\ M_{12}^*(\theta) & N(\theta) + C^*(\theta) V C(\theta) \end{bmatrix} \begin{bmatrix} E^{-1}(\theta) A(\theta) \\ I \end{bmatrix} \prec 0. \quad (2.24)$$

where $M_{11}(\theta) = \sum_{j=1}^L \theta_j M_{11j}$, $M_{12}(\theta) = \sum_{j=1}^L \theta_j M_{12j}$, $N(\theta) = \sum_{j=1}^L \theta_j N_j$, $A(\theta) = \sum_{j=1}^L \theta_j A_j$ *and* $C(\theta) = \sum_{j=1}^L \theta_j C_j$.

This lemma can be seen as the extension of Theorem 2.4 to a general formulation. The proof follows the same lines as that of Theorem 2.4.

Proof The first step is, thanks to Lemma 2.3, to notice that (2.23) is equivalent to assessing that for all $\theta \in \mathbb{E}^L$ one has

$$\begin{bmatrix} M_{11}(\theta) & M_{12}(\theta) \\ M_{12}^*(\theta) & N(\theta) + C^*(\theta) V C(\theta) \end{bmatrix} + \mathrm{He}\left\{ \begin{bmatrix} F_1 \\ F_2 \end{bmatrix} \begin{bmatrix} E(\theta) & -A(\theta) \end{bmatrix} \right\} \prec 0$$

Finally (2.24) is obtained by post- and premultiplying by $\begin{bmatrix} E^{-1}(\theta) A(\theta) \\ I \end{bmatrix}$ and its transpose, respectively. $\qquad\square$

By analogy to Theorem 2.4 one expects some results such as Theorem 2.5. It indeed holds and is based on the properties (b) and (c) of Lemma 2.2.

Lemma 2.5 *Under the same setting as Lemma 2.4 except for* $M_{11j} = M_{11}$, $M_{12j} = M_{12}$, $N_j = N$, *and* $E_j = E$ $(j \in \mathscr{I}_L)$, *suppose* $V \succeq 0$, E *is invertible, and* $M_{11} \succeq 0$. *Then the following three conditions are equivalent:*

(i) The following LMI holds for all $\theta \in \mathbb{E}^L$

$$\begin{bmatrix} E^{-1} A(\theta) \\ I \end{bmatrix}^* \begin{bmatrix} M_{11} & M_{12} \\ M_{12}^* & N + C^*(\theta) V C(\theta) \end{bmatrix} \begin{bmatrix} E^{-1} A(\theta) \\ I \end{bmatrix} \prec 0. \quad (2.25)$$

(ii) The following LMI holds for all $j \in \mathscr{I}_L$

$$\begin{bmatrix} E^{-1} A_j \\ I \end{bmatrix}^* \begin{bmatrix} M_{11} & M_{12} \\ M_{12}^* & N + C_j^* V C_j \end{bmatrix} \begin{bmatrix} E^{-1} A_j \\ I \end{bmatrix} \prec 0. \quad (2.26)$$

(iii) There exists F_1 and F_2 such that for all $j \in \mathscr{I}_L$

$$\begin{bmatrix} M_{11} & M_{12} \\ M_{12}^* & N + C_j^* V C_j \end{bmatrix} + He\left\{ \begin{bmatrix} F_1 \\ F_2 \end{bmatrix} \begin{bmatrix} E & -A_j \end{bmatrix} \right\} \prec 0 \qquad (2.27)$$

Moreover, if these equivalent inequality conditions hold, then the S-variables in (2.27) can be chosen as $F_1 = -(M_{11} + \varepsilon W)E^{-1}$ and $F_2 = -M_{12}^* E^{-1}$ where W is a free-positive definite matrix $W \succ 0$ and $\varepsilon > 0$ is sufficiently small. If $M_{11} \succ 0$ one can chose $\varepsilon = 0$.

Proof Equivalence between (i) and (ii) is due to Lemma 2.3. Equivalence between (ii) and (iii) is direct by the properties (b) and (c) of Lemma 2.2. □

Notice that Lemma 2.4 provides a finite dimensional LMI test for rationally dependent matrices $E^{-1}(\theta)A(\theta)$. This feature is important for the analysis of more elaborated uncertain systems than LTI polytopic systems. This topic is also closely related to the analysis of uncertain descriptor systems. These issues are treated in detail in Chap. 3.

2.5 SV-LMI Results for Robust Performance Analysis Problems

Let us consider the LTI system G described by

$$G : \begin{cases} \dot{x}(t) = Ax(t) + Bw(t), \ x(0) = 0, \\ z(t) = Cx(t) + Dw(t) \end{cases} \qquad (2.28)$$

where $x \in \mathbb{R}^n$ is the state, $w \in \mathbb{R}^{n_w}$ the disturbance input, $z \in \mathbb{R}^{n_z}$ the performance output, and $A \in \mathbb{R}^{n \times n}$, $B \in \mathbb{R}^{n \times n_w}$, $C \in \mathbb{R}^{n_z \times n}$, $D \in \mathbb{R}^{n_z \times n_w}$. In this section, we derive various SV-LMIs characterizing time- and frequency-domain performances of the system (2.28). The performances of interest includes pole location, H_2 performance, H_∞ performance, and impulse-to-peak performances. Then, we apply those SV-LMIs to robust performance analysis of polytopic uncertain LTI system G_θ described by

$$G_\theta : \begin{cases} \dot{x}(t) = A(\theta)x(t) + B(\theta)w(t), \ x(0) = 0, \\ z(t) = C(\theta)x(t) + D(\theta)w(t) \end{cases},$$
$$\theta \in \mathbb{E}^L, \quad \begin{bmatrix} A(\theta) & B(\theta) \\ C(\theta) & D(\theta) \end{bmatrix} = \sum_{j=1}^L \theta_j \begin{bmatrix} A_j & B_j \\ C_j & D_j \end{bmatrix}. \qquad (2.29)$$

Here, A_j ($j \in \mathscr{I}_L$), etc., are known matrices.

Some of the results are derived for the dual system described by

$$G_\theta^* : \begin{cases} \dot{x}_d(t) = A^T(\theta)x_d(t) + C^T(\theta)w_d(t), \ x_d(0) = 0, \\ z_d(t) = B^T(\theta)x_d(t) + D^T(\theta)w_d(t) \end{cases}. \qquad (2.30)$$

2.5.1 Robust Regional Pole Location Analysis

In this subsection, we analyze the location of eigenvalues of a given matrix A in the complex plane. Namely, we will discuss whether $\lambda(A) \subset \mathbb{O}$ holds for given $\mathbb{O} \subset \mathbb{C}$. This analysis could be considered as a preliminary step for regional pole placement by feedback control.

2.5.1.1   QMI Regions

For the analysis of regional pole location, we focus on the following Quadratic Matrix Inequality (QMI) regions defined by $R \in \mathbb{H}^{2d}$

$$\mathbb{O}(R) := \left\{ \lambda \in \mathbb{C} : \begin{bmatrix} \lambda^* I_d & I_d \end{bmatrix} R \begin{bmatrix} \lambda I_d \\ I_d \end{bmatrix} = R_{11}\lambda^*\lambda + R_{12}^*\lambda + R_{12}\lambda^* + R_{22} \prec 0 \right\}.$$

(2.31)

This definition of regions of the complex plane has been introduced in [4]. Such regions are convex if $R_{11} \succeq 0$. In that case, take L such that $R_{11} = L^*L$. A Schur complement argument indicates that the region is also described by the LMI

$$\begin{bmatrix} R_{12}^*\lambda + R_{12}\lambda^* + R_{22} & \lambda^* L^* \\ \lambda L & -I \end{bmatrix} =$$
$$\begin{bmatrix} R_{12}^* & 0 \\ L & 0 \end{bmatrix}\lambda + \begin{bmatrix} R_{12} & L^* \\ 0 & 0 \end{bmatrix}\lambda^* + \begin{bmatrix} R_{22} & 0 \\ 0 & -I \end{bmatrix} \prec 0.$$

(2.32)

This fact illustrates that QMI regions are more general than LMI regions defined in [5]. The subclass of QMI regions with R_{11} positive semi-definite is exactly identical to the LMI regions.

Before giving examples of QMI regions, recall that for real-valued LTI system (2.28), if $\lambda_{\pm} = -\zeta\omega_n \pm j\omega_d \subset \lambda(A)$ is a pair of poles, then for the associated mode $0 < \zeta < 1$ is the damping ratio, $\omega_d = \omega_n\sqrt{1-\zeta^2}$ is the damped natural frequency, $\omega_n (= |\lambda_{\pm}|)$ is the undamped natural frequency, and $\tau = \frac{1}{\zeta\omega_n}$ is the time constant.

For $d = 1$, there are two types of regions: half planes (if $R_{11} = 0$) and discs (otherwise). Examples of such regions follow.

- $R_{11} = 0$, $R_{12} = 1$, $R_{22} = 0$ then $\mathbb{O}(R) = \mathbb{C}_-$ the left-hand side of the complex plane. Proving that poles in this region are equivalent to Hurwitz stability.
- $R_{11} = 0$, $R_{12} = -1$, $R_{22} = 0$ then $\mathbb{O}(R) = \mathbb{C}_+$ the right-hand side of the complex plane. A system having its poles in such region is said to be anti-stable.
- $R_{11} = 0$, $R_{12} = 1$, $R_{22} = -2\alpha$ then $\mathbb{O}(R) = \{\lambda \in \mathbb{C} : \text{Re}(\lambda) < \alpha\}$. If poles are in such region the modes of the system have time constants smaller than $-\frac{1}{\alpha}$ for $\alpha < 0$. This property is also called α-stability. It is illustrated in Fig. 2.2a.

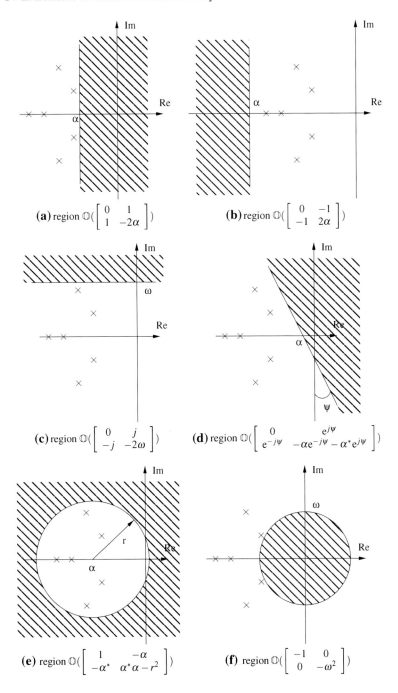

Fig. 2.2 Examples of pole location regions. The striped region is the exterior of the QMI region

- $R_{11} = 0$, $R_{12} = -1$, $R_{22} = 2\alpha$ then $\mathbb{O}(R) = \{\lambda \in \mathbb{C}: \quad \text{Re}(\lambda) > \alpha\}$. If poles are in such region the modes of the system have time constants greater than $-\frac{1}{\alpha}$ for $\alpha < 0$. Such region is illustrated in Fig. 2.2b.
- $R_{11} = 0$, $R_{12} = j$, $R_{22} = -2\omega$ then $\mathbb{O}(R) = \{\lambda \in \mathbb{C}: \quad \text{Im}(\lambda) < \omega\}$. If poles are in such region the modes of a real-valued system have damped natural frequencies smaller than ω: $\text{Im}(-\zeta\omega_n \pm j\omega_d) \leq \omega$ if and only if $|\omega_d| \leq \omega$. Such region is illustrated in Fig. 2.2c.
- $R_{11} = 0$, $R_{12} = e^{j\psi}$, $R_{22} = -\alpha e^{-j\psi} - \alpha^* e^{j\psi}$ then $\mathbb{O}(R)$ is the half plane limited by a line that crosses the values α and orthogonal to $e^{j\psi}$. Such region is illustrated in Fig. 2.2d. For $\alpha = 0$ and $\psi = \arcsin \zeta$, if poles are in such region the modes of a real-valued system have a damping ratio greater than ζ. If $\psi = 0$ and α is real it corresponds to the set $\mathbb{O}(R) = \{\lambda \in \mathbb{C}: \quad \text{Re}(\lambda) < \alpha\}$. If $\psi = \pi/2$ and $\alpha = j\omega$ it corresponds to the set $\mathbb{O}(R) = \{\lambda \in \mathbb{C}: \quad \text{Im}(\lambda) < \omega\}$.
- $R_{11} = 1$, $R_{12} = 0$, $R_{22} = -1$ then $\mathbb{O}(R) = \mathbb{D}$ is the open unit disc. Proving that poles are in this region is equivalent to Schur stability of discrete-time systems.
- $R_{11} = -1$, $R_{12} = 0$, $R_{22} = 1$ then $\mathbb{O}(R)$ is the exterior of the unit disc. A discrete-time system having its poles in such region is said to be anti-stable.
- $R_{11} = 1$, $R_{12} = -\alpha$, $R_{22} = \alpha^*\alpha - r^2$ then $\mathbb{O}(R)$ is the open disc centered at $\alpha \in \mathbb{C}$ with radius r. Such region is illustrated in Fig. 2.2e. For $\alpha = 0$ and $r = \omega$, if poles are in such region the modes of the system have undamped natural frequencies ω_n smaller than ω.
- $R_{11} = -1$, $R_{12} = \alpha$, $R_{22} = r^2 - \alpha^*\alpha$ then $\mathbb{O}(R)$ is the exterior of the disc centered at $\alpha \in \mathbb{C}$ with radius r. For $\alpha = 0$ and $r = \omega$, if poles are in such region the modes of the system have undamped natural frequencies ω_n greater than ω. Such region is illustrated in Fig. 2.2f.

Pole location in QMI regions can therefore without difficulty specify characteristics on the modes of linear systems. These are frequently used in the regional pole placement designs [5, 6].

In the definition of QMI regions, (2.31) the matrix R has $2d$ rows and $2d$ columns. But, the regions that have been described above are those with $d = 1$ only. Known examples from the literature of regions with $d > 1$ are intersections of regions of smaller dimensions. It is indeed quite easy to see that

$$\mathbb{O}\left(\begin{bmatrix} R_{11} & R_{12} \\ R_{12}^* & R_{22} \end{bmatrix}\right) \cap \mathbb{O}\left(\begin{bmatrix} \tilde{R}_{11} & \tilde{R}_{12} \\ \tilde{R}_{12}^* & \tilde{R}_{22} \end{bmatrix}\right) = \mathbb{O}\left(\begin{bmatrix} R_{11} & 0 & R_{12} & 0 \\ 0 & \tilde{R}_{11} & 0 & \tilde{R}_{12} \\ R_{12}^* & 0 & R_{22} & 0 \\ 0 & \tilde{R}_{12}^* & 0 & \tilde{R}_{22} \end{bmatrix}\right).$$

This feature is most useful for design problems. In analysis it is not. Indeed, an alternative is to prove first that the poles belong to one region and, separately to prove they belong to the other region. It amounts to solving two separate LMI problems of smaller size than the LMI problem defined for the intersection of regions. Yet, a useful application of this property is to convert a region defined by a complex valued R into a region defined by a real-valued matrix.

To illustrate this feature consider the region defined by $R_{11} = 0$, $R_{12} = e^{j\psi}$, $R_{22} = 0$ that amounts to poles located below a line crossing the origin and that makes an angle ψ with the imaginary axis. Eigenvalues of a real matrix are symmetric with respect to the real axis. Hence, proving that a real-valued matrix has its eigenvalues in $\mathbb{O}(R)$ implies that the poles are as well in the conjugate of that region. Thus, the eigenvalues are in the intersection of the two regions defined by:

$$\mathbb{O}\left(\left[\begin{array}{cc|cc} 0 & 0 & e^{j\psi} & 0 \\ 0 & 0 & 0 & e^{-j\psi} \\ \hline e^{-j\psi} & 0 & 0 & 0 \\ 0 & e^{j\psi} & 0 & 0 \end{array}\right]\right) \tag{2.33}$$

Since any change of basis applied simultaneously to all blocs of R does not modify the feasibility of the inequality in (2.31), this conic region is equivalently defined by:

$$\mathbb{O}\left(\left[\begin{array}{cc|cc} 0 & 0 & \cos(\psi) & \sin(\psi) \\ 0 & 0 & -\sin(\psi) & \cos(\psi) \\ \hline \cos(\psi) & -\sin(\psi) & 0 & 0 \\ \sin(\psi) & \cos(\psi) & 0 & 0 \end{array}\right]\right). \tag{2.34}$$

This is how conic regions are defined in [5].

2.5.1.2 LMI Tests for Pole Location in QMI Regions

Let us recall that how the condition $\lambda(A) \subset \mathbb{O}(R)$ can be written in terms of LMIs.

Lemma 2.6 *For given matrices $A \in \mathbb{C}^{n \times n}$, $R \in \mathbb{H}^{2d}$, the following conditions are equivalent:*

(i) $\lambda(A) \subset \mathbb{O}(R)$.
(ii) There exists $P \in \mathbb{H}^n$ such that

$$\begin{bmatrix} I_d \otimes A^* & I \end{bmatrix} \begin{bmatrix} R_{11} \otimes P & R_{12} \otimes P \\ R_{12}^* \otimes P & R_{22} \otimes P \end{bmatrix} \begin{bmatrix} I_d \otimes A \\ I \end{bmatrix} \prec 0. \tag{2.35}$$

By convention, we call P the Lyapunov (matrix) variable in (2.35).

Proof In this lemma, the implication (ii) $\Rightarrow$ (i) is straightforward to see. Indeed, if (2.35) holds, then for any eigenvalue-eigenvector pair $(\lambda, \xi) \in \mathbb{C} \times \mathbb{C}^n$ of A we have

$$\begin{aligned}
&(I_d \otimes \xi)^* \begin{bmatrix} I_d \otimes A^* & I \end{bmatrix} \begin{bmatrix} R_{11} \otimes P & R_{12} \otimes P \\ R_{12}^* \otimes P & R_{22} \otimes P \end{bmatrix} \begin{bmatrix} I_d \otimes A \\ I \end{bmatrix} (I_d \otimes \xi) \\
&= \begin{bmatrix} I_d \otimes (\lambda\xi)^* & I_d \otimes \xi^* \end{bmatrix} \begin{bmatrix} R_{11} \otimes P & R_{12} \otimes P \\ R_{12}^* \otimes P & R_{22} \otimes P \end{bmatrix} \begin{bmatrix} I_d \otimes (\lambda\xi) \\ I_d \otimes \xi \end{bmatrix} \\
&= (\xi^* P \xi) \begin{bmatrix} \lambda^* I_d & I_d \end{bmatrix} R \begin{bmatrix} \lambda I_d \\ I_d \end{bmatrix} \prec 0.
\end{aligned}$$

Since $P \succ 0$, we have $\xi^* P \xi > 0$, and hence the above inequality implies

$$\begin{bmatrix} \lambda^* I_d & I_d \end{bmatrix} R \begin{bmatrix} \lambda I_d \\ I_d \end{bmatrix} \prec 0.$$

It follows that $\lambda \in \mathbb{O}(R)$.

The implication (i) $\Rightarrow$ (ii) is more technical. It follows exactly the lines of the proof given in appendix of [5]. $\square$

The LMI condition (2.35) is complex valued in general and P is complex as well. The LMI should be interpreted as follows: a complex Hermitian matrix is negative definite $X \prec 0$ if and only if

$$\begin{bmatrix} \mathrm{Re}(X) & \mathrm{Im}(X) \\ -\mathrm{Im}(X) & \mathrm{Re}(X) \end{bmatrix} \prec 0. \tag{2.36}$$

A special case of the LMI condition (2.35) is when A is real valued. If R is real valued as well, then one can without conservatism look for real-valued P. If A is real valued and R is not, P should be considered complex valued. An alternative is to double the size of the matrix by including the conjugate region (as it is done for the cone in (2.33)) and, by a change of basis to convert it to a region defined by a real-valued matrix (as it is done for the cone in (2.34)).

2.5.1.3 LMI Tests for Robust Pole Location Analysis

Based on Lemma 2.6 one gets the following robust pole location result based on S-variables.

Theorem 2.8 *The uncertain LTI system (2.29) has all its poles located in $\mathbb{O}(R)$ for all $\theta \in \mathbb{E}^L$ if there exist $P_j \in \mathbb{H}^n$ ($j \in \mathscr{I}_L$) and $F_i \in \mathbb{C}^{dn \times dn}$ ($i = 1, 2$) such that*

$$P_j \succ 0,$$
$$\begin{bmatrix} R_{11} \otimes P_j & R_{12} \otimes P_j \\ R_{12}^* \otimes P_j & R_{22} \otimes P_j \end{bmatrix} + \mathrm{He}\left\{ \begin{bmatrix} F_1 \\ F_2 \end{bmatrix} \begin{bmatrix} I & -I_d \otimes A_j \end{bmatrix} \right\} \prec 0 \ (\forall j \in \mathscr{I}_L). \tag{2.37}$$

Proof The result is a direct application of Lemma 2.4. $\square$

By analogy to Corollary 2.4, an SV-LMI-based pole location result can be produced for the dual system $\dot{x}_d = A^T(\theta) x_d$.

Corollary 2.5 *The uncertain LTI system (2.29) has all its poles located in $\mathbb{O}(R)$ for all $\theta \in \mathbb{E}^L$ if there exist $X_j \in \mathbb{H}^n$ ($j \in \mathscr{I}_L$) and $F_i \in \mathbb{C}^{dn \times dn}$ ($i = 1, 2$) such that*

$$X_j \succ 0,$$
$$\begin{bmatrix} R_{11} \otimes X_j & R_{12} \otimes X_j \\ R_{12}^* \otimes X_j & R_{22} \otimes X_j \end{bmatrix} + \mathrm{He}\left\{ \begin{bmatrix} I \\ -I_d \otimes A_j \end{bmatrix} \begin{bmatrix} F_1 & F_2 \end{bmatrix} \right\} \prec 0 \ (\forall j \in \mathscr{I}_L).$$
$$\tag{2.38}$$

It is shown on examples that these two results are not equivalent. The reason for it, as for Hurwitz stability, is that the underlying parameter-dependent Lyapunov matrices are not of the same type (see Theorem 2.6).

The two S-variable results are valid for all regions, even for nonconvex regions for which R_{11} is not positive semi-definite. If such assumption is added, then one gets the following result.

Theorem 2.9 *If $R_{11} \succeq 0$ the uncertain LTI system (2.29) has all its poles located in $\mathbb{O}(R)$ for all $\theta \in \mathbb{E}^L$ if there exist $\Pi \in \mathbb{H}^n$ such that*

$$\Pi \succ 0, \quad \begin{bmatrix} I_d \otimes A_j^* & I \end{bmatrix} \begin{bmatrix} R_{11} \otimes \Pi & R_{12} \otimes \Pi \\ R_{12}^* \otimes \Pi & R_{22} \otimes \Pi \end{bmatrix} \begin{bmatrix} I_d \otimes A_j \\ I \end{bmatrix} \prec 0. \qquad (2.39)$$

Moreover, if (2.39) holds, then (2.37) also holds. A possible choice of variables is $P_j = \Pi$, $F_1 = -(R_{11} + \varepsilon I) \otimes \Pi$, and $F_2 = -R_{12}^* \otimes \Pi$ where $\varepsilon > 0$ is small enough. In case $R_{11} \succ 0$ the choice $\varepsilon = 0$ is admissible.

Proof The results readily follow from Lemma 2.5 with the choice $\mathsf{M}_{11} = R_{11} \otimes \Pi$, $\mathsf{M}_{12} = R_{12} \otimes \Pi$, $\mathsf{M}_{22} = R_{22} \otimes \Pi$, $\mathsf{E} = I$, $\mathsf{A}_j = I_d \otimes A_j$. The S-variables can be obtained if we let $W = \varepsilon I \otimes \Pi \succ 0$ in Lemma 2.5. □

Notice that for $R_{11} = 0$, $R_{12} = 1$ and $R_{22} = 0$ one has $\mathbb{O}(R) = \mathbb{C}_-$ and one recovers the LMI conditions obtained for Hurwitz stability. Theorem 2.9 is the pole location extension of Theorem 2.2. For that reason it is called a "quadratic stability" result. Theorem 2.8 is the pole location extension of Theorem 2.4. Corollary 2.5 is the pole location extension of Corollary 2.4. As for the case of Hurwitz stability, it is guaranteed that Theorem 2.8 and Corollary 2.5 are less conservative than Theorem 2.9.

On page 26, it is shown that identical QMI regions may be described by different choices of the R matrix. Under some conditions, one can have $\mathbb{O}(R) = \mathbb{O}(\tilde{R})$ with $R \neq \tilde{R}$. Moreover, these matrices may be of different dimensions. For results of Lemma 2.6, modifying the matrix that defines the regions has no effect since the LMI conditions are necessary and sufficient. For the S-variables tests of Theorem 2.8 and Corollary 2.5, describing regions with matrices R of enlarged dimensions increases the dimensions of the S-variables, thus the number of decision variables and therefore potentially reduces the conservatism. However, in several systematic ways for modifying the matrix defining the pole location region, such modification does not bring any improvement as illustrated below:

- For all square nonsingular $\check{R}$ one has

$$\mathbb{O}\left(\begin{bmatrix} R_{11} & R_{12} \\ R_{12}^* & R_{22} \end{bmatrix} \right) = \mathbb{O}\left(\begin{bmatrix} \check{R}^* R_{11} \check{R} & \check{R}^* R_{12} \check{R} \\ \check{R}^* R_{12}^* \check{R} & \check{R}^* R_{22} \check{R} \end{bmatrix} \right).$$

This modification is exactly what is applied when going from (2.33) to (2.34) (for the choice $\check{R} = \frac{1}{\sqrt{2}} \begin{bmatrix} 1 & -j \\ j & -1 \end{bmatrix}$). Modifying in this way the matrix that defines the region has no effect on the conservatism of Theorem 2.8. To prove this fact, write (2.37) for the modified matrix:

$$
\begin{bmatrix} \check{R}^* R_{11} \check{R} \otimes P_j & \check{R}^* R_{12} \check{R} \otimes P_j \\ \check{R}^* R_{12}^* \check{R} \otimes P_j & \check{R}^* R_{22} \check{R} \otimes P_j \end{bmatrix} + \mathrm{He} \left\{ \begin{bmatrix} F_1 \\ F_2 \end{bmatrix} \begin{bmatrix} I_{dn} & -I_d \otimes A_j \end{bmatrix} \right\} \prec 0.
$$

(2.40)

Pre- and postmultiply this inequality by $\begin{bmatrix} \check{R}^{-*} \otimes I_n & 0 \\ 0 & \check{R}^{-*} \otimes I_n \end{bmatrix}$ and its transpose conjugate, respectively, due to properties of the Krönecker product it gives

$$
\begin{bmatrix} R_{11} \otimes P_j & R_{12} \otimes P_j \\ R_{12}^* \otimes P_j & R_{22} \otimes P_j \end{bmatrix} + \mathrm{He} \left\{ \begin{bmatrix} \check{F}_1 \\ \check{F}_2 \end{bmatrix} \begin{bmatrix} I_{dn} & -I_d \otimes A_j \end{bmatrix} \right\} \prec 0
$$

(2.41)

with $\check{F}_i = (\check{R}^{-*} \otimes I_n) F_i (\check{R}^{-1} \otimes I_n)$, for $i = 1, 2$. It is exactly the same as the conditions in Theorem 2.8.

- For all positive definite $\hat{R} \succ 0$ one has

$$
\mathbb{O}\left(\begin{bmatrix} R_{11} & R_{12} \\ R_{12}^* & R_{22} \end{bmatrix} \right) = \mathbb{O}\left(\begin{bmatrix} \hat{R} \otimes R_{11} & \hat{R} \otimes R_{12} \\ \hat{R} \otimes R_{12}^* & \hat{R} \otimes R_{22} \end{bmatrix} \right).
$$

(2.42)

The choice of $\hat{R}$ may seem a critical issue. Indeed, one cannot search simultaneously for the P_j matrices and $\hat{R}$. Fortunately, one can always choose $\hat{R} = I_r$ equal to the identity matrix without conservatism. To prove this fact write (2.37) for the increased dimension region with $\hat{R} \in \mathbb{H}^r$:

$$
\begin{bmatrix} \hat{R} \otimes R_{11} \otimes P_j & \hat{R} \otimes R_{12} \otimes P_j \\ \hat{R} \otimes R_{12}^* \otimes P_j & \hat{R} \otimes R_{22} \otimes P_j \end{bmatrix} + \mathrm{He} \left\{ \begin{bmatrix} F_1 \\ F_2 \end{bmatrix} \begin{bmatrix} I_{rdn} & -I_{rd} \otimes A_j \end{bmatrix} \right\} \prec 0.
$$

(2.43)

Pre- and postmultiply this inequality by $\begin{bmatrix} \hat{R}^{-1/2} \otimes I_{dn} & 0 \\ 0 & \hat{R}^{-1/2} \otimes I_{dn} \end{bmatrix}$, due to properties of the Krönecker product it gives:

$$
\begin{bmatrix} I_r \otimes R_{11} \otimes P_j & I_r \otimes R_{12} \otimes P_j \\ I_r \otimes R_{12}^* \otimes P_j & I_r \otimes R_{22} \otimes P_j \end{bmatrix} + \mathrm{He} \left\{ \begin{bmatrix} \hat{F}_1 \\ \hat{F}_2 \end{bmatrix} \begin{bmatrix} I_{rdn} & -I_{rd} \otimes A_j \end{bmatrix} \right\} \prec 0
$$

(2.44)

where $\hat{F}_i = (R^{-1/2} \otimes I_{dn}) F_i (R^{-1/2} \otimes I_{dn})$ for $i = 1, 2$. The LMIs hence hold for $\hat{R}$ replaced by I_r. There is no conservatism reduction brought by the degrees of freedom in $\hat{R}$.

When increasing r it is trivial to see that conditions are at least no more conservative (if the LMIs for a value $\hat{R} = I_{r_1}$ then they imply that these also hold for $\hat{R} = I_{r_2}$ if $r_2 \geq r_1$). Conservatism reduction may be expected thanks to the increased number of decision variables in the S-variables $\hat{F}_i$. Unfortunately it is not. The inequality (2.44) may as well be written (when reordering the rows and columns) as

$$I_r \otimes \begin{bmatrix} R_{11} \otimes P_j & R_{12} \otimes P_j \\ R_{12}^* \otimes P_j & R_{22} \otimes P_j \end{bmatrix} + \mathrm{He}\left\{ \check{F}\left(I_r \otimes \begin{bmatrix} I_{dn} & -I_d \otimes A_j \end{bmatrix}\right)\right\} \prec 0 \quad (2.45)$$

where $\check{F}$ contains the same coefficients as $\begin{bmatrix} \hat{F}_1^T & \hat{F}_2^T \end{bmatrix}$ but reordered accordingly. The upper left block of size $2n \times 2n$ of inequality (2.45) needs to be negative-definite which reads as

$$\begin{bmatrix} R_{11} \otimes P_j & R_{12} \otimes P_j \\ R_{12}^* \otimes P_j & R_{22} \otimes P_j \end{bmatrix} + \mathrm{He}\left\{\begin{bmatrix} \check{F}_{11} \\ \check{F}_{21} \end{bmatrix} \begin{bmatrix} I_{dn} & -I_d \otimes A_j \end{bmatrix}\right\} \prec 0.$$

It is exactly the same as the conditions in Theorem 2.8. There is thus no improvement to be expected when enlarging the dimensions of the matrix defining the region in this way.

- Suppose R_{11} contains one or more positive eigenvalue. It can then be decomposed in $R_{11} = \acute{R}_{11} + L^*L$ and by Schur complement argument one gets

$$\mathbb{O}\left(\begin{bmatrix} R_{11} & R_{12} \\ R_{12}^* & R_{22} \end{bmatrix}\right) = \mathbb{O}\left(\begin{bmatrix} \acute{R}_{11} & 0 & R_{12} & L^* \\ 0 & 0 & 0 & 0 \\ R_{12}^* & 0 & R_{22} & 0 \\ L & 0 & 0 & -I \end{bmatrix}\right). \quad (2.46)$$

This way of modifying the matrix defining the region is exactly what is applied to get the LMI description of convex QMI regions in (2.32). Again, this way of modifying the matrix defining the region does not bring any reduction of conservatism. To observe this fact, consider the condition (2.37) for the modified the matrix defining the region:

$$\begin{bmatrix} \acute{R}_{11} \otimes P_j & 0 & R_{12} \otimes P_j & L^* \otimes P_j \\ 0 & 0 & 0 & 0 \\ R_{12}^* \otimes P_j & 0 & R_{22} \otimes P_j & 0 \\ L \otimes P_j & 0 & 0 & -I \otimes P_j \end{bmatrix}$$

$$+ \mathrm{He}\left\{\begin{bmatrix} F_{11} & F_{12} \\ F_{21} & F_{22} \\ F_{31} & F_{32} \\ F_{41} & F_{42} \end{bmatrix} \begin{bmatrix} I_{dn} & 0 & -I_d \otimes A_j & 0 \\ 0 & I_{dn} & 0 & -I_d \otimes A_j \end{bmatrix}\right\} \prec 0. \quad (2.47)$$

Define the following matrices

$$Z_1^* = (F_{12} + (L^* \otimes I)F_{42})F_{22}^{-1}, \quad Z^* = \begin{bmatrix} I_{dn} & -Z_1^* & 0 & L^* \otimes I_n \\ 0 & -Z_2^* & I_{dn} & 0 \end{bmatrix}.$$
$$Z_2^* = F_{32}F_{22}^{-1}$$

and notice the following facts

$$Z^* \begin{bmatrix} \acute{R}_{11} \otimes P_j & 0 & R_{12} \otimes P_j & L^* \otimes P_j \\ 0 & 0 & 0 & 0 \\ \hline R_{12}^* \otimes P_j & 0 & R_{22} \otimes P_j & 0 \\ L \otimes P_j & 0 & 0 & -I \otimes P_j \end{bmatrix} Z = \begin{bmatrix} R_{11} \otimes P_j & R_{12} \otimes P_j \\ R_{12}^* \otimes P_j & R_{22} \otimes P_j \end{bmatrix}$$

$$Z^* \begin{bmatrix} F_{11} & F_{12} \\ F_{21} & F_{22} \\ F_{31} & F_{32} \\ F_{41} & F_{42} \end{bmatrix} = \begin{bmatrix} F_{11} - Z_1^* F_{21} + (L^* \otimes I_n) F_{41} & 0 \\ -Z_2^* F_{21} + F_{31} & 0 \end{bmatrix} = \begin{bmatrix} \acute{F}_1 & 0 \\ \acute{F}_2 & 0 \end{bmatrix}$$

$$\begin{bmatrix} I_{dn} & 0 & -I_d \otimes A_j & 0 \\ 0 & I_{dn} & 0 & -I_d \otimes A_j \end{bmatrix} Z = \begin{bmatrix} I_{dn} & -I_d \otimes A_j \\ -Z_1 - L \otimes A_j & -Z_2 \end{bmatrix}.$$

Hence pre- and postmultiplying (2.47) by Z^* and Z, respectively, implies the existence of $\acute{F}_{11}$ and $\acute{F}_{21}$ such that

$$\begin{bmatrix} R_{11} \otimes P_j & R_{12} \otimes P_j \\ R_{12}^* \otimes P_j & R_{22} \otimes P_j \end{bmatrix} + \text{He} \left\{ \begin{bmatrix} \acute{F}_1 \\ \acute{F}_2 \end{bmatrix} \begin{bmatrix} I_{dn} & -I_d \otimes A_j \end{bmatrix} \right\} \prec 0.$$

It is exactly the same as the conditions in Theorem 2.8. There is again no improvement in terms of conservatism brought by increasing artificially the dimensions of the matrix defining the region in this way.

2.5.2 Robust H₂ Performance Analysis

Let us consider the LTI system G described by (2.28). Suppose G is stable and strictly proper, i.e., the matrix A is Hurwitz stable and $D = 0$. Then, the H_2 norm of the system G is defined by

$$\|G\|_2 := \sqrt{\frac{1}{2\pi} \int_{-\infty}^{\infty} \text{trace}\{G(j\omega)^* G(j\omega)\} d\omega}. \tag{2.48}$$

From the Parseval relation, this can be rewritten as

$$\|G\|_2 = \sqrt{\int_{-\infty}^{\infty} \text{trace}\{g(t)^T g(t)\} dt} \tag{2.49}$$

where $g(\cdot) : \mathbb{R} \to \mathbb{R}^{n_z \times n_w}$ is the impulse response matrix of the system G defined by

$$g(t) := \begin{cases} C \exp(At)B & t \geq 0, \\ 0 & t < 0. \end{cases} \tag{2.50}$$

It is well known that the H_2 norm is useful if we evaluate the size of the system G in terms of stochastic disturbance input w. Indeed, if the input signal w is a standard white noise, then the squared H_2 norm $\|G\|_2^2$ coincides with the asymptotic mean of the squared norm of the output z [7, pp.237–239].

From (2.48) and (2.50), we can derive an algebraic characterization of the H_2 norm $\|G\|_2$. Indeed, from (2.49) and (2.50), we have

$$\|G\|_2^2 = \int_0^\infty \mathrm{trace}\left\{ B^T \exp(A^T t)C^T C \exp(At)B \right\} = \mathrm{trace}(B^T P_o B)$$

where P_o is the observability gramian of the pair (C, A) defined by

$$P_o := \int_0^\infty \exp(A^T t)C^T C \exp(At)dt.$$

As is well known, the observability gramian P_o can also be characterized as a unique solution for the Lyapunov equation

$$P_o A + A^T P_o + C^T C = 0.$$

It follows that the next lemma holds.

Lemma 2.7 *For the LTI system G described by (2.28), suppose A is Hurwitz stable and $D = 0$. Then, we have*

$$\|G\|_2^2 = \mathrm{trace}(B^T P_o B) \tag{2.51}$$

where P_o is a unique solution for the Lyapunov equation

$$P_o A + A^T P_o + C^T C = 0. \tag{2.52}$$

We next characterize the H_2 norm in terms of LMIs.

Theorem 2.10 *For the LTI system G described by (2.28), suppose $D = 0$. Then, for a given scalar $\bar{\gamma}_2 > 0$, the following conditions are equivalent:*

(i) The matrix A is Hurwitz stable and $\|G\|_2 < \bar{\gamma}_2$.
(ii) There exists $P \in \mathbb{S}_{++}^n$ and $Z \in \mathbb{S}_{++}^{n_w}$ such that

$$PA + A^T P + C^T C \prec 0, \tag{2.53a}$$

$$B^T P B - Z \prec 0 \tag{2.53b}$$

$$\overline{\gamma}_2^2 > \mathrm{trace}(Z). \tag{2.53c}$$

Proof of Theorem 2.10 (i) $\Rightarrow$ (ii) Suppose (i) holds. Then, we have

$$\|G\|_2^2 = \mathrm{trace}(B^T P_o B) < \overline{\gamma}_2^2$$

where P_o is the observability gramian obtained from (2.52). To complete the proof, let us define W as a unique solution for the Lyapunov equation

$$WA + A^T W + I = 0.$$

Since A is Hurwitz, we see that $W \succ 0$. Moreover, let us define a scalar $\beta > 0$ by

$$\beta = \frac{\overline{\gamma}_2^2 - \|G\|_2^2}{2\mathrm{trace}(B^T W B)}.$$

Then, we can confirm

$$\begin{aligned}
(P_o + \beta W)A + A^T (P_o + \beta W) + C^T C &= -\beta I, \\
\mathrm{trace}\left(B^T (P_o + \beta W)B\right) = \|G\|_2^2 + \frac{\overline{\gamma}_2^2 - \|G\|_2^2}{2} &= \frac{\overline{\gamma}_2^2 + \|G\|_2}{2} < \overline{\gamma}_2^2.
\end{aligned} \tag{2.54}$$

It follows that (2.53a)–(2.53c) holds with $P = P_o + \beta W$ and $Z = B^T (P_o + \beta W)B + \varepsilon I$ for $\varepsilon \in (0, \frac{\overline{\gamma}_2^2 - \|G\|_2^2}{2})$.

(ii) $\Rightarrow$ (i) Suppose (2.53a) holds. Then, it is clear that A is Hurwitz stable form Theorem 2.1. Moreover, if we define

$$Q := -(PA + A^T P + C^T C) \succ 0,$$

it is obvious that

$$PA + A^T P + C^T C + Q = 0.$$

Subtracting (2.52) from the above equality, we have

$$(P - P_o)A + A^T (P - P_o) + Q = 0.$$

Since A is stable as noted, the above Lyapunov equality implies $P \succ P_o$. Therefore, we readily conclude

$$\overline{\gamma}_2^2 > \text{trace}(Z) > \text{trace}(B^T P B) \geq \text{trace}(B^T P_o B) = \|G\|_2^2.$$

This completes the proof. □

Based on (2.53a)–(2.53c), one gets the following result on SV-LMI-based robust H_2 performance analysis.

Theorem 2.11 *The uncertain LTI system (2.29) is robustly stable and satisfies the robust H_2 performance $\|G_\theta\|_2 < \overline{\gamma}_2$ ($\forall \theta \in \mathbb{E}^L$) if there exist $P_j \in \mathbb{S}_{++}^n$, $Z_j \in \mathbb{S}_{++}^{n_w}$ ($j \in \mathscr{I}_L$) and F_i ($i = 1, 2, 3, 4$) such that*

$$\begin{bmatrix} 0 & P_j \\ P_j & C_j^T C_j \end{bmatrix} + \text{He}\left\{\begin{bmatrix} F_1 \\ F_2 \end{bmatrix}\begin{bmatrix} I & -A_j \end{bmatrix}\right\} \prec 0 \ (j \in \mathscr{I}_L), \tag{2.55a}$$

$$\begin{bmatrix} P_j & 0 \\ 0 & -Z_j \end{bmatrix} + \text{He}\left\{\begin{bmatrix} F_3 \\ F_4 \end{bmatrix}\begin{bmatrix} I & -B_j \end{bmatrix}\right\} \prec 0 \ (j \in \mathscr{I}_L), \tag{2.55b}$$

$$\overline{\gamma}_2^2 > \text{trace}(Z_j) \ (j \in \mathscr{I}_L). \tag{2.55c}$$

Proof The result is a direct application of Lemma 2.4. Indeed, if we identify (2.55a) and (2.55b) with (2.23), respectively, the LMIs corresponding to (2.24) read as

$$P(\theta)A(\theta) + A(\theta)^T P(\theta) + C(\theta)^T C(\theta) \prec 0, \tag{2.56a}$$

$$B(\theta)^T P(\theta)B(\theta) - Z(\theta) \prec 0, \tag{2.56b}$$

$$\overline{\gamma}_2^2 > \text{trace}(Z(\theta)), \quad \forall \theta \in \mathbb{E}^L \tag{2.56c}$$

where $P(\theta) := \sum_{j=1}^L \theta_j P_j$, $Z(\theta) := \sum_{j=1}^L \theta_j Z_j$. From Theorem 2.10, the above condition (2.56a)–(2.56b) clearly shows that the uncertain LTI system (2.29) is robustly stable and satisfies the robust H_2 performance $\|G_\theta\|_2 < \overline{\gamma}_2$ ($\forall \theta \in \mathbb{E}^L$). □

By analogy to Corollary 2.4 and noticing that $\|G_\theta\|_2 = \|G_\theta^*\|_2$, a second robust S-variable H_2 performance result can be produced for the dual system.

Corollary 2.6 *The uncertain LTI system (2.29) is robustly stable and satisfies the robust H_2 performance $\|G_\theta\|_2 < \overline{\gamma}_2$ ($\forall \theta \in \mathbb{E}^L$) if there exist $X_j \in \mathbb{S}^n$ ($j \in \mathscr{I}_L$) and F_i ($i = 1, 2, 3, 4$) such that*

$$\begin{bmatrix} 0 & X_j \\ X_j & B_j B_j^T \end{bmatrix} + \text{He}\left\{\begin{bmatrix} I \\ -A_j \end{bmatrix}\begin{bmatrix} F_1 & F_2 \end{bmatrix}\right\} \prec 0 \ (j \in \mathscr{I}_L), \tag{2.57a}$$

$$\begin{bmatrix} X_j & 0 \\ 0 & -Z_j \end{bmatrix} + \text{He} \left\{ \begin{bmatrix} I \\ -C_j \end{bmatrix} \begin{bmatrix} F_3 & F_4 \end{bmatrix} \right\} \prec 0 \ (j \in \mathscr{I}_L), \tag{2.57b}$$

$$\overline{\gamma}_2^2 > \text{trace}(Z_j) \ (j \in \mathscr{I}_L). \tag{2.57c}$$

It is shown on examples that these two results are not equivalent. One reason for it is, similarly to the Hurwitz stability cases, that the underlying parameter-dependent Lyapunov matrices are not of the same type (see Theorem 2.6). Another reason comes from the fact that the "quadratic stability" counterparts of these results are not equivalent either.

Theorem 2.12 *The uncertain LTI system (2.29) is robustly stable and satisfies the robust H_2 performance $\|G_\theta\|_2 < \overline{\gamma}_2$ ($\forall \theta \in \mathbb{E}^L$) if there exist $\Pi \in \mathbb{S}^n$ such that*

$$\Pi A_j + A_j^T \Pi + C_j^T C_j \prec 0 \ (j \in \mathscr{I}_L), \tag{2.58a}$$

$$B_j^T \Pi B_j - Z_j \prec 0 \ (j \in \mathscr{I}_L), \tag{2.58b}$$

$$\overline{\gamma}_2^2 > \text{trace}(Z_j) \ (j \in \mathscr{I}_L). \tag{2.58c}$$

Moreover, if (2.58a)–(2.58c) holds, then (2.55a)–(2.55c) also holds. A possible choice of variables is $P_j = \Pi$, $F_1 = -\varepsilon \Pi$, $F_2 = -\Pi$, $F_3 = -\Pi$, and $F_4 = 0$ where $\varepsilon > 0$ is small enough.

Corollary 2.7 *The uncertain LTI system (2.29) is robustly stable and satisfies the robust H_2 performance $\|G_\theta\|_2 < \overline{\gamma}_2$ ($\forall \theta \in \mathbb{E}^L$) if there exist $\Xi \in \mathbb{S}^n$ such that*

$$\Xi A_j^T + A_j \Xi + B_j B_j^T \prec 0 \ (j \in \mathscr{I}_L), \tag{2.59a}$$

$$C_j \Xi C_j^T - Z_j \prec 0 \ (j \in \mathscr{I}_L), \tag{2.59b}$$

$$\overline{\gamma}_2^2 > \text{trace}(Z_j) \ (j \in \mathscr{I}_L). \tag{2.59c}$$

Moreover, if (2.59a)–(2.59c) holds, then (2.57a)–(2.57c) also holds. A possible choice of variables is $X_j = \Xi$, $F_1 = -\varepsilon \Xi$, $F_2 = -\Xi$, $F_3 = -\Xi$, and $F_4 = 0$ where $\varepsilon > 0$ is small enough.

The LMI in Theorem 2.12 readily follows from Theorem 2.10 with simple convexity arguments. The latter assertion of this theorem is a direct consequence of Lemma 2.5. Note that the S-variable LMI in Theorem 2.11 is no more conservative than the LMI in Theorem 2.12 due to this assertion. Similar comments apply also to the LMIs in Corollary 2.7–2.6.

2.5.3 Robust H_∞ Performance Analysis

Suppose that the LTI system G described by (2.28) is stable (i.e., the matrix A is Hurwitz stable). Then, the H_∞ norm of the system G is defined by

$$\|G\|_\infty := \sup_{\omega \in \mathbb{R}} \|G(j\omega)\|. \tag{2.60}$$

The H_∞ norm plays an important role in robust control theory. The cerebrated small gain theorem ensures that the feedback connection with stable LTI systems G_1 and G_2 is stable if $\|G_1\|_\infty \|G_2\|_\infty < 1$ holds. It follows that if $\|G\|_\infty < \gamma_\infty$, then the feedback connection with G and an uncertain stable LTI system Δ is stable whenever $\|\Delta\|_\infty < 1/\gamma_\infty$. In this way, the H_∞ norm serves as a reasonable measure for robust stability of feedback systems against unstructured uncertainties. Another important interpretation of the H_∞ norm is that it coincides with the L_2 induced norm of the system G. Namely, we have

$$\|G\|_\infty = \sup_{w \in L_2, \|w\|_2 \neq 0} \frac{\|z\|_2}{\|w\|_2}. \tag{2.61}$$

Usually, when evaluating the H_∞ norm of the system G, the disturbance input w and the performance output z are chosen so that the good performance of the system G can be translated to the insensitiveness of z with respect to w. It follows that we can ensure good performance if $\|G\|_\infty$ is small enough.

From the definition (2.60), we see that $\|G\|_\infty < \overline{\gamma}_\infty$ holds for given $\overline{\gamma}_\infty > 0$ if and only if

$$G(j\omega)^* G(j\omega) \prec \overline{\gamma}_\infty^2 I \ (\forall \omega \in (\mathbb{R} \cup \{\infty\}))$$

or equivalently,

$$\begin{bmatrix} (j\omega I - A)^{-1}B \\ I \end{bmatrix}^* \hat{\Theta} \begin{bmatrix} (j\omega I - A)^{-1}B \\ I \end{bmatrix} \prec 0 \ (\forall \omega \in (\mathbb{R} \cup \{\infty\})), \tag{2.62}$$

$$\hat{\Theta} = \begin{bmatrix} C & D \\ 0 & I_{n_w} \end{bmatrix}^T \Theta_{\mathrm{BD}} \begin{bmatrix} C & D \\ 0 & I_{n_w} \end{bmatrix}, \quad \Theta_{\mathrm{BD}} := \begin{bmatrix} I & 0 \\ 0 & -\overline{\gamma}_\infty^2 I \end{bmatrix} \tag{2.63}$$

It is well recognized that numerous performances of LTI systems defined in the frequency domain can be characterized in the form of (2.62). For example, if we set

$$\hat{\Theta} := \begin{bmatrix} C & D \\ 0 & I_{n_w} \end{bmatrix}^T \Theta_{\mathrm{PR}} \begin{bmatrix} C & D \\ 0 & I_{n_w} \end{bmatrix}, \quad \Theta_{\mathrm{PR}} := \begin{bmatrix} 0 & -I \\ -I & 0 \end{bmatrix}$$

then the corresponding condition becomes

$$G(j\omega) + G(j\omega)^* \succ 0 \ (\forall \omega \in (\mathbb{R} \cup \{\infty\})).$$

This is nothing but the positive realness condition for the system G.

The condition (2.62) is often called *frequency domain inequality (FDI)*. The FDI condition is apparently hard to check since it involves semi-infinite constraint (i.e., we have to take care of all ω belonging to $\mathbb{R} \cup \{\infty\}$). The celebrated Kalman–Yakubovich–Popov (KYP) lemma successfully remove this semi-infinite constraint and translate the FDI into a finite-dimensional LMI.

Lemma 2.8 (KYP Lemma)[8] *For given $A \in \mathbb{R}^{n \times n}$, $B \in \mathbb{R}^{n \times n_w}$ and*

$$\hat{\Theta} = \begin{bmatrix} \hat{\Theta}_{11} & \hat{\Theta}_{12} \\ \hat{\Theta}_{12}^T & \hat{\Theta}_{22} \end{bmatrix} \in \mathbb{S}_{n+n_w}, \quad \hat{\Theta}_{11} \in \mathbb{S}_n, \quad \hat{\Theta}_{22} \in \mathbb{S}_{n_w},$$

suppose $\hat{\Theta}_{11} \succeq 0$. Then, the following conditions are equivalent.

(i) *The matrix A is Hurwitz stable and the FDI (2.62) holds.*
(ii) *There exists $P \in \mathbb{S}_{++}^n$ such that*

$$\begin{bmatrix} PA + A^T P & PB \\ B^T P & 0 \end{bmatrix} + \hat{\Theta} \prec 0. \tag{2.64}$$

The condition (ii) is easy to check numerically since (2.64) is a finite-dimensional LMI. For the proof of (i) $\Rightarrow$ (ii), we need some mathematical preliminaries [8]. However, the proof of (ii) $\Rightarrow$ (i) is fairly easy. Indeed, if (2.64) holds, then we have $PA + A^T P \prec 0$ since $\hat{\Theta}_{11} \succeq 0$. Since $P \in \mathbb{S}_{++}^n$, this implies that A is Hurwitz stable. On the other hand, (2.64) can be rewritten equivalently as

$$\text{He}\left\{ \begin{bmatrix} P \\ 0 \end{bmatrix} \begin{bmatrix} (A - j\omega I) & B \end{bmatrix} \right\} + \hat{\Theta} \prec 0.$$

where $\omega \in \mathbb{R}$ is arbitrary. Since A is Hurwitz stable, we see $j\omega I - A$ is nonsingular, and hence we have

$$\begin{bmatrix} (j\omega I - A)^{-1} B \\ I \end{bmatrix}^* \left(\text{He}\left\{ \begin{bmatrix} P \\ 0 \end{bmatrix} \begin{bmatrix} (A - j\omega I) & B \end{bmatrix} \right\} + \hat{\Theta} \right) \begin{bmatrix} (j\omega I - A)^{-1} B \\ I \end{bmatrix}$$
$$= \begin{bmatrix} (j\omega I - A)^{-1} B \\ I \end{bmatrix}^* \hat{\Theta} \begin{bmatrix} (j\omega I - A)^{-1} B \\ I \end{bmatrix} \prec 0 \ (\forall \omega \in \mathbb{R}).$$

Finally, if (2.64) holds, then it is obvious that $\hat{\Theta}_{22} \prec 0$ and hence the FDI (2.62) holds for $\omega = \infty$.

We next provides a robust SV-LMI corresponding to (2.64) for the case when $\hat{\Theta}$ is structured as follows:

$$\hat{\Theta} = \begin{bmatrix} C & D \\ 0 & I_{n_w} \end{bmatrix}^T \Theta \begin{bmatrix} C & D \\ 0 & I_{n_w} \end{bmatrix}, \quad \Theta = \begin{bmatrix} \Theta_{11} & \Theta_{12} \\ \Theta_{12}^T & \Theta_{22} \end{bmatrix}, \quad \Theta_{11} \succeq 0. \quad (2.65)$$

It trivially includes the case when $\hat{\Theta}$ has no structure since it corresponds to $C = I$ and $D = 0$.

Theorem 2.13 *The uncertain LTI system* (2.29) *is robustly stable and the FDI* (2.62) *holds for all* $\theta \in \mathbb{E}^L$ *if there exist* $P_j \in \mathbb{S}^n$ ($j \in \mathcal{I}_L$) *and* F_i ($i = 1, 2, 3$) *such that*

$$\begin{bmatrix} 0 & P_j & 0 \\ P_j & \hat{\Theta}_{11j} & \hat{\Theta}_{12j} \\ 0 & \hat{\Theta}_{12j}^T & \hat{\Theta}_{22j} \end{bmatrix} + \mathrm{He}\left\{ \begin{bmatrix} F_1 \\ F_2 \\ F_3 \end{bmatrix} \begin{bmatrix} I & -A_j & -B_j \end{bmatrix} \right\} \prec 0 \ (j \in \mathcal{I}_L) \quad (2.66)$$

where $\hat{\Theta}_j := \begin{bmatrix} \hat{\Theta}_{11j} & \hat{\Theta}_{12j} \\ \hat{\Theta}_{12j}^T & \hat{\Theta}_{22j} \end{bmatrix} = \begin{bmatrix} C_j & D_j \\ 0 & I \end{bmatrix}^T \Theta \begin{bmatrix} C_j & D_j \\ 0 & I \end{bmatrix}.$

Proof The result is a direct application of Lemma 2.4. □

The "quadratic stability" counterpart of this result can be given as follows.

Theorem 2.14 *The uncertain LTI system* (2.29) *is robustly stable and the FDI* (2.62) *holds for all* $\theta \in \mathbb{E}^L$ *if there exist* $\Pi \in \mathbb{S}^n$ *such that for all* $j \in \mathcal{I}_L$

$$\begin{bmatrix} \Pi A_j + A_j^T \Pi & \Pi B_j \\ B_j^T \Pi & 0 \end{bmatrix} + \hat{\Theta}_j \prec 0. \quad (2.67)$$

Moreover, if (2.67) *holds, then* (2.66) *also holds. A possible choice of variables is* $P_j = \Pi$, $F_1 = -\varepsilon \Pi$, $F_2 = -\Pi$, *and* $F_3 = 0$ *where* $\varepsilon > 0$ *is small enough.*

Proof The LMI (2.67) readily follows from (2.64) with simple convexity arguments. The latter result is a direct application of Lemma 2.5. □

Dual versions of the above two theorems are accessible thanks to the following technical result. That applies assuming that diagonal blocks of Θ satisfy certain conditions.

Lemma 2.9 *The two following conditions on G are equivalent*

$$\begin{bmatrix} G \\ I \end{bmatrix}^* \begin{bmatrix} \Theta_{11} & \Theta_{12} \\ \Theta_{12}^* & \Theta_{22} \end{bmatrix} \begin{bmatrix} G \\ I \end{bmatrix} \prec 0 \quad (2.68)$$

$$\begin{bmatrix} G^* \\ I \end{bmatrix}^* \begin{bmatrix} \Theta_{11d} & \Theta_{12d} \\ \Theta_{12d}^* & \Theta_{22d} \end{bmatrix} \begin{bmatrix} G^* \\ I \end{bmatrix} \prec 0 \quad (2.69)$$

where

(i) *If* $\Theta_{11} \succ 0$

$$\Theta_{11d} = (\Theta_{12}^*\Theta_{11}^{-1}\Theta_{12} - \Theta_{22})^{-1}$$

$$\Theta_d = \begin{bmatrix} \Theta_{11d} & \Theta_{12d} \\ \Theta_{12d}^* & \Theta_{22d} \end{bmatrix} = \begin{bmatrix} \Theta_{11d} & \Theta_{11d}\Theta_{12}^*\Theta_{11}^{-1} \\ \Theta_{11}^{-1}\Theta_{12}\Theta_{11d} & \Theta_{11}^{-1}\Theta_{12}\Theta_{11d}\Theta_{12}^*\Theta_{11}^{-1} - \Theta_{11}^{-1} \end{bmatrix}$$

(ii) *If* $\Theta_{11} = 0$ *and* Θ_{12} *is square non singular:*

$$\Theta_d = \begin{bmatrix} \Theta_{11d} & \Theta_{12d} \\ \Theta_{12d}^* & \Theta_{22d} \end{bmatrix} = \begin{bmatrix} 0 & \Theta_{12}^{-1} \\ \Theta_{12}^{-*} & \Theta_{12}^{-1}\Theta_{22}\Theta_{12}^{-*} \end{bmatrix}$$

Proof Let us start with the simplest case (ii). Condition (2.68) also reads as

$$G^*\Theta_{12} + \Theta_{12}^*G + \Theta_{22} \prec 0.$$

Pre- and postmultiply by the invertible matrix Θ_{12}^{-*} and its transpose, respectively. It gives exactly (2.69) with the given Θ_d matrix.

Now let us consider the case (i). Condition (2.68) reads as

$$G^*\Theta_{11}G + G^*\Theta_{12} + \Theta_{12}^*G + \Theta_{22} \prec 0.$$

Since Θ_{11} is assumed to positive definite, it is quite simple to show that this inequality can also be written as

$$(G^*\Theta_{11} + \Theta_{12}^*)\Theta_{11}^{-1}(\Theta_{11}G + \Theta_{12}) + \Theta_{22} - \Theta_{12}^*\Theta_{11}^{-1}\Theta_{12} \prec 0.$$

Since again Θ_{11} is positive definite, we can apply Schur complement argument to the above inequality to get

$$\begin{bmatrix} \Theta_{22} - \Theta_{12}^*\Theta_{11}^{-1}\Theta_{12} & G^*\Theta_{11} + \Theta_{12}^* \\ \Theta_{11}G + \Theta_{12} & -\Theta_{11} \end{bmatrix} \prec 0.$$

A converse Schur complement argument applied to the upper left block gives

$$-\Theta_{11} - (\Theta_{11}G + \Theta_{12})(\Theta_{22} - \Theta_{12}^*\Theta_{11}^{-1}\Theta_{12})^{-1}(G^*\Theta_{11} + \Theta_{12}^*) \prec 0.$$

Pre- and postmultiply this inequality by Θ_{11}^{-1} and develop the expression to get (2.69) with the given Θ_d matrix.   □

In case one of the two conditions in Lemma 2.9 is applicable, then the following dual versions of Theorem 2.13 and Theorem 2.14 are obtained.

Corollary 2.8 *The uncertain LTI system (2.29) is robustly stable and the FDI (2.62) holds for all* $\theta \in \mathbb{E}^L$ *if there exist* $X_j \in \mathbb{S}^n$ ($j \in \mathscr{I}_L$) *and* F_i ($i = 1, 2, 3$) *such that*

$$\begin{bmatrix} 0 & X_j & 0 \\ X_j & \hat{\Theta}_{11j} & \hat{\Theta}_{12j} \\ 0 & \hat{\Theta}_{12j}^T & \hat{\Theta}_{22j} \end{bmatrix} + \text{He} \left\{ \begin{bmatrix} I \\ -A_j \\ -C_j \end{bmatrix} [F_1 \; F_2 \; F_3] \right\} \prec 0 \; (j \in \mathscr{I}_L) \qquad (2.70)$$

$$\text{where } \begin{bmatrix} \hat{\Theta}_{11j} & \hat{\Theta}_{12j} \\ \hat{\Theta}_{12j}^T & \hat{\Theta}_{22j} \end{bmatrix} = \begin{bmatrix} B_j & 0 \\ D_j & I \end{bmatrix} \Theta_d \begin{bmatrix} B_j & 0 \\ D_j & I \end{bmatrix}^T.$$

Corollary 2.9 *The uncertain LTI system* (2.29) *is robustly stable and the FDI* (2.62) *holds for all* $\theta \in \mathbb{E}^L$ *if there exist* $\Xi \in \mathbb{S}^n$ *such that for all* $j \in \mathscr{I}_L$

$$\begin{bmatrix} \Xi A_j^T + A_j \Xi & \Xi C_j^T \\ C_j \Xi & 0 \end{bmatrix} + \hat{\Theta}_j \prec 0, \quad \hat{\Theta}_j = \begin{bmatrix} B_j & 0 \\ D_j & I \end{bmatrix} \Theta_d \begin{bmatrix} B_j & 0 \\ D_j & I \end{bmatrix}^T. \qquad (2.71)$$

Moreover, if (2.71) *holds, then* (2.70) *also holds. A possible choice of variables is* $X_j = \Xi$, $F_1 = -\varepsilon \Xi$, $F_2 = -\Xi$, *and* $F_3 = 0$ *where* $\varepsilon > 0$ *is small enough.*

2.5.4   Robust Impulse-To-Peak Performance Analysis

Suppose the LTI system G described by (2.28) is stable (i.e., the matrix A is Hurwitz stable) and $D = 0$. In this subsection, we focus on the value

$$\gamma_{IP} := \sup_{t \geq 0, \|\alpha\| = 1} \|z(t)\| \qquad (2.72)$$

where $z(t)$ is the response of (2.28) to an impulse input $w = \delta(t)\alpha$, $\alpha \in \mathbb{R}^{n_w}$. The value γ_{IP} defined by (2.72) is the peak (supremum) of the performance output measured in Euclidean norm with respect to all unitary impulse inputs. It should be noted that z can also be characterized as the output of the input-free system described by

$$\begin{cases} \dot{x}(t) = A x(t), \; x(0) = x_0 = B\alpha, \\ z(t) = C x(t). \end{cases} \qquad (2.73)$$

For given $\overline{\gamma}_{IP} > 0$, a sufficient condition to guarantee $\gamma_{IP} < \overline{\gamma}_{IP}$ can be given in terms of LMIs [2].

Theorem 2.15 *For the LTI system* G *described by* (2.28), *suppose* $D = 0$. *The matrix* A *is stable and* $\gamma_{IP} < \overline{\gamma}_{IP}$ *holds if there exists* $P \in \mathbb{S}_{++}^n$ *such that*

$$PA + A^T P \prec 0, \qquad (2.74a)$$

$$B^T P B - \overline{\gamma}_{IP}^2 I_{n_w} \prec 0, \qquad (2.74b)$$

$$C^T C - P \prec 0. \qquad (2.74c)$$

Proof Suppose (2.74a)–(2.74c) holds with $P \in \mathbb{S}^n_{++}$. From (2.74a), the matrix A is Hurwitz stable. Moreover, if define $g_P(t) := P^{\frac{1}{2}} \exp(At) B\alpha$ for given $\alpha \in \mathbb{R}^{n_w}$ with $\|\alpha\| = 1$, we can readily see from (2.74c) that

$$\|z(t)\| \leq \|g_P(t)\| \quad (\forall t \geq 0). \tag{2.75}$$

With this in mind, let us focus on g_P. From (2.74b), we first obtain

$$\|g_P(0)\| = \|P^{\frac{1}{2}} B\alpha\| < \overline{\gamma}_{IP} \|\alpha\| = \overline{\gamma}_{IP}. \tag{2.76}$$

Moreover, for all $t > 0$, we have

$$\begin{aligned}
\frac{d\|g_P(t)\|^2}{dt} &= \frac{d(\alpha^T B^T \exp(A^T t) P \exp(At) B\alpha)}{dt} \\
&= \alpha^T B^T \exp(A^T t)(PA + A^T P) \exp(At) B\alpha \\
&\leq 0
\end{aligned} \tag{2.77}$$

since (2.74a) holds. From (2.76) and (2.77), we have $\|g_P(t)\| < \overline{\gamma}_{IP}$ ($\forall t \geq 0$). Since (2.75) holds, we can conclude that $\|z(t)\| < \overline{\gamma}_{IP}$ ($\forall t \geq 0$) holds as well. □

In contrast with other performances discussed in the preceding subsections, the LMI characterization in Theorem 2.15 is sufficient for $\gamma_{IP} < \overline{\gamma}_{IP}$. The LMI condition (2.74a)–(2.74c) is somewhat complicated, but the corresponding SV-LMIs can be derived without any further ado. This is done in the following. Before that we illustrate the usefulness of impulse-to-peak performance in characterizing time responses for bounded initial conditions.

Ensuring adequate impulse-to-peak performance (i.e., keeping γ_{IP} small enough) is often very important. In some practical applications, the state vector must be kept "small" to avoid undesirable phenomena such as saturation, etc. Since the magnitude of the state vector of course depends on the initial condition, it is reasonable to analyze whether

$$z(t)^T W_z z(t) \leq 1 \ \forall (t, x_0) \in (\mathbb{R}_+, \mathbb{X}_0), \quad \mathbb{X}_0 := \{x_0 \in \mathbb{R}^n : x_0^T W_0 x_0 \leq 1\} \tag{2.78}$$

holds for (2.73), where $W_z \in \mathbb{S}^{n_z}_{++}$ and $W_0 \in \mathbb{S}^n_{++}$ are appropriately chosen weighting matrices. A sufficient condition for (2.78) readily follows from Theorem 2.15.

Corollary 2.10 *Suppose $W_z \in \mathbb{S}^{n_z}_{++}$ and $W_0 \in \mathbb{S}^n_{++}$ are given. Then, for the LTI system (2.73), the matrix A is stable and the condition (2.78) holds if there exists $P \in \mathbb{S}^n_{++}$ such that*

$$PA + A^T P \prec 0, \tag{2.79a}$$

$$P - W_0 \prec 0, \tag{2.79b}$$

$$C^T W_z C - P \prec 0. \tag{2.79c}$$

Proof of Corollary 2.10 Again, the condition (2.79a) implies the Hurwitz stability of A. As for the condition (2.78), we first note that $x_0 \in \mathbb{X}_0$ can always be represented as $x_0 = W_0^{-\frac{1}{2}} v$ for some $\|v\| \leq 1$. If $x_0 = 0$, the condition (2.78) is obviously satisfied, and therefore we consider the case where $x_0 \neq 0$ (and hence $v \neq 0$). From (2.79b) we readily obtain

$$x_0^T W_0 x_0 > x_0^T P x_0 \quad \Leftrightarrow \quad v^T v > x_0^T P x_0 \quad \Rightarrow \quad 1 > x_0^T P x_0 \; (\forall x_0 \in \mathbb{X}_0).$$

From (2.79a), the above last inequality and (2.79c), the assertion of the corollary readily follows by Theorem 2.15.   □

We next consider briefly the robust impulse-to-peak performance analysis for the polytopic uncertain LTI system (2.29). The issue is to determine whether the following two robustness properties hold or not:

(A) $\lambda(A(\theta)) \subset \mathbb{C}_- \; (\forall \theta \in \mathbb{E}^L)$,
(B) $\|z_\theta(t)\| < \overline{\gamma}_{\text{IP}} \; (\forall (t, \theta) \in \mathbb{R}_+ \times \mathbb{E}^L)$.

Here, z_θ is the response matrix to an impulse input $w = \delta(t)\alpha, \alpha \in \mathbb{R}^{n_w}$ with $\|\alpha\| = 1$ of the plant (2.29).

Theorem 2.16 *The uncertain LTI system (2.29) is robustly stable and the robust impulse-to-peak performance* (B) *holds if there exist* $P_j \in \mathbb{S}^n \; (j \in \mathscr{I}_L)$ *and* $F_i \; (i = 1, 2, 3, 4)$ *such that*

$$\begin{bmatrix} 0 & P_j \\ P_j & 0 \end{bmatrix} + \text{He} \left\{ \begin{bmatrix} F_1 \\ F_2 \end{bmatrix} \begin{bmatrix} I & -A_j \end{bmatrix} \right\} \prec 0 \; (j \in \mathscr{I}_L). \tag{2.80a}$$

$$\begin{bmatrix} P_j & 0 \\ 0 & -\overline{\gamma}_{IP}^2 I_{n_w} \end{bmatrix} + \text{He} \left\{ \begin{bmatrix} F_3 \\ F_4 \end{bmatrix} \begin{bmatrix} I & -B_j \end{bmatrix} \right\} \prec 0 \; (j \in \mathscr{I}_L) \tag{2.80b}$$

$$C_j^T C_j - P_j \prec 0 \; (j \in \mathscr{I}_L). \tag{2.80c}$$

Proof The result is a direct application of Lemma 2.4.   □

The "quadratic stability" counterpart of this result can be given in the following. The latter assertion of the next theorem follows from Lemma 2.5.

Theorem 2.17 *The uncertain LTI system (2.29) is robustly stable and the robust impulse-to-peak performance* (B) *holds if there exist* $\Pi \in \mathbb{S}^n$ *such that for all* $j \in \mathscr{I}_L$

$$\Pi A_j + A_j^T \Pi \prec 0, \tag{2.81a}$$

$$B_j^T \Pi B_j - \overline{\gamma}_{IP}^2 I_{n_w} \prec 0, \tag{2.81b}$$

$$C_j^T C_j - \Pi \prec 0. \tag{2.81c}$$

Moreover, if (2.81a)–(2.81c) holds, then (2.80a)–(2.80c) also holds. A possible choice of variables is $P_j = \Pi$, $F_1 = -\varepsilon\Pi$, $F_2 = -\Pi$, $F_3 = -\Pi$, and $F_4 = 0$ where $\varepsilon > 0$ is small enough.

We shall next consider the results for the dual system.

Lemma 2.10 *For the LTI system G described by (2.28), suppose $D = 0$. The matrix A is stable and $\gamma_{IP} < \overline{\gamma}_{IP}$ holds if there exists $X \in \mathbb{S}^n_{++}$ such that*

$$XA^T + AX \prec 0, \tag{2.82a}$$

$$CXC^T - \overline{\gamma}^2_{IP}I_{n_z} \prec 0, \tag{2.82b}$$

$$BB^T - X \prec 0. \tag{2.82c}$$

Moreover, if (2.74a)–(2.74c) hold for some matrix P, then (2.82a)–(2.82c) also hold and $X = \overline{\gamma}^2_{IP}P^{-1}$ is a possible choice.

Proof We shall prove that (2.82a)–(2.82c) with $X = \overline{\gamma}^2_{IP}P^{-1}$ follows immediately from (2.74a)–(2.74c). Pre- and postmultiply (2.74a) by $X = \overline{\gamma}^2_{IP}P^{-1}$ gives exactly (2.82a). Divide (2.74b) and (2.74c) by $\overline{\gamma}^{-2}_{IP}$, it gives

$$B^T X^{-1}B - I_{n_w} \prec 0, \qquad C^T(\overline{\gamma}^{-2}_{IP}I_{n_z})^{-1}C - X^{-1} \prec 0.$$

Apply a Schur complement argument to these inequalities to get

$$\begin{bmatrix} -X & B \\ B^T & -I_{n_w} \end{bmatrix} \prec 0, \qquad \begin{bmatrix} -X^{-1} & C^T \\ C & -\overline{\gamma}^2_{IP}I_{n_z} \end{bmatrix} \prec 0$$

Converse Schur complement argument concludes that (2.82c) and (2.82b) hold. □

Based on Lemma 2.10 the dual counterparts of Theorem 2.16 and Theorem 2.17 are derived.

Theorem 2.18 *The uncertain LTI system (2.29) is robustly stable and the robust impulse-to-peak performance (B) holds if there exist $X_j \in \mathbb{S}^n$ ($j \in \mathcal{I}_L$) and F_i ($i = 1, 2, 3, 4$) such that*

$$\begin{bmatrix} 0 & X_j \\ X_j & 0 \end{bmatrix} + \mathrm{He}\left\{ \begin{bmatrix} I \\ -A_j \end{bmatrix} \begin{bmatrix} F_1 & F_2 \end{bmatrix} \right\} \prec 0 \, (j \in \mathcal{I}_L). \tag{2.83a}$$

$$\begin{bmatrix} X_j & 0 \\ 0 & -\overline{\gamma}^2_{IP}I_{n_z} \end{bmatrix} + \mathrm{He}\left\{ \begin{bmatrix} I \\ -C_j \end{bmatrix} \begin{bmatrix} F_3 & F_4 \end{bmatrix} \right\} \prec 0 \, (j \in \mathcal{I}_L) \tag{2.83b}$$

$$B_j B_j^T - X_j \prec 0 \, (j \in \mathcal{I}_L). \tag{2.83c}$$

Theorem 2.19 *The uncertain LTI system* (2.29) *is robustly stable and the robust impulse-to-peak performance* (B) *holds if there exist* $\Pi \in \mathbb{S}^n$ *such that for all* $j \in \mathscr{I}_L$

$$\Xi A_j^T + A_j \Xi \prec 0, \tag{2.84a}$$

$$C_j \Xi C_j - \bar{\gamma}_{IP}^2 I_{n_z} \prec 0, \tag{2.84b}$$

$$B_j B_j^T - \Xi \prec 0. \tag{2.84c}$$

Moreover, if (2.84a)–(2.84c) *holds, then* (2.83a)–(2.83c) *also holds. A possible choice of variables is* $X_j = \Xi$, $F_1 = -\varepsilon\Xi$, $F_2 = -\Xi$, $F_3 = -\Xi$, $F_4 = 0$ *where* $\varepsilon > 0$ *is small enough.*

Once again, there are no implications in either direction between Theorem 2.16, Theorem 2.17 and their dual counterparts, Theorem 2.18 and Theorem 2.19. The only implications are that the S-variable results of Theorem 2.16 and Theorem 2.18 are less conservative (no more conservative) than the "quadratic stability" counterparts of Theorem 2.17 and Theorem 2.19, respectively.

2.6 Numerical Examples

In this section, we illustrate the effectiveness of SV-LMIs for robust performance analysis problems for polytopic uncertain LTI systems.

2.6.1 Quarter-Car Suspension Example

2.6.1.1 Definition of the Uncertain System

Let us revisit the quarter-car suspension model discussed in Sect. 2.2. By following [9], we let the following nominal values of the parameters

$$M_{nom} = 320\,\text{Kg}, \; m = 40\,\text{Kg},$$
$$k_1 = 180\,\text{KN/m}, \; k_{2,nom} = 200\,\text{KN/m}, \; c = 1000\,\text{Ns/m}$$

The main objectives for the suspension system design are to achieve (a) good ride comfort, (b) good road-holding ability, and (c) small suspension deflection. To fulfill these specifications, the following generalized plant is defined:

$$\begin{cases} \dot{x}(t) = A(M, k_2)x(t) + B_1(k_2)w(t) + B_2(M)u(t), \\ z(t) = C_1(M)x(t) \qquad\qquad\qquad + D_{12}(M)u(t). \end{cases}$$

Here, $A(M, k_2)$ and $B_2(M)$ are as in (2.7)

$$
A(M, k_2) := \begin{bmatrix} 0 & 1 & 0 & 0 \\ -\dfrac{k_1}{M} & -\dfrac{c}{M} & \dfrac{k_1}{M} & \dfrac{c}{M} \\ 0 & 0 & 0 & 1 \\ \dfrac{k_1}{m} & \dfrac{c}{m} & -\dfrac{k_1+k_2}{m} & -\dfrac{c}{m} \end{bmatrix}, \quad B_2(M) := \begin{bmatrix} 0 \\ \dfrac{1}{M} \\ 0 \\ -\dfrac{1}{m} \end{bmatrix}
$$

and input/output performance signals are defined by

$$
B_1(k_2) := \begin{bmatrix} 0 \\ 0 \\ 0 \\ \dfrac{k_2}{m} \end{bmatrix}, \quad C_1(M) := \begin{bmatrix} -\dfrac{k_1}{M} & -\dfrac{c}{M} & \dfrac{k_1}{M} & \dfrac{c}{M} \\ W_{z_c} & 0 & 0 & 0 \\ 0 & 0 & 0 & 0 \end{bmatrix}, \quad D_{12}(M) := \begin{bmatrix} \dfrac{1}{M} \\ 0 \\ W_u \end{bmatrix}.
$$

In this generalized plant, the disturbance input w corresponds to the deviation of the ground height, whereas the performance output z is nothing but $z = [\ddot{z}_c \; W_{z_c} z_c \; W_u u]^T$. Here, $W_{z_c} = 2.5 > 0$ and $W_u = 0.15 > 0$ are weightings for the position z_c and the input u. Therefore, the H_∞ norm of the transfer $w \to z$ characterizes a tradeoff between the three required specifications (a), (b), and (c).

The considered uncertainties are as follows. The weight of the chassis can be increased by 20% from its nominal value. The wheel stiffness can have up to 10% discrepancy around the nominal value.

$$
M_{min} = M_{nom} = 320, \qquad M_{max} = 1.2 \cdot M_{nom} = 384,
$$
$$
k_{2,min} = 0.9 \cdot k_{2,nom} = 180, \; k_{2,max} = 1.1 \cdot k_{2,nom} = 220.
$$

2.6.1.2 H_∞ Performance

First, we study the open-loop system. The nominal system is stable and its H_∞ norm is 13.080. The H_∞ norm computed on the vertices of the polytope (the four extremal combinations of the uncertainties) gives the following values 12.433, 13.814, 11.172, and 12.464. The value 13.814 obtained for $M = M_{min}$ and $k_2 = k_{2,min}$ may be a worst-case value of the H_∞ norm of the open-loop system, but some combination of uncertainties may as well give some larger value of the H_∞ norm, or even destabilize the system. To guarantee robustly some upper-bound on the H_∞ norm we apply results of Theorem 2.14, that is, we seek for a unique quadratic Lyapunov function for all uncertainties. The obtained optimum of this LMI problem is 88.978. It is an upper bound on the robust H_∞ performance. Still, there is clearly a huge gap between this upper bound and the values computed on vertices. In order to reduce the gap, we apply SV-LMI results of Theorem 2.13 which are guaranteed to be no more conservative. The obtained optimum of the SV-LMI problem is 13.814. Since it coincides with the H_∞ norm computed for $M = M_{min}$ and $k_2 = k_{2,min}$, we can

Table 2.1 H_∞ performance of the quarter-car suspension

	Nominal	Worst Vertex	SV-LMI (2.66)	Dual-SV-LMI (2.70)	Quad-LMI (2.67)	Dual-quad-LMI (2.71)
Open-loop	13.080	13.814	13.814	13.814	88.978	88.990
Closed-loop	9.710	9.859	9.859	9.859	18.497	18.497

conclude that

- 13.814 is the worst-case H_∞ norm and
- the SV-LMIs of Theorem 2.13 are nonconservative on this example.

The same study is now performed for the closed-loop system with the following state-feedback control

$$u = \begin{bmatrix} 7 & 21 & -8 & 3 \end{bmatrix} x.$$

For the nominal values of the parameters the H_∞ norm is 9.710. This state-feedback control improves the performance of the system. Computed on the vertices the H_∞ norm takes the following values 9.027, 9.539, 9.162, 9.859. The value 9.859 obtained for $M = M_{min}$ and $k_2 = k_{2,max}$ may be a worst-case value of the H_∞ norm of the closed-loop system. However, as above, this cannot be ensured at this stage. The LMIs of Theorem 2.14 are solved and give the guaranteed upper bound on the worst-case H_∞ norm 18.497. The SV-LMIs of Theorem 2.13 are solved; at the optimum, these give the improved upper-bound 9.859. It follows that the same conclusions apply as for the open-loop case. The SV-LMIs are nonconservative for this example.

Recall that results of Theorem 2.14 and Theorem 2.13 have dual counterparts which are, respectively, Corollary 2.9 and Corollary 2.8. The LMIs of these corollaries happen to produce the same values as for the SV-LMIs. All results for the open-loop and closed-loop cases are summarized in Table 2.1.

2.6.1.3 H_2 Performance

The same tests are now performed but assuming that the performance is measured according to an H_2 norm. Results are summarized in Table 2.2. The state-feedback gain is tuned in order to attenuate the H_∞ norm. It happens to have a (small) negative effect on the H_2 norm. Again the results show that, for this example, worst-case values of the uncertainties are at vertices and the SV-LMIs are sometimes exact (nonconservative). More precisely, notice that primal and dual LMIs produce different results both in the SV-LMI case or in the standard "quadratic stability" case. For this example, the dual versions are less conservative and the dual SV-LMIs are exact. The primal SV-LMIs are conservative with a small conservatism gap while the classical "quadratic stability" conditions are all highly conservative.

Table 2.2 H_2 performance of the quarter-car suspension

	Nominal	Worst vertex	Dual-SV-LMI (2.57a)–(2.57c)	SV-LMI (2.55a)–(2.55c)	Dual-quad-LMI (2.59a)–(2.59c)	Quad-LMI (2.58a)–(2.58c)
Open-loop	4.136	4.251	4.251	4.401	9.972	16.568
Closed-loop	4.143	4.388	4.388	4.396	5.469	8.058

Table 2.3 Impulse-to-peak performance of the quarter-car suspension

	Nominal	Worst vertex	Dual-SV-LMI (2.83a)–(2.83c)	SV-LMI (2.80a)–(2.80c)	Dual-quad-LMI (2.84a)–(2.84c)	Quad-LMI (2.81a)–(2.81c)
Open-loop	15.625	17.187	17.330	17.331	17.523	17.522
Closed-loop	15.578	17.136	17.339	17.339	17.382	17.382

2.6.1.4 Impulse-To-Peak Performance

Similar tests are done assuming performance is measured according to impulse-to-peak properties. Since the system has scalar perturbation input $w \in \mathbb{R}$, computation of the impulse-to-peak performance of a given system, let say the nominal one for example, is possible by simulation. In practice, we applied the `impulse` Matlab function. Results are summarized in Table 2.3.

2.6.2 Stability of Randomly Generated Examples

The tests that follow are done on artificial randomly generated uncertain models. Generating models of increasing sizes (order of the model, number of vertices of the polytope) allows to test the conservatism and the numerical complexity of the exposed LMI results.

2.6.2.1 Procedure for Random Generation of Polytopes

Relevance of randomly generated examples is often debatable. Indeed, although mathematically correct, the generated examples may not be illustrative for engineering issues. In order to have randomly generated models that fit some of the features of models issued from modeling of engineering problems, we have tried to fulfill the following criteria:

- Sparsity: Usual state-space models issued from real-life problems are sparse and the sparse structure is preserved by the uncertainties. A zero coefficient usually means that there is no information transiting between components. It cannot be considered to be uncertain.

- Few uncertain coefficients: The number of uncertain coefficients should be of the same order of magnitude as the number of vertices. Typically, for real-life models, the uncertainties are related to coefficients that occur a few times in the matrices of the model, all other coefficients being assumed to be known.
- Not all critical vertices: In order to test conservatism with respect to stability, one should generate examples in which worst-case poles are close to be unstable. Since only vertices are generated and worst cases are not at vertices, this issue cannot be handled exactly. A heuristic approach is to generate vertices close to be unstable. But, to be realistic, not all vertices should have this feature.

In order to comply with these specifications, we have developed a heuristic method combining randomized and line search features.

- Inputs: The inputs are the sizes of the system to be generated, that is, n the order and L the number of vertices; a positive coefficient d specifying the density of the sparse matrices; a positive coefficient α specifying the distance to instability; a positive coefficient p specifying the maximal amplitude of uncertainties.
- Initialization: The initialization step aims at generating a nominal A matrix with specified density d and with poles such that the distance to instability is clearly satisfied. Namely the matrix should be such that for all eigenvalues λ of A one has $\mathrm{Re}(\lambda) < -1.3\alpha$. The way this initialization is done is as follows:

  0.1. Generate a random matrix A_0 with density d and a second random matrix $\tilde{A}_0$ with the same sparsity pattern (should have same zero coefficients as A_0).
  0.2. Perform a line search over the scalar decision variable x, starting with $x = 0$, with the constraints

  $$x^* = \max x : \quad \max(\mathrm{Re}(\lambda(A_0 + x\tilde{A}_0))) \leq -1.3\alpha, \ 0 \leq x \leq p.$$

  If p is sufficiently large and $\tilde{A}_0$ is a destabilizing direction, the matrix $A_0 + x^*\tilde{A}_0$ has one (or more) eigenvalue with real part equal to -1.3α, that is, at the specified distance to instability. If p is small or if $A_0 + x\tilde{A}_0$ has all eigenvalues with real part smaller than -1.3α whatever positive x then the matrix $A_0 + x^*\tilde{A}_0$ has all eigenvalues with real part smaller than -1.3α. Both these cases are satisfactory.
  0.3. If all real parts of eigenvalues of $A_0 + x^*\tilde{A}_0$ are smaller than -1.3α take $A = A_0 + x^*\tilde{A}_0$. If not start again at step 0.1.

- Generate vertices: Almost the same procedure is applied to generate the vertices but this time with perturbation matrices with smaller density and by requiring the strict distance to instability α. For each vertex $j \in \mathscr{I}_L$ the exact procedure is as follows:

  j.1 Generate a random matrix $\tilde{A}_j$ with low density (typically one or two non zero coefficients) and fitting the sparsity pattern of A (no non zero coefficients at row-column positions where the coefficients of A are zero).

j.2 Perform a line search over the scalar decision variable x_j, starting with $x_j = 0$, with the constraints

$$x_j^* = \max x_j : \quad \max(\mathrm{Re}(\lambda(A + x_j \tilde{A}_j))) \leq -\alpha , \; 0 \leq x_j \leq \mathsf{p}.$$

If p is sufficiently large and $\tilde{A}_j$ is a destabilizing direction, the matrix $A + x_j^* \tilde{A}_j$ has one (or more) eigenvalue with real part equal to $-\alpha$, that is, at the specified distance to instability. If p is small or if $A + x_j \tilde{A}_j$ has all eigenvalues with real part smaller than $-\alpha$ whatever positive x_j then the matrix $A + x_j^* \tilde{A}_j$ has all eigenvalues with real part smaller than $-\alpha$. Since A is chosen to have poles with real part strictly smaller than -1.3α and by continuity arguments one of these two cases holds.

j.3 Take $A_j = A + x_j^* \tilde{A}_j$ to be a vertex of the polytope and proceed to the next one.

For illustration, the procedure is applied twice with $n = 4$, $L = 2$, $\mathsf{d} = 0.7$, $\alpha = 0.1, \mathsf{s}$ and $\mathsf{p} = 1$. One (random) result gives the following two vertices

$$A_1 = \begin{bmatrix} 0 & 1.0851 & 0 & -1.3040 \\ 0 & -0.0883 & -0.5298 & 0 \\ 0.6548 & 0 & -1.0984 & 1.2803 \\ 1.0987 & 0 & 0 & -1.5144 \end{bmatrix},$$

$$A_2 = \begin{bmatrix} 0 & 1.0851 & 0 & -1.3040 \\ 0 & -0.0883 & -0.5943 & 0 \\ 0.6548 & 0 & -1.2043 & 1.2725 \\ 1.0987 & 0 & 0 & -1.4806 \end{bmatrix}$$

with the following eigenvalues

$$\lambda(A_1) = \begin{matrix} -0.1000 \pm 0.8430j \\ -1.2505 \pm 0.7363j \end{matrix}, \quad \lambda(A_2) = \begin{matrix} -0.1000 \pm 0.8758j \\ -1.2865 \pm 0.7112j \end{matrix}$$

In this example, both vertices have poles at the specified distance to instability. Both matrices have the same sparsity pattern and only four coefficients out of nine are different.

Another (random) result of the procedure gives the following two vertices

$$A_1 = \begin{bmatrix} 0.8117 & 0 & 0.1146 & -1.3448 \\ 0.9845 & -0.6881 & 1.1752 & 0 \\ 0 & 0 & -0.2259 & 0 \\ 0.6236 & 0 & 0 & -1.0199 \end{bmatrix},$$

$$A_2 = \begin{bmatrix} -0.1851 & 0 & 0.1146 & -1.3448 \\ 0.9845 & -2.3668 & 1.1752 & 0 \\ 0 & 0 & -0.2259 & 0 \\ 0.6236 & 0 & 0 & -1.0199 \end{bmatrix}$$

with the following eigenvalues

$$\lambda(A_1) = \begin{matrix} -0.6881 \\ -0.1041 \pm 0.0045j, \\ -0.2259 \end{matrix} \quad \lambda(A_2) = \begin{matrix} -2.3668 \\ -0.6025 \pm 0.8152j \\ -0.2259 \end{matrix}$$

In this second example, none of the vertices are exactly at the specified distance to instability (although one is quite close). Sparsity and low number of uncertain elements criteria are satisfied.

Note that the procedure for generating randomly these systems has a numerical cost that is not negligible due to the possibly many trials for having a good initial guess of the matrix A and due to the many line search optimizations. The computation time for this random generation is large compared to the LMI tests to be evaluated as illustrated in the following. This is a price to pay to have interesting, reality-like, illustrative tests.

2.6.2.2 Robust Stability Testing

Series of tests are done for different values of the order n and the number of vertices L. The generated random models are according to the above-described procedure with parameters $\alpha = 0.05$, d $= 0.6$, and p $= 10$. The results are given in Table 2.4. The values are the number of samples that comply the tests and the values in curly braces are the mean computation times.

Computation time of LMI conditions that is given in the table includes both the parser YALMIP time (the time used to build the LMI problem and to convert it into the appropriate SDP-like format acceptable for the solver) and the SDP solver computation time (SDPT3 was used for these tests). Both these computations times grow with the size of the problem but not quite in the same way. The parser time mainly depends on the number of LMIs to declare, that is the number of vertices L, while the solver time grows polynomially with respect to the overall size of the problem characterized by $n^2 L$. This difference explains why for small size problems the quad-LMI tests and SV-tests have almost identical computation time and the difference increases rapidly with the order of the system n.

When reading the table it is clear that SV results are much less conservative than classical quadratic stability type results. Moreover, the reduction of conservatism increases as the size of the problems grow (in n and L). The improvement brought by SV results is of 40 % for $n = 3$, $L = 3$. It is of 1,000 % for $n = 5$, $L = 6$. This reduction of conservatism is not at the expense of large increase in computation time.

Table 2.4 Robust stability tests for 100 random systems (for each row of the table), generated with parameters $\alpha = 0.05$, $d = 0.6$ and $p = 10$

n	L	rand gen	quad-LMI (2.9)	SV-LMI (2.13)	Dual-SV-LMI (2.17)	SV different (Theorem 2.6)	Unstable	Unknown
3	3	{ 2.47s}	70 {0.40s}	98 {0.47s}	96 {0.50s}	4	0	1
4	3	{ 3.18s}	48 {0.44s}	97 {0.54s}	98 {0.51s}	1	1	1
5	3	{10.90s}	32 {0.49s}	92 {0.65s}	94 {0.63s}	2	2	4
3	4	{ 2.11s}	53 {0.45s}	96 {0.52s}	96 {0.52s}	4	0	2
4	4	{ 4.28s}	32 {0.54s}	91 {0.69s}	95 {0.64s}	6	0	4
5	4	{10.57s}	28 {0.53s}	90 {0.79s}	90 {0.79s}	0	3	7
6	4	{32.36s}	27 {0.58s}	90 {0.95s}	91 {0.92s}	1	2	7
3	5	{ 2.45s}	54 {0.46s}	92 {0.62s}	91 {0.62s}	5	0	6
4	5	{ 3.77s}	22 {0.51s}	85 {0.77s}	84 {0.78s}	3	2	12
5	5	{10.78s}	23 {0.64s}	87 {1.05s}	87 {1.02s}	0	4	9
3	6	{ 2.50s}	42 {0.49s}	87 {0.72s}	87 {0.73s}	12	1	6
4	6	{ 3.67s}	17 {0.55s}	76 {0.91s}	76 {0.95s}	8	6	14
5	6	{11.02s}	7 {0.58s}	79 {1.18s}	78 {1.18s}	5	4	15
3	7	{ 2.32s}	47 {0.48s}	88 {0.77s}	91 {0.76s}	7	0	7
4	7	{ 4.38s}	16 {0.59s}	72 {1.16s}	78 {1.12s}	10	1	19
5	7	{11.40s}	8 {0.60s}	68 {1.40s}	66 {1.47s}	8	2	27
3	8	{ 2.93s}	32 {0.58s}	78 {0.93s}	83 {0.87s}	12	1	12
4	8	{ 4.65s}	15 {0.70s}	63 {1.37s}	65 {1.35s}	6	5	28
5	8	{10.06s}	6 {0.66s}	67 {1.59s}	67 {1.61s}	6	2	28
6	8	{32.62s}	5 {0.69s}	66 {2.10s}	66 {2.08s}	10	1	28

The value given in curly braces is the mean cpu time for generating the examples and for solving the LMIs

As mentioned in Theorem 2.6, the two primal and dual SV-LMI results may not be feasible simultaneously. This fact is illustrated by the value in the column "SV different." Considering, for example, the row corresponding to $n = 3$, $L = 6$, 87 systems are proved robustly stable by each SV LMI conditions. Among the 87 cases proved robustly stable with the SV-LMI condition (2.13), six give unfeasible dual SV-LMIs (2.17). Conversely, among the 87 cases proved robustly stable with the dual SV-LMI condition (2.17), six give unfeasible primal SV-LMIs (2.13). Overall, combining the two results, $87 + 6 = 93$ of the 100 tested systems are proved to be robustly stable by either one of the SV-LMI results. Among the remaining seven cases, one is proved to correspond to a nonrobustly stable system. This fact is obtained by finding a "worst case" value of the uncertainties for which the system is unstable. The way to find such value of the uncertainties is described in Chap. 3. Since all vertices of the generated polytopes are stable, these "worst case" values are not at vertices.

The last column of Table 2.4 indicates the number of generated random examples for which we were not able to conclude with neither robust stability nor instability.

The fact that these are not equal to zero illustrates the conservatism of the SV results (and the imperfectness of the worst-case uncertainty extraction scheme). Although much less conservative than the quadratic stability results, these results do have some conservatism. Further reduction of this conservatism is considered in Chap. 3.

2.6.2.3 Robust Discrete-Time Stability Testing

A same series of tests is performed for discrete-time systems. Stability corresponds to pole location in the unit disc. The system generation is identical except for the maximal real part of eigenvalues test replaced by maximal modulus test. Results are given in Table 2.5. The values are the number of samples that comply the tests and the values in curly braces are the mean computation times.

Conclusions that can be drawn from these results are quite similar to those in the continuous-time case. The main difference is in the very low number of samples for

Table 2.5 Robust stability tests for 100 random discrete-time systems (for each row of the table), generated with parameters $\alpha = 0.05$, $d = 0.6$ and $p = 10$

n	L	Rand gen	Quad-LMI (2.9)	SV-LMI (2.37)	Dual-SV-LMI (2.38)	SV different (see Theorem 2.6)	Unstable	Unknown
3	3	{0.70s}	83 {0.41s}	97 {0.49s}	96 {0.44s}	1	4	0
4	3	{0.86s}	61 {0.48s}	96 {0.51s}	96 {0.50s}	0	4	0
5	3	{0.89s}	49 {0.50s}	92 {0.67s}	92 {0.61s}	0	8	0
3	4	{0.84s}	58 {0.45s}	90 {0.57s}	88 {0.53s}	2	12	0
4	4	{0.92s}	42 {0.53s}	92 {0.68s}	90 {0.64s}	2	10	0
5	4	{1.13s}	39 {0.61s}	87 {0.82s}	87 {0.73s}	0	13	0
6	4	{1.62s}	28 {0.58s}	97 {0.91s}	96 {0.90s}	1	4	0
3	5	{0.90s}	67 {0.45s}	89 {0.57s}	87 {0.54s}	2	13	0
4	5	{0.95s}	34 {0.48s}	78 {0.80s}	77 {0.66s}	1	23	0
5	5	{1.11s}	33 {0.53s}	81 {0.97s}	80 {0.82s}	1	20	0
3	6	{1.04s}	54 {0.50s}	83 {0.68s}	81 {0.61s}	2	18	1
4	6	{1.12s}	38 {0.52s}	83 {0.94s}	78 {0.85s}	5	22	0
5	6	{1.41s}	22 {0.58s}	77 {1.26s}	77 {1.04s}	0	22	1
3	7	{1.14s}	62 {0.51s}	82 {0.75s}	82 {0.65s}	0	18	0
4	7	{1.23s}	34 {0.52s}	86 {0.98s}	83 {0.90s}	3	17	0
5	7	{1.38s}	13 {0.57s}	69 {1.36s}	67 {1.11s}	2	33	0
3	8	{1.40s}	58 {0.60s}	80 {0.90s}	78 {0.71s}	2	22	0
4	8	{1.45s}	26 {0.56s}	69 {1.15s}	68 {0.96s}	1	32	0
5	8	{1.69s}	6 {0.68s}	66 {1.64s}	26 {1.39s}	4	38	0
6	8	{2.10s}	4 {0.68s}	73 {1.98s}	70 {1.78s}	3	30	0

The value given in curly braces is the mean cpu time for generating the examples and for solving the LMIs

which we are not able to state whether these are robustly stable systems or not. It gives the feeling that these SV-LMI results are close to be nonconservative. Compared to the continuous-time case this is rather surprising but is maybe due to the random system generation.

2.6.2.4 Airbus Provided Models

In 2009–2012 a study took place involving the Directorate General for Civil Aviation, LAAS-CNRS and the aircraft manufacturer Airbus through convention NGCI F/20 334/DA PPUJ. Among many studied topics, one of them was the evaluation of LMI methods for robust performance analysis of closed-loop longitudinal dynamics of a civil aircraft. The results that are described in the following are results of joint work on this topic involving Denis Arzelier, Guilherme De Calazans Chevarria, Dimitri Peaucelle at LAAS-CNRS and Guilhem Puyou at Airbus. These are also published in [10]. The numerical results of the book are recomputed values based on updated codes.

In the study, the certification is lead for a control law of the aircraft motion in the vertical plane (longitudinal). All the closed-loop components have been modeled. It includes:

- the actuators, namely the elevators, that have been modeled by a first order transfer function;
- the flight mechanics which has been restricted to only short period dynamics (angle of attack x_α and pitch rate x_q);
- the sensors delay and filtering have been reduced to first order transfer functions;
- the flight control law is mostly static but includes a scheduling part with respect to the speed (V_c), the Mach number (M_a) and the center of gravity (c_g).

The closed-loop model has been linearized on 633 flight points that map the parametric domain with respect to the weight (3 values [min medium max]), balance (center of gravity position [max forward, max aft]), speed V_c and Mach number (see Fig. 2.3). A regular spacing is used for V_c points generation. Nevertheless, minimum speed depends on the weight parameter value so that original points which are below the minimal bound are always brought back to the closest limit (see Fig. 2.3 in the lower part of the V_c range). A non regular spacing is used for Mach points generation. It comes from the aerodynamic coefficient study that highlights nonlinearity in the high Mach range. A tightest grid is used for high Mach number values.

For each of the $j = 1, \ldots, 633$ points, the linearized model is given by:

$$\begin{bmatrix} \dot{x}(t) \\ z(t) \end{bmatrix} = \begin{bmatrix} A_j & B_j \\ C_j & D_j \end{bmatrix} \begin{bmatrix} x(t) \\ w(t) \end{bmatrix} \tag{2.85}$$

where $x \in \mathbb{R}^9$ is the state vector, w the vector of exogenous inputs, z the vector of controlled outputs. The channels w and z will be defined in the following for several performance criteria.

Fig. 2.3 Flight Domain in the (V_c, M_a) plane

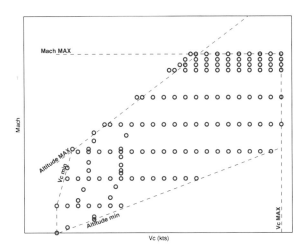

Based on the knowledge of the 633 linearized models and of the true model dependency upon the flight conditions, it could be possible to derive an uncertain model with respect to flight condition parameters (load factor, Mach number etc.). Such uncertain model would depend on the 6 parameters and should fit exactly the given linearized models on the 633 points. In between the 633 it would correspond to one possible interpolation, possibly not fitting the real values. But the major drawback of such model would be its size. Some attempts, handling the interpolation issue and based on the physics of the plant are presented in [11]. The LFT models provided have uncertainty blocks of size 250–300. These are not for the longitudinal axis problem but based on these results one can expect an uncertainty block of size 150 for the problem we consider and maybe only for a part of the flight envelope. Such dimension are incompatible with the tools we aim at testing.

It is therefore proposed here to build several small size uncertain models, each of which are valid in the neighborhood of a given fight condition point. The collection of these uncertain models for the 633 flight conditions therefore cover the flight envelope. Analysis of each of these gives for each individual fight condition the expected level of robust performances valid in the close neighborhood.

A heuristic algorithm has been designed for choosing neighboring flight points. Given a selected flight point j, it selects among the 632 others the set $N(j) \subset \{1 \ldots 633\}$ of neighboring points by using a criteria that combines the Euclidian distance in the 6-dimensional space of flight condition parameters and the search along parametric directions. Define $L(j)$ the number of elements of $N(j)$. The algorithm was tuned specifically for the given problem to generate sets $N(j)$ with acceptable number of elements. Namely, $L(j)$ goes from 8 in the corners of the flight envelope to 13 and has a mean value of 12.2.

As indicated above, the chosen uncertain modeling process leads to describing the system at a point j taking into account the neighboring points $N(j)$. We have chosen to do so in the natural affine polytopic way considered in this book. It needs no

assumption on the physics of the problem. Each uncertain model at point j is defined
as the linear combination of the $L(j)$ neighboring models with indexes in $N(j)$:

$$\begin{bmatrix} A_j(\theta_j) & B_j(\theta_j) \\ C_j(\theta_j) & D_j(\theta_j) \end{bmatrix} = \sum_{j \in N(j)} \theta_{j,j} \begin{bmatrix} A_j & B_j \\ C_j & D_j \end{bmatrix}, \qquad \theta_j \in \mathbb{E}^{L(j)} \qquad (2.86)$$

2.6.2.5 Robust Performance Tests

The first main issue in analyzing feedback control systems is closed-loop robust
stability of the models (2.86). For those flight conditions at which robust stability
is assessed, 5 robust criteria are evaluated to assess different performances of the
flight control system. Three of these criteria are input–output type performances.
The inputs w and the outputs z for each of these depend on the criteria and are
explained below.

- Robust pole placement in the left-half plane such that $\mathrm{Re}(\lambda) \leq \alpha$ to evaluate
  settling time.
- Robust pole placement in conic sectors with an angle ψ with respect to the imaginary axis to evaluate damping of modes.
- H_2 norm usually helps to evaluate the structural loads in turbulence and can also
  be used as an indirect measure of consumption. In our study, we evaluate robust
  upper bounds on the H_2-norm assuming the performance input w is a noise on the
  measured signals: pitch rate x_q and load factor x_{nz}, while the performance output
  $z = u \in \mathbb{R}$ is the control signal. The H_2 norm thus gives an information about the
  effect of measurement noise on the control effort.
- H_∞ norm may be used for measuring several features: comfort against turbulence,
  actuator activity or stability margins. In our study z is composed of the following
  three signals: the angle of attack x_α, the pitch rate x_q and the load factor x_{nz},
  while $w \in \mathbb{R}$ is an additional disturbance signal applied on the control inputs. The
  H_∞-norm thus measures a stability margin.
- The impulse-to-peak performance is a way to evaluate the peak of the time response
  of the system to nonzero initial conditions. The issue of saturation of actuators
  signals can be analyzed using this criteria. Here, it is used to assess the behavior of
  the control signals to nonzero initial conditions in the aircraft. Namely, $w \in \mathbb{R}^2$ is
  chosen to define a unit ball of initial conditions and $z \in \mathbb{R}$ measures the maximal
  deflection of the elevator actuators.

We have done tests for all the 633 flight points. For each flight point $j = 1 \ldots 633$
the steps were the following:

1. Generate the polytope of neighbors.
2. Test if all vertices $j \in N(j)$ are stable. If not, the polytope is proved not to be
   robustly stable. If all vertices are stable the quad-LMI test (2.9) is performed. In
   case it fails to prove robust stability, SV-LMI test (2.13) is performed. Robust
   stability is undetermined if all vertices are stable and SV-LMI test fails.

3. For robustly stable systems, upper bounds on the 5 robust criteria are evaluated with available methods and compared.

Robust stability results.

For robust stability assessment, it turned out that 3 of the 633 flight points are unstable. These are involved as neighbors in 15 other polytopes. A total of 18 flight points were thus proved not to be robustly stable. Of the 615 remaining polytopes, 612 were proved to be robustly stable using the quad-LMI test (2.9). The remaining 3 are proved to be robustly stable using SV-LMI test (2.13).

The size of the LMIs (number of variables and number of rows of the constraints) depends on the number of vertices. Since the number of vertices is not the same for all flight points, Table 2.6 gives the mean values of these dimensions over the 633 points. The table also gives the mean computation time for testing feasibility of the LMIs.

Settling time.

Finding an upper bound on the real part of the poles of the uncertain system cannot be done by LMI optimization. A bisection algorithm is applied in which LMI tests are run at each step to test a possible upper bound α. The initial interval of the bisection algorithm is $[\alpha_m(j)\ 0]$ where $\alpha_m(j)$ is the largest real part of all poles at the vertices of the robustly stable polytope of neighbors around flight point j:

$$\alpha_m(j) = \max_{j \in N(j)} \max \text{Real}(\lambda(A_j)).$$

Bisection stops when 1 % tolerance is reached. The optimal upper bound is denoted $\alpha^*(j)$. For comparison of the methods over all flight points, the following ratio is computed $\alpha_\%(j) = (\alpha^*(j) - \alpha_m(j))/|\alpha_m(j)|$. Its mean value in percentage is given in Table 2.7 for the different methods.

The SV-LMI methods happen to be much less conservative than the quad-LMI method. Because of that, the number of iterations in the bisection is much lower for SV-LMI methods, thus making these not only less conservative but also of comparable bisection convergence time.

Damping.

Exactly the same procedure as above is applied to compare methods for the damping criterion. Results are given in Table 2.8.

Again, the SV-LMI method overcomes the quad-LMI results in terms of conservatism. It is at the expense of an increase in computation time. Compared to the tests

Table 2.6 LMI sizes and times for stability tests

	Mean nb of variables	Mean nb of rows	Mean time (s)
Quad-LMI (2.9)	45.0	127.7	0.49
SV-LMI (2.13)	683.3	204.3	1.53

Table 2.7 Results for settling time criterion

	$\alpha_\%$	Mean time per LMIs (s)	Mean number of bisection iterates	Mean bisection time (s)
Quad-LMI (2.39)	11.69 %	0.85	8.0	6.82
SV-LMI (2.37)	2.09 %	2.21	4.33	9.57

Table 2.8 Results for damping criterion

	$\psi_\%$	Mean time per LMIs (s)	Mean number of bisection iterates	Mean bisection time (s)
Quad-LMI (2.39)	9.45 %	1.78	8.0	14.27
SV-LMI (2.37)	1.69 %	7.78	4.43	34.48

Table 2.9 Results for robust H_2 cost

	γ_2 %	Mean nb of variables	Mean nb of rows	Mean time (s)
Quad-LMI (2.58a)–(2.58a)	30.82 %	46.0	136.7	0.89
SV-LMI (2.55a)–(2.55c)	0.01 %	728.3	257.1	3.16

Table 2.10 Results for robust H_∞ cost

	γ_∞%	Mean nb of variables	Mean nb of rows	Mean time (s)
Quad-LMI (2.67)	31.21 %	46.0	154.1	0.89
SV-LMI (2.66)	0.14 %	694.1	231.3	2.83

for settling-time criterion, the size of the LMIs happen to be double because pole location regions such as sectors used to prove damping are complex valued regions. These need double-sized LMI to be coded, see equation (2.36).

Input–Output performances.

Computation of the upper bounds on the three input–output criteria are done by direct LMI optimization. Again, a ratio $\gamma_\%$(j) between the worst performance over all vertices of the polytope at point j and the obtained upper bound is computed. For the impulse-to-peak performance, $\gamma_{IP,m}$ is computed by simulation.

Results for the H_2, H_∞ and impulse-to-peak performances are given, respectively, in Tables 2.9, 2.10 and 2.11.

The reduction of conservatism due to the SV-LMI methods is significant and the comparisons with vertices indicate that SV-LMI results are close to being nonconservative. Moreover, globally the computation time does not increase dramatically.

Table 2.11 Results for robust impulse-to-peak performance

	$\gamma_{IP}\%$	Mean nb of variables	Mean nb of rows	Mean time (s)
Quad-LMI (2.81a)–(2.81c)	42.27 %	46.0	281.8	1.23
SV-LMI (2.80a)–(2.80c)	27.91 %	688.7	375.8	2.87

2.7 Conclusions

This chapter exposes the SV-LMI technique applied for a set of classical robust control system analysis problems. The considered problems are stability, pole location in convex sets of the complex plane, H_∞, H_2 and impulse-to-peak performances. For all these problems, classical "quadratic stability" based methods are recalled and compared in terms of conservatism and numerical complexity. SV-LMIs are proved to be always less conservative and as illustrated on examples the conservatism reduction is most significant. Noticeably, the conservatism reduction is not at the expense of huge increase of the computation time. Nevertheless, there is still some conservatism in SV-LMIs. The reasons for this conservatism are precisely explained and noticed on examples. The aim of the chapter that follows is to build with the same SV-LMI technique conditions with further reduced conservatism.

References

1. Zhou K, Doyle JC (1997) Essentials of robust control. Prentice Hall, Englewood Cliffs
2. Boyd S, El Ghaoui L, Feron E, Balakrishnan V (1994) Linear matrix inequalities in system and control theory. SIAM Studies in Applied Mathematics, Philadelphia
3. Skelton RE, Iwasaki T, Grigoriadis K (1998) A unified approach to linear control design., Systems and controlTaylor and Francis, London
4. Peaucelle D, Arzelier D, Bachelier O, Bernussou J (2000) A new robust D-stability condition for real convex polytopic uncertainty. Syst Control Lett 40(1):21–30
5. Chilali M (1996) Méthodes LMI pour l'Analyse et la Synthèse Multi-Critère. PhD thesis, Paris IX, Dauphine
6. Haddad WM, Bernstein DS (1992) Controller design with regional pole constraints. IEEE Trans Autom Control 37:54–61
7. Trentelman HL, Stoorvogel AA, Hutus M (2001) Control theory for linear systems. Springer, London
8. Rantzer A (1996) On the Kalman-Yakubovitch-Popov lemma. Syst Control Lett 28:7–10
9. Fallah MS, Bhat R, Xie W (2009) H_∞ robust control of active suspensions: a practical point of view. In: American control conference, pp 1385–1389, St. Louis, 2009
10. Chevarria G, Peaucelle D, Arzelier D, Puyou G (2010) Robust analysis of the longitudinal control of a civil aircraft using RoMulOC. In: IEEE conference on computer aided control system design, Yokohama, Septembre 2010
11. Hecker S, Pfifer H (2010) Generation of lfrs for the cofcluo nonlinear aircraft model. In: 2nd workshop on clearance of flight control laws, Stockholm, 2010

Chapter 3
Descriptor Case and System Augmentation

3.1 Robust Stability of Systems in Descriptor Form

In this chapter we present S-Variable approach to the analysis of uncertain systems represented in descriptor form. Moreover, we show that the descriptor representation leads to the idea of system augmentation, which is useful in reducing the conservatism of S-Variable approach to robustness analysis of uncertain LTI systems affected by polytopic-type uncertainties.

3.1.1 Systems in Descriptor Form

Let us consider the following descriptor representation of systems

$$E_{xx}\dot{x}(t) + E_{x\pi}\pi(t) = Ax(t) + Bu(t) \tag{3.1}$$

where $x \in \mathbb{R}^{n_x}$ is the state of the system, $u \in \mathbb{R}^{n_u}$ is the control input and $\pi \in \mathbb{R}^{n_\pi}$ is an auxiliary signal linearly constrained by the equation. The matrices defining the system may be square or not: $E_{xx} \in \mathbb{R}^{n \times n_x}$, $E_{x\pi} \in \mathbb{R}^{n \times n_\pi}$, $A \in \mathbb{R}^{n \times n_x}$, $B \in \mathbb{R}^{n \times n_u}$.

As exposed in [1] such descriptor modeling is convenient and natural for most physical processes. For mechanical systems it is the natural form of Lagrange equations. It is most useful for modeling interconnected processes as in robotics, electrical circuit network as well as for social, biological and economic systems.

An example of system described by such equations is the quarter-car suspension model considered in Chap. 2. The equation of motion are

$$\begin{aligned} M\ddot{z}_c &= -\pi_{k_1} - \pi_c + u \\ m\ddot{z}_w &= \pi_{k_1} - \pi_{k_2} + \pi_c - u \end{aligned} \tag{3.2}$$

© Springer-Verlag London 2015
Y. Ebihara et al., *S-Variable Approach to LMI-Based Robust Control*,
Communications and Control Engineering, DOI 10.1007/978-1-4471-6606-1_3

where $\pi_{k_1} = k_1(z_c - z_w)$ and $\pi_{k_2} = k_2 z_w$ are the stiffness forces of the suspension and the wheel, $\pi_c = c(\dot{z}_c - \dot{z}_w)$ is the dumper force, u (= f in the notation of Chap. 2) is the control input for the active suspension. A natural descriptor model for this system that preserves the structure of the definition of the system is as follows:

$$
\begin{bmatrix} m & 0 & 0 & 0 \\ 0 & 1 & 0 & 0 \\ 0 & 0 & M & 0 \\ 0 & 0 & 0 & 1 \\ 0 & 0 & 0 & 0 \\ 0 & 0 & 0 & 0 \\ 0 & 0 & 0 & 0 \end{bmatrix} \dot{x} +
\begin{bmatrix} -1 & 1 & -1 \\ 0 & 0 & 0 \\ 1 & 0 & 1 \\ 0 & 0 & 0 \\ 1 & 0 & 0 \\ 0 & 1 & 0 \\ 0 & 0 & 1 \end{bmatrix} \pi
$$

$$
= \begin{bmatrix} 0 & 0 & 0 & 0 \\ 1 & 0 & 0 & 0 \\ 0 & 0 & 0 & 0 \\ 0 & 0 & 1 & 0 \\ 0 & -k_1 & 0 & k_1 \\ 0 & k_2 & 0 & 0 \\ -c & 0 & c & 0 \end{bmatrix} x +
\begin{bmatrix} -1 \\ 0 \\ 1 \\ 0 \\ 0 \\ 0 \\ 0 \end{bmatrix} u
$$

(3.3)

where $x^T = \begin{bmatrix} \dot{z}_w & z_w & \dot{z}_c & z_c \end{bmatrix}^T$ and $\pi^T = \begin{bmatrix} \pi_{k_1} & \pi_{k_2} & \pi_c \end{bmatrix}^T$.

As seen in Chap. 2 this model can be reformulated as a non-descriptor model $\dot{x} = \hat{A}x + \hat{B}u$. The operation for building $\hat{A}$ and $\hat{B}$ amounts to multiplying from the left hand side the equality by $\widehat{E}^{-1} = \begin{bmatrix} E_{xx} & E_{x\pi} \end{bmatrix}^{-1}$ (the matrix is indeed square and invertible in the considered case) as in

$$
\underbrace{\widehat{E}^{-1} \left[E_{xx} \dot{x}(t) + E_{x\pi} \pi(t) \right]}_{=\begin{bmatrix} \dot{x} \\ \pi \end{bmatrix}} = \underbrace{\widehat{E}^{-1} A}_{=\begin{bmatrix} \hat{A} \\ \hat{A}_\pi \end{bmatrix}} x(t) + \underbrace{\widehat{E}^{-1} B}_{=\begin{bmatrix} \hat{B} \\ \hat{B}_\pi \end{bmatrix}} u(t)
$$

and removing rows involving π. Unfortunately, this procedure has drawbacks (or cannot be carried out) if, for example, $\begin{bmatrix} E_{xx} & E_{x\pi} \end{bmatrix}$ is non-invertible, or badly conditioned, or contains uncertainties.

In the literature (see [1–3] for example), descriptor systems are more often described by equations of the following type

$$
\tilde{E}\dot{\tilde{x}}(t) = \tilde{A}\tilde{x}(t) + Bu(t) \tag{3.4}
$$

where $\tilde{E}$ and $\tilde{A}$ have same number of columns. Our modeling (3.1) is not contradictory. One can indeed always recover (3.4) by choosing

$$\tilde{x}(t) = \begin{bmatrix} x(t) \\ \int_0^t \pi(\tau)d\tau \end{bmatrix}, \quad \begin{array}{l} \tilde{E} = \begin{bmatrix} E_{xx} & E_{x\pi} \end{bmatrix} \in \mathbb{R}^{n \times (n_x + n_\pi)}, \\ \tilde{A} = \begin{bmatrix} A & 0 \end{bmatrix} \in \mathbb{R}^{n \times (n_x + n_\pi)} \end{array}.$$

Although non conservative, this transformation is not recommended since it increases artificially the order of the model from n_x to $n_x + n_\pi$.

3.1.2 Stability

As explained in detail in [4], the analysis of descriptor systems is slightly more complex than that of usual, non-descriptor, systems. In addition to exponential modes (which can be converging or diverging), linear descriptor systems may contain so-called impulsive modes. See [1–3] for detailed discussions about these impulsive modes and their interpretations in terms of matrix pencil and Laplace transform. Here we propose a time domain description. Impulsive modes are impulsive phenomena in response to non-zero initial conditions. Their L_2 norm is unbounded and thus considered as unstable modes.

Assume $u(t) = 0$ ($t \geq 0$) in (3.1) and non-zero initial condition $x(0^-) \neq 0$ such that there exists a solution x_{o1} to

$$E_{xx}x_{o1} + E_{x\pi}\pi_{o1} = 0, \quad x_{o1} = -x(0^-) \tag{3.5}$$

In that case, $x(t) = x_{o1}\delta(t)$ is a solution to the equations. Such behavior is called a *non-dynamic mode* [5]. The impulse component works as a reset to zero of the involved components of the state.

Assume now that for that same x_{o1} the product $Ax_{o1} \neq 0$ and there exists a solution (x_{o2}, π_{o2}) to the following linear equation

$$E_{xx}x_{o2} + E_{x\pi}\pi_{o2} = Ax_{o1}. \tag{3.6}$$

Then $x(t) = x_{o1}\delta(t) + x_{o2}\delta^{(1)}(t)$ is yet another solution to the differential equation (3.1). This solution is an *impulsive mode*. Other impulsive modes are possible with increasing derivatives of the Dirac distribution if the following recursive equations have nonzero solutions $(x_{o\iota}, \pi_{o\iota})$, $\iota > 1$:

$$E_{xx}x_{o\iota} + E_{x\pi}\pi_{o\iota} = Ax_{o(\iota-1)}.$$

These impulsive modes exist due to propagations of the initial conditions x_{o1} through the system equations. These modes are to be avoided in practice. A system without such impulsive modes (all solutions to (3.5) are such that $Ax_{o1} = 0$) is said *impulse free*.

A linear descriptor system as considered in this book is said to be *stable* if it is impulse free and the state converges to zero for zero input $u = 0$ and all initial conditions admitted by the system:

$$\lim_{t \to \infty} x(t) \to 0 \quad (\forall\, x(0) \in \mathbb{R}^{n_x}) \tag{3.7}$$

In the literature this property is sometimes referred to as *admissibility* and contains the properties that Eq. (3.1) is regular (defines a unique solution), impulsive free, and the state converges to zero for zero input. Following the example of [6] we choose to keep the terminology *stable* as for usual linear systems to make the notions as simple as possible.

Theorem 3.1 *Let the following factorization* $\begin{bmatrix} E_{xx} & E_{x\pi} \end{bmatrix} = E_1 \begin{bmatrix} E_{2xx} & E_{2x\pi} \end{bmatrix}$ *where E_1 is full column rank and let $E_2 = E_{2x\pi}^{\perp} E_{2xx}$. The system (3.1) is stable if there exist matrices $P \in \mathbb{S}^{n_x}$ and Y of appropriate dimensions such that the following LMI conditions hold:*

$$(E_2 E_2^{\circ})^T P (E_2 E_2^{\circ}) \succ 0,$$
$$\begin{bmatrix} E_1 & -A \end{bmatrix}^{T\perp} \begin{bmatrix} 0 & P_e^T \\ P_e & 0 \end{bmatrix} \begin{bmatrix} E_1 & -A \end{bmatrix}^{T\perp T} \prec 0, \tag{3.8}$$
$$P_e = (E_2^T P + Y^T E_2^{\perp}) E_{2x\pi}^{\perp}.$$

Proof First notice that by definition of $N^{\perp}$ the following conditions are equivalent

$$\begin{aligned} N^{T\perp} M N^{T\perp T} \prec 0 &\Leftrightarrow y^T M y < 0, \quad \forall\, y \neq 0 : y^T N^T = 0 \\ &\Leftrightarrow y^T M y < 0, \quad \forall\, y \neq 0 : N y = 0. \end{aligned} \tag{3.9}$$

Assume the conditions (3.8) hold. The second inequality reads as

$$2\varpi^T P_e^T x = 2\varpi^T E_{2x\pi}^{\perp T} (P E_2 + E_2^{\perp T} Y) x < 0, \quad \forall \begin{bmatrix} \varpi \\ x \end{bmatrix} \neq 0 : E_1 \varpi - A x = 0.$$

Take $\varpi = E_{2xx} \dot{x} + E_{2x\pi} \pi$ (and hence $E_1 \varpi = E_{xx} \dot{x} + E_{x\pi} \pi$), the condition is

$$2\dot{x}^T E_{2xx}^T E_{2x\pi}^{\perp T} (P E_2 + E_2^{\perp T} Y) x < 0, \quad \forall \begin{bmatrix} \dot{x} \\ \pi \\ x \end{bmatrix} \neq 0 : E_{xx} \dot{x} + E_{x\pi} \pi = A x.$$

By definition of E_2 this condition also reads as

$$2\dot{x}^T E_2^T (P E_2 + E_2^{\perp T} Y) x < 0, \quad \forall \begin{bmatrix} \dot{x}^T & \pi^T & x^T \end{bmatrix}^T \neq 0 : E_{xx} \dot{x} + E_{x\pi} \pi = A x.$$

Since $E_2^T E_2^{\perp T} = 0$, one gets

$$\dot{V}(x) = 2\dot{x}^T E_2^T P E_2 x < 0, \quad \forall \begin{bmatrix} \dot{x}^T & \pi^T & x^T \end{bmatrix}^T \neq 0 : E_{xx}\dot{x} + E_{x\pi}\pi = Ax \quad (3.10)$$

where $V(x) = x^T E_2^T P E_2 x$ is a quadratic function of the state. Its derivative is strictly negative along the trajectories of the system. The condition $(E_2 E_2^\circ)^T P(E_2 E_2^\circ) \succ 0$, by definition of E_2°, is equivalent to $x^T E_2^T P E_2 x > 0$ for all x such that $E_2 x \neq 0$. The function V is hence positive. It is equal to zero only on the null space of E_2. By Lyapunov theory one concludes that all initial conditions that are not in the kernel of E_2 produce converging to zero trajectories.

Consider now solutions to $E_2 x_{o1} = 0$. These are exactly the solutions to $E_{xx} x_{o1} + E_{x\pi}\pi_{o1} = 0$ corresponding to the reset behavior of descriptor systems. Assume this reset behavior is source of an impulsive mode $x(t) = x_{o2}\delta(t) \neq 0$, $\dot{x}(t) = x_{o1}\delta(t) + x_{o2}\delta^{(2)}(t)$. This signal is solution to the system equation and would produce the following derivative of the Lyapunov function $\dot{V}(x) = \delta\delta^{(2)}2x_{o2}^T E_2^T P E_2 x_{o2}$ which cannot be negative. But it should satisfy condition (3.10). It is thus zero, $x_{o2} = 0$. There is no impulsive mode.   □

The result presented in Theorem 3.1 is a variation on results presented in [7–10]. It may look complex but its structure is defined in order to cope with uncertain systems in the next subsection. Before that, let us apply the conditions to special cases.

- Assume $E_{x\pi} = 0$ and $E_{xx} = I$, that is, the system is in usual non-descriptor form. In that case $E_{x\pi}^{\perp} = I$, $E_2 = E_2^\circ = I$, $E_2^{\perp}$ is empty and

$$\begin{bmatrix} E_1 & -A \end{bmatrix}^{T\perp} = \begin{bmatrix} I & -A \end{bmatrix}^{T\perp T} = \begin{bmatrix} A^T & I \end{bmatrix}.$$

The LMI conditions (3.8) are then exactly

$$P \succ 0, \quad A^T P + PA \prec 0$$

which is the classical Lyapunov condition of Theorem 2.1.
- Assume $E_{x\pi} = 0$ and $E_{xx} = E = E_2$ is square (that is for the choice $E_1 = I$). In that case the LMI conditions (3.8) read as

$$(EE^\circ)^T P(EE^\circ) \succ 0, \quad A^T(PE + E^{\perp T}Y) + (E^T P + Y^T E^{\perp})A \prec 0$$

which is equivalent to the conditions in [7, 8].

Compared to these existing conditions from the literature, the result of Theorem 3.1 provides an extension to non-square matrices E and A. Moreover, it exploits the structure $E = \begin{bmatrix} E_{xx} & E_{x\pi} \end{bmatrix}$.

3.1.3 Uncertain Descriptor Systems

As discussed in Sect. 3.1.1, descriptor modeling is suitable and natural for describing many processes. But another advantage of descriptor modeling is with respect to uncertainties. Any model that depends rationally on uncertain parameters can be converted to an affine polytopic one at the expense of adding descriptor type algebraic constraints. This feature is discussed in [11, 12] for the case of polynomially non-linear systems that are converted to quadratically non-linear models. It is also exposed in [13] for the case of parameter-dependent systems. The result of these papers can be further generalized as follows.

Theorem 3.2 *Assume a parameter-dependent descriptor model*

$$\bar{E}_{xx}(\theta)\dot{x} + \bar{E}_{x\pi}(\theta)\bar{\pi} = \bar{A}(\theta)x + \bar{B}(\theta)u, \quad \theta \in \mathbb{R}^L \tag{3.11}$$

in which the θ-dependent matrices are rational with respect to the components of θ, then, there always exists another parameter-dependent descriptor model

$$E_{xx}(\theta)\dot{x} + E_{x\pi}(\theta)\pi = A(\theta)x + B(\theta)u, \quad \theta \in \mathbb{R}^L \tag{3.12}$$

in which the θ-dependent matrices are affine functions of θ.

Before giving a formal proof, let us illustrate this fact on a simple example. Take the following system rationally dependent on one parameter $\theta \in \mathbb{R}$

$$\dot{x} = \begin{bmatrix} -1 + 2\theta \ (1-\theta)^{-1} \\ 0 \quad\quad -2 - \theta^2 \end{bmatrix} x.$$

Since descriptor forms are allowed, a first step is to remove all denominators form this representation and get only polynomial coefficients:

$$\begin{bmatrix} 1 - \theta & 0 \\ 0 & 1 \end{bmatrix} \dot{x} = \begin{bmatrix} (-1 + 2\theta)(1-\theta) & 1 \\ 0 & -2 - \theta^2 \end{bmatrix} x. \tag{3.13}$$

To obtain only affine coefficients the method is then to introduce fictive auxiliary signals that decompose products of parameter-dependent coefficients. For example one can decompose powers of coefficients as $\theta^2 x = \theta\pi$ for $\pi = \theta x$. The resulting affine descriptor model is

$$\begin{bmatrix} 1 - \theta & 0 \\ 0 & 1 \\ 0 & 0 \\ 0 & 0 \end{bmatrix} \dot{x} + \begin{bmatrix} 2\theta & 0 \\ 0 & \theta \\ 1 & 0 \\ 0 & 1 \end{bmatrix} \pi = \begin{bmatrix} -1 + 3\theta & 1 \\ 0 & -2 \\ \theta & 0 \\ 0 & \theta \end{bmatrix} x.$$

Another choice to decompose (3.13) is based on the formula

$$\begin{bmatrix} (-1+2\theta)(1-\theta) & 0 \\ 0 & -\theta^2 \end{bmatrix} x = \begin{bmatrix} -1+2\theta & 0 \\ 0 & -\theta \end{bmatrix} \hat{\pi}, \quad \hat{\pi} = \begin{bmatrix} 1-\theta & 0 \\ 0 & \theta \end{bmatrix} x.$$

In this latter case the resulting, equivalent, affine descriptor model is

$$\begin{bmatrix} 1-\theta & 0 \\ 0 & 1 \\ 0 & 0 \\ 0 & 0 \end{bmatrix} \dot{x} + \begin{bmatrix} 1-2\theta & 0 \\ 0 & \theta \\ 1 & 0 \\ 0 & 1 \end{bmatrix} \hat{\pi} = \begin{bmatrix} 0 & 1 \\ 0 & -2 \\ 1-\theta & 0 \\ 0 & \theta \end{bmatrix} x.$$

The example illustrates that affine descriptor models may not be unique. The choice of π can be seen as a problem of choosing monomials in which the θ-dependent polynomials are expressed.

Proof of Theorem 3.2 For a start let us rewrite (3.11) with a unique rationally dependent matrix:

$$\underbrace{\left[-\bar{E}_{xx}(\theta) \ -\bar{E}_{x\pi}(\theta) \ \bar{A}(\theta) \ \bar{B}(\theta) \right]}_{M(\theta)} \lambda = 0, \quad \lambda = \begin{bmatrix} \dot{x} \\ \bar{\pi} \\ x \\ u \end{bmatrix}.$$

Here, $M(\theta)$ is rationally dependent on θ and it can be represented via a linear-fractional transform (LFT) [14, 15] as

$$M(\theta) = M_{11} + M_{12}\Delta(\theta)(I - M_{22}\Delta(\theta))^{-1} M_{21}$$

where $\Delta(\theta)$ is linear with respect to θ (can even be taken diagonal). As it is done in the above exposed example the affine dependent model is obtained by introducing additional fictive signals $\bar{\pi}_2 = (I - M_{22}\Delta(\theta))^{-1} M_{21}\lambda$. This notation gives the affine representation

$$\begin{bmatrix} M_{11} \\ M_{21} \end{bmatrix} \lambda + \begin{bmatrix} M_{12}\Delta(\theta) \\ M_{22}\Delta(\theta) - I \end{bmatrix} \bar{\pi}_2 = 0 \tag{3.14}$$

which can be converted to (3.12) by taking $\pi = \begin{bmatrix} \bar{\pi}^T & \bar{\pi}_2^T \end{bmatrix}^T$. The system matrices are such that

$$\begin{bmatrix} M_{11} \\ M_{21} \end{bmatrix} = \begin{bmatrix} -E_{xx} & -E_{x\pi 1} & A & B \end{bmatrix}, \quad E_{x\pi}(\theta) = \begin{bmatrix} E_{x\pi 1} \begin{bmatrix} -M_{12}\Delta(\theta) \\ I - M_{22}\Delta(\theta) \end{bmatrix} \end{bmatrix}.$$

$\square$

As illustrated above, the affine modeling (3.12) is not unique. This is not surprising since LFT representations are not unique either. Thus, applying the method exposed in the proof produces different results depending on the LFT representation.

Moreover, one should remark that affine descriptor modeling can have smaller dimensions than LFTs (and usually do have much smaller dimensions). To illustrate this fact consider the following LFT

$$\dot{x} = \left[-1 + \begin{bmatrix} -1 & -1 \end{bmatrix} \Delta \left[I - \begin{bmatrix} -1 & -1 \\ 0 & 0 \end{bmatrix} \Delta \right]^{-1} \begin{bmatrix} -1 \\ 1 \end{bmatrix} \right] x$$

with $\Delta = \begin{bmatrix} \delta_1 & 0 \\ 0 & \delta_2 \end{bmatrix}$. This LFT happens to be minimal and corresponds to a feedback loop $w = \Delta z$ on a linear system in which $w \in \mathbb{R}^2$ and $z \in \mathbb{R}^2$. It involves four scalar exogenous signals. After some tedious computation that system also writes as $\dot{x} = -\frac{1+\delta_1}{1+\delta_2}x$. Its minimal affine descriptor model is $(1 + \delta_1)\dot{x} = -(1 + \delta_2)x$. This model is clearly of smaller size than the LFT (no exogenous signal at all). It is also smaller than the descriptor model that would be obtained via the procedure in the proof of Theorem 3.2.

3.1.4 S-Variable Results for Robust Stability

We shall now give the S-Variable type results for the stability analysis of uncertain descriptor systems. Assume without loss of generality that a rationally parameter-dependent system has been converted to an affine descriptor model

$$E_{xx}(\theta)\dot{x} + E_{x\pi}(\theta)\pi = A(\theta)x, \quad \theta \in \mathbb{E}^L \subset \mathbb{R}^L \tag{3.15}$$

where the parameter-dependent matrices are linear and given by

$$E_{xx}(\theta) := \sum_{j=1}^{L} \theta_j E_{xxj}, \quad E_{x\pi}(\theta) := \sum_{j=1}^{L} \theta_j E_{x\pi j}, \quad A(\theta) := \sum_{j=1}^{L} \theta_j A_j. \tag{3.16}$$

Here, $E_{xxj} \in \mathbb{R}^{n \times n_x}$, $E_{x\pi j} \in \mathbb{R}^{n \times n_\pi}$, $A_j \in \mathbb{R}^{n \times n_x}$ ($j \in \mathcal{I}_L$) are known matrices. On the other hand, the set $\mathbb{E}^L$ is a standard simplex in $\mathbb{R}^L$ defined by

$$\mathbb{E}^L := \left\{ \theta \in \mathbb{R}^L : \theta \geq 0, \; \mathbf{1}^T\theta = 1 \right\}.$$

As in Chap. 2, $\theta \in \mathbb{E}^L$ is assumed to be a time-invariant uncertainty. This model is a *polytopic descriptor uncertain model* defined by the vertices $E_{xxj}, E_{x\pi j}, A_j$.

Assumption 3.1 It is assumed that

$$\begin{bmatrix} E_{xx}(\theta) & E_{x\pi}(\theta) \end{bmatrix} = E_{1x}(\theta) \begin{bmatrix} E_{2xx} & E_{2x\pi} \end{bmatrix} \tag{3.17}$$

where $E_{1x}(\theta) = \sum_{j=1}^{L} \theta_j E_{1xj}$ is full column rank for all $\theta \in \mathbb{E}^L$.

Recalling that $\begin{bmatrix} E_{2xx} & E_{2x\pi} \end{bmatrix}$ reflects the descriptor structure that is source of non-dynamic and impulsive modes, the assumption states that the non-dynamic and eventual impulsive modes are independent of the uncertainty. The assumption is legitimate. More precisely, it is intimately related to the well-posedness assumption of LFTs. This fact can be seen from (3.14). The equivalent LFT model is well-posed if for all uncertainties $M_{22}\Delta(\theta) - I$ is non-singular. If it is the case then the column rank of $\begin{bmatrix} E_{xx}(\theta) & E_{x\pi}(\theta) \end{bmatrix}$ is independent of $\theta \in \mathbb{E}^L$.

Theorem 3.3 *If Assumption 3.1 holds, let* $E_2 = E_{2x\pi}^{\perp} E_{2xx}$. *The system* (3.15) *is robustly stable (that is, stable for all* $\theta \in \mathbb{E}^L$) *if there exist* $P_j = P_j^T$, Y_j $(j \in \mathcal{I}_L)$, F_1 *and* F_2 *such that for all* $j \in \mathcal{I}_L$:

$$(E_2 E_2^{\circ})^T P_j (E_2 E_2^{\circ}) \succ 0,$$

$$\begin{bmatrix} 0 & P_{ej}^T \\ P_{ej} & 0 \end{bmatrix} + \mathrm{He}\left\{ \begin{bmatrix} F_1 \\ F_2 \end{bmatrix} \begin{bmatrix} E_{1xj} & -A_j \end{bmatrix} \right\} \prec 0, \qquad (3.18)$$

$$P_{ej} = (E_2^T P_j + Y_j^T E_2^{\perp}) E_{2x\pi}^{\perp}.$$

This theorem is the generalization of Theorem 2.4 to descriptor systems. Its proof follows exactly the lines of Lemma 2.4. More precisely, Lemma 2.4 corresponds to the special case when E_1 is square. With this additional assumption one gets that

$$\begin{bmatrix} E_1(\theta) & -A(\theta) \end{bmatrix}^{T\perp} = \begin{bmatrix} E_{1x}^{-1}(\theta)A(\theta) \\ I \end{bmatrix}^T.$$

The stability proved by Theorem 3.3 corresponds hence also to the stability of the rationally dependent system

$$E_{2xx}\dot{x} + E_{2x\pi}\pi = E_{1x}^{-1}(\theta)A(\theta)x.$$

This property is not surprising since, thanks to Theorem 3.2, all rationally dependent systems can be converted to affine ones.

For non descriptor systems the SV-LMIs of Theorem 2.4 are proved to be less conservative than "quadratic stability" type results of Theorem 2.2. Links between the two types of results are demonstrated to be related to Lemma 2.5. By analogy, a "quadratic stability" type result for descriptor systems is obtained applying the results of Lemma 2.5. The quadratic stability result needs a stronger assumption on the system matrices.

Assumption 3.2 It is assumed that there are no uncertainties in the matrices E_{xx} and $E_{x\pi}$ and

$$\begin{bmatrix} E_{xx}(\theta) & E_{x\pi}(\theta) \end{bmatrix} = E_{1x} \begin{bmatrix} E_{2xx} & E_{2x\pi} \end{bmatrix} \qquad (3.19)$$

where E_{1x} is square non-singular.

Theorem 3.4 *If Assumption 3.2 holds, let* $E_2 = E_{2x\pi}^{\perp} E_{2xx}$. *The system* (3.15) *is robustly stable (that is stable for all* $\theta \in \mathbb{E}^L$*) if there exist* $P = P^T, Y$ *such that for all* $j \in \mathcal{I}_L$:

$$(E_2 E_2^{\circ})^T P(E_2 E_2^{\circ}) \succ 0,$$

$$\begin{bmatrix} E_{1x}^{-1} A_j \\ I \end{bmatrix}^T \begin{bmatrix} 0 & P_e^T \\ P_e & 0 \end{bmatrix} \begin{bmatrix} E_{1x}^{-1} A_j \\ I \end{bmatrix} \prec 0, \tag{3.20}$$

$$P_e = (E_2^T P + Y^T E_2^{\perp}) E_{2x\pi}^{\perp}.$$

Moreover, if (3.20) *holds then there always exists a solution to* (3.18). *A possible choice of S-Variables is* $F_1 = -\varepsilon W E_{1x}^{-1}$, $F_2 = -(E_2^T P + Y^T E_2^{\perp}) E_{2x\pi}^{\perp} E_{1x}^{-1}$ *with W any positive definite matrix and* $\varepsilon > 0$ *sufficiently small.*

The comparison of Theorems 3.3 and 3.4 indicates that the S-Variable type result is the most general of the two. The "quadratic stability" type result is a sub case for which the number of decision variables can be reduced. Ways for reduction of the number of decision variables is further explored in a systematic way in Sect. 3.2.

3.1.5 Performances

In Chap. 2, in addition to stability, several performance criteria are considered. The same criteria may be considered for descriptor systems. Rather than giving all the derivations of such results, we only expose the robust SV-LMI tests for two of these performance analysis problems, namely: regional pole location and L_2 induced norm. Full derivation of all results equivalent to those in Chap. 2 are possible following the exposed methodology. They are not included in order to concentrate on important conservatism reduction and exactness verification issues in the following sections.

3.1.5.1 Regional Finite Pole Location

Theorem 3.5 *Suppose Assumption 3.1 holds and let* $E_2 = E_{2x\pi}^{\perp} E_{2xx}$. *The system* (3.15) *has finite poles robustly located in* $\mathbb{O}(R)$ *(that is, for all* $\theta \in \mathbb{E}^L$*) if there exist* $P_j = P_j^T$, Y_j *(* $j \in \mathcal{I}_L$ *) and* **F** *such that for all* $j \in \mathcal{I}_L$:

$$(E_2 E_2^{\circ})^T P_j (E_2 E_2^{\circ}) \succ 0$$

$$\begin{bmatrix} R_{11} \otimes (E_{2x\pi}^{\perp T} P_j E_{2x\pi}^{\perp}) & R_{12} \otimes P_{ej}^* \\ R_{12}^* \otimes P_{ej} & R_{22} \otimes (E_2^T P_j E_2) \end{bmatrix}$$
$$+ \mathrm{He} \left\{ \mathbf{F} \left[I_d \otimes E_{1xj} - I_d \otimes A_j \right] \right\} \prec 0 \tag{3.21}$$
$$P_{ej} = (E_2^T P_j + Y_j^T E_2^{\perp}) E_{2x\pi}^{\perp}.$$

Moreover, if R_{11} has one or more non-negative eigenvalue, then the system is robustly impulse free.

Proof The first stage of the proof is to apply the method of Lemma 2.4: by convexity, if the conditions (3.21) hold for all vertices $j \in \mathscr{I}_L$, then they also hold for all values in the convex hull when taking $P(\theta) = \sum_{j=1}^{L} \theta_j P_j$ and $Y(\theta) = \sum_{j=1}^{L} \theta_j Y_j$. The first inequality of (3.21) states that $x^* E_2^T P(\theta) E_2 x > 0$ for all $E_2 x \neq 0$. Then, pre and post multiply the remaining inequality by $\left[I_d \otimes \varpi^T \; I_d \otimes x^T \right]$ and its transpose, respectively, where

$$\varpi = E_{2xx}\dot{x} + E_{2x\pi}\pi. \tag{3.22}$$

Following the methodology of the proof Theorem 3.1 it conduces to the property

$$
\begin{aligned}
&R_{11}(\dot{x}^* E_2^T P(\theta) E_2 \dot{x}) \\
&+R_{12}(\dot{x}^* E_2^T P(\theta) E_2 x) \\
&+R_{12}^*(x^* E_2^T P(\theta) E_2 \dot{x}) \\
&+R_{22}(x^* E_2^T P(\theta) E_2 x)
\end{aligned}
\prec 0, \quad \forall \begin{bmatrix} \dot{x} \\ \pi \\ x \end{bmatrix} \neq 0 : E_{xx}(\theta)\dot{x} + E_{x\pi}(\theta)\pi = A(\theta)x
$$

Assume now that $\lambda(\theta)$ is a finite pole of the system, that is: there exists a non zero x such that $E_{xx}(\theta)\lambda(\theta)x + E_{x\pi}(\theta)\pi = A(\theta)x$ holds for some π. For that choice of x and $\dot{x} = \lambda(\theta)x$ the above inequality implies

$$(R_{11}\lambda^*(\theta)\lambda(\theta) + R_{12}\lambda(\theta) + R_{12}^*\lambda(\theta) + R_{22})(x^* E_2^T P(\theta) E_2 x) \prec 0.$$

The right-hand side term of this product is strictly positive if $E_2 x \neq 0$. Hence for all non-impulsive modes the poles belong to the region $\mathbb{O}(R)$ as defined in (2.31).

Consider now the reset behavior $E_2 x_{1o} = 0$. Assume an eventual impulsive mode that it could generate: $E_{xx}x_{2o} + E_{x\pi}\pi_{2o} = Ax_{1o}$. For that pair (x_{1o}, x_{2o}) one gets that $R_{11}(x_{2o}^* E_2^T P(\theta) E_2 x_{2o}) \prec 0$ for all $E_2 x_{2o} \neq 0$. If R_{11} is not definite negative one concludes that $E_2 x_{2o} = 0$. There is no impulsive mode. $\square$

3.1.5.2 L_2-Induced Norm Performance

Let us now consider a system with performance input/output vectors

$$
\begin{aligned}
E_{xx}(\theta)\dot{x} + E_{x\pi}(\theta)\pi &= A(\theta)x + B(\theta)w \\
E_{zx}(\theta)\dot{x} + E_{z\pi}(\theta)\pi + z &= C(\theta)x + D(\theta)w
\end{aligned}, \quad \theta \in \mathbb{E}^L \subset \mathbb{R}^L \tag{3.23}
$$

where the parameter-dependent matrices are supposed polytopic. Parameter dependency is described by (2.29), (3.16) and

$$E_{zx}(\theta) := \sum_{j=1}^{L} \theta_j E_{zxj}, \quad E_{z\pi}(\theta) := \sum_{j=1}^{L} \theta_j E_{z\pi j} \tag{3.24}$$

where $E_{zxj} \in \mathbb{R}^{n_z \times n_x}$ and $E_{z\pi j} \in \mathbb{R}^{n_z \times n_\pi}$ are again known vertex matrices. For the system (3.23), a stronger assumption is made in replacement to Assumption 3.1.

Assumption 3.3 It is assumed that

$$
\begin{bmatrix} E_{xx}(\theta) & E_{x\pi}(\theta) \\ E_{zx}(\theta) & E_{z\pi}(\theta) \end{bmatrix} = \begin{bmatrix} E_{1x}(\theta) \\ E_{1z}(\theta) \end{bmatrix} \begin{bmatrix} E_{2x} & E_{2\pi} \end{bmatrix} \tag{3.25}
$$

where $\begin{bmatrix} E_{1x}(\theta) \\ E_{1z}(\theta) \end{bmatrix} = \sum_{j=1}^{L} \theta_j \begin{bmatrix} E_{1xj} \\ E_{1zj} \end{bmatrix}$ is full column rank for all $\theta \in \mathbb{E}^L$.

Under this assumption, we can obtain the next theorem for the robust L_2 gain (L_2 induced norm) performance of the uncertain system (3.23).

Theorem 3.6 *Suppose Assumption 3.3 holds and let* $E_2 = E_{2\pi}^{\perp} E_{2x}$. *Then, the system* (3.23) *is robustly stable and satisfies the* L_2 *induced norm property* $\|z\|_2 \le \gamma_\infty \|w\|_2$ *for all* $\theta \in \mathbb{E}^L$ *if there exist* $P_j = P_j^T$, Y_j ($j \in \mathscr{I}_L$) *and* F *such that for all* $j \in \mathscr{I}_L$:

$$
(E_2 E_2^\circ)^T P_j (E_2 E_2^\circ) \succ 0
$$

$$
\begin{bmatrix} 0 & 0 & P_{ej}^T & 0 \\ 0 & I & 0 & 0 \\ P_{ej} & 0 & 0 & 0 \\ 0 & 0 & 0 & -\overline{\gamma}_\infty^2 I \end{bmatrix} + \mathrm{He} \left\{ \mathsf{F} \begin{bmatrix} E_{1xj} & 0 & -A_j & -B_j \\ E_{1zj} & I & -C_j & -D_j \end{bmatrix} \right\} \prec 0 \tag{3.26}
$$

$$
P_{ej} = (E_2^T P_j + Y_j^T E_2^{\perp}) E_{2\pi}^{\perp}.
$$

Proof The first stage of the proof is to apply the method of Lemma 2.4: by convexity, if the conditions hold for all vertices $j \in \mathscr{I}_L$, then they also hold for all values in the convex hull when taking $P(\theta) = \sum_{j=1}^{L} \theta_j P_j$ and $Y(\theta) = \sum_{j=1}^{L} \theta_j Y_j$. The first inequality of (3.26) implies that $x^* E_2^T P(\theta) E_{2x} > 0$ for all $E_{2x} \ne 0$. Then, pre and post multiply the remaining inequality by the following vector

$$
\begin{bmatrix} \varpi^T & z^T & x^T & w^T \end{bmatrix}
$$

and its transpose, respectively, where ϖ is defined by (3.22). In view of the fact that

$$
\begin{bmatrix} E_{1x}(\theta) & 0 \\ E_{1z}(\theta) & I \end{bmatrix} \begin{bmatrix} \varpi \\ z \end{bmatrix} = \begin{bmatrix} A(\theta) & B(\theta) \\ C(\theta) & D(\theta) \end{bmatrix} \begin{bmatrix} x \\ w \end{bmatrix}
$$

holds along the trajectory of (3.23), the resulting condition is exactly

$$
z^T z - \overline{\gamma}_\infty^2 w^T w + 2\dot{x}^T E_2^T P(\theta) E_{2x} < 0
$$

for all non-zero trajectory of (3.23). In case of zero disturbances ($w = 0$) it proves stability as in the proof of Theorem 3.1. For zero initial conditions one gets

$$\int_0^t z^T(\tau)z(\tau)d\tau < \overline{\gamma}_\infty^2 \int_0^t w^T(\tau)w(\tau)d\tau - x^T(t)E_2^T P(\theta)E_2 x(t)$$

$$\leq \overline{\gamma}_\infty^2 \int_0^t w^T(\tau)w(\tau)d\tau$$

which, when t goes to infinity, reads as $\|z\|_2 \leq \overline{\gamma}_\infty \|w\|_2$. $\qquad\qquad\Box$

3.2 Reducing Size of SV-LMIs

In this section, we are slightly apart from the system analysis using SV-LMIs and concentrate our attention on the treatment of SV-LMIs themselves, i.e., how to reduce the size of SV-LMIs for computational efficiency.

All SV-LMIs build up to now are of the type

$$\mathsf{M}_j(\mathsf{P}_j) + \mathrm{He}\left\{\mathsf{FN}_j\right\} \prec 0 \quad (j \in \mathscr{I}_L). \tag{3.27}$$

Their computational complexity is both related to the size of decision variables $(\mathsf{P}_j, \mathsf{F})$, and to the size of the constraints. A nice feature when $L = 1$ is that the size of the constraints can be reduced to

$$\mathsf{N}^{T\perp}\mathsf{M}(\mathsf{P})\mathsf{N}^{T\perp T} \prec 0$$

where the S-Variables are completely removed. The issue considered in this section is how to perform such numerical complexity reduction when $L > 1$, keeping the nice feature of SV-LMIs (e.g., employing parameter-dependent Lyapunov matrices for robust performance analysis of uncertain LTI systems so that less conservative results can be obtained). It is not expected to be possible in general but particular structures of the problem allow it. The results in the following sections are sophisticated versions of [16].

3.2.1 Removing Parameter Independent Rows

Let us consider the example of the quarter-car suspension model whose descriptor form is given by (3.3). When considering stability of this system the SV-LMIs involve the following matrix

$$N_j = \begin{bmatrix} E_{xxj} & E_{x\pi j} & -A_j \end{bmatrix}$$

$$= \begin{bmatrix}
m_j & 0 & 0 & 0 & -1 & 1 & -1 & 0 & 0 & 0 & 0 \\
0 & 1 & 0 & 0 & 0 & 0 & 0 & -1 & 0 & 0 & 0 \\
0 & 0 & M_j & 0 & 1 & 0 & 1 & 0 & 0 & 0 & 0 \\
0 & 0 & 0 & 1 & 0 & 0 & 0 & 0 & 0 & -1 & 0 \\
0 & 0 & 0 & 0 & 1 & 0 & 0 & 0 & k_{1j} & 0 & -k_{1j} \\
0 & 0 & 0 & 0 & 0 & 1 & 0 & 0 & -k_{2j} & 0 & 0 \\
0 & 0 & 0 & 0 & 0 & 0 & 1 & c_j & 0 & -c_j & 0
\end{bmatrix}.$$

Assuming M and m are non zero, the factorization of Assumption 3.1 can be done with $E_{1x}(\theta) = \begin{bmatrix} E_{xx}(\theta) & E_{x\pi}(\theta) \end{bmatrix}$, $\begin{bmatrix} E_{2xx} & E_{2x\pi} \end{bmatrix} = I_7$, $E_2 = I_4$ and $E_{2x\pi}^\perp = \begin{bmatrix} I_4 & 0_{4,3} \end{bmatrix}$. Therefore, if we view the SV-LMI (3.18) in Theorem 3.3 in the form of (3.27), the matrices N_j become $N_j = \begin{bmatrix} E_{xxj} & E_{x\pi j} & -A_j \end{bmatrix}$ and have 7 rows and 11 columns. The related S-Variables F would have $11 \times 7 = 77$ decisions variables which is already a large number although the system is of order 4. One way to reduce this number of decision variables is to notice that some rows are parameter independent. Even if all parameters M_j, m_j, k_{1j}, k_{2j} and c_j are uncertain this matrix contains two rows (rows 2 and 4) that are parameter independent. The following lemma allows to take into account this information to reduce without conservatism the size of the S-Variables.

Lemma 3.1 *If there exists an invertible* T *such that for all* $j = 1, \ldots, L$

$$N_j = T \begin{bmatrix} N_{1j} \\ N_2 \end{bmatrix}, \tag{3.28}$$

then the following two LMI problems are equivalent

(i) *There exist* (P_j, F) *solution of* (3.27).
(ii) *There exist* $(P_j, \hat{F})$ *solution of*

$$\hat{M}_j(P_j) + He\{\hat{F}\hat{N}_j\} < 0, \quad j = 1, \ldots, L \tag{3.29}$$

where $\hat{M}_j(P_j) = N_2^{T\perp} M_j(P_j) N_2^{T\perp T}$ *and* $\hat{N}_j = N_{1j} N_2^{T\perp T}$.

Conditions with variables $(P_j, \hat{F})$ are both of smaller dimensions ($N_2^{T\perp}$ has less rows than columns) and have less decision variables ($\hat{F}$ has less rows than F).

Before proving the result, it is applied for illustration on the quarter-car suspension model. The rows 2 and 4 are parameter independent. The N_2 matrix that can be chosen as

$$N_2 = \begin{bmatrix}
0 & 1 & 0 & 0 & 0 & 0 & 0 & -1 & 0 & 0 & 0 \\
0 & 0 & 0 & 1 & 0 & 0 & 0 & 0 & 0 & -1 & 0
\end{bmatrix}$$

which gives

$$
N_2^{T\perp} =
\begin{bmatrix}
1\,0\,0\,0 & 0\,0\,0 & 0\,0\,0\,0 \\
0\,1\,0\,0 & 0\,0\,0 & 1\,0\,0\,0 \\
0\,0\,1\,0 & 0\,0\,0 & 0\,0\,0\,0 \\
0\,0\,0\,1 & 0\,0\,0 & 0\,0\,1\,0 \\
0\,0\,0\,0 & 1\,0\,0 & 0\,0\,0\,0 \\
0\,0\,0\,0 & 0\,1\,0 & 0\,0\,0\,0 \\
0\,0\,0\,0 & 0\,0\,1 & 0\,0\,0\,0 \\
0\,0\,0\,0 & 0\,0\,0 & 0\,1\,0\,0 \\
0\,0\,0\,0 & 0\,0\,0 & 0\,0\,0\,1
\end{bmatrix}
$$

and hence

$$
N_{1j}N_2^{T\perp T} =
\begin{bmatrix}
m_j & 0 & 0 & 0 & -1 & 1 & -1 & 0 & 0 \\
0 & 0 & M_j & 0 & 1 & 0 & 1 & 0 & 0 \\
0 & 0 & 0 & 0 & 1 & 0 & 0 & k_{1j} & -k_{1j} \\
0 & 0 & 0 & 0 & 0 & 1 & 0 & -k_{2j} & 0 \\
0 & c_j & 0 & -c_j & 0 & 0 & 1 & 0 & 0
\end{bmatrix}
$$

The resulting number of decision variables in $\hat{F}$ is $5 \cdot 9 = 45$. A 40 % reduction!

Proof of Lemma 3.1 Assume that (3.28) holds and decompose the S-Variable as $FT^{-1} = [\, F_1 \; F_2 \,]$. The SV-LMIs (3.27) then read as

$$
M_j(P_j) + \text{He}\{F_1 N_{1j}\} + \text{He}\{F_2 N_2\} < 0, \quad j = 1, \ldots, L \qquad (3.30)
$$

By elimination lemma [17] the condition is equivalent to (3.29) where $\hat{F} = N_2^{T\perp} F_1$. □

3.2.2 Removing Some Parameter Independent Columns

Let us consider again the stability test of Theorem 3.3 applied to the quarter-car suspension model. The condition (3.18) reads as

$$
\begin{aligned}
&P_j \succ 0_{4,4} \\
&\begin{bmatrix}
0_{4,4} & 0_{4,3} & P_j \\
0_{3,4} & 0_{3,3} & 0_{3,4} \\
P_j & 0_{4,3} & 0_{4,4}
\end{bmatrix}
+ \text{He}\{F [\, E_{xxj} \; E_{x\pi j} \; -A_j \,]\} < 0.
\end{aligned}
$$

Applying Lemma 3.1 the LMIs become the following

$$P_j \succ 0_{4,4}$$

$$
\begin{bmatrix}
K_1^T P_j + P_j K_1 & 0_{4,3} & P_j K_2 \\
0_{3,4} & 0_{3,3} & 0_{3,2} \\
K_2^T P_j & & 0_{2,3} & 0_{2,2}
\end{bmatrix}
$$

$$
+ \mathrm{He} \left\{ \hat{F}
\begin{bmatrix}
m_j & 0 & 0 & 0 & -1 & 1 & -1 & 0 & 0 \\
0 & 0 & M_j & 0 & 1 & 0 & 1 & 0 & 0 \\
0 & 0 & 0 & 0 & 1 & 0 & 0 & k_{1j} & -k_{1j} \\
0 & 0 & 0 & 0 & 0 & 1 & 0 & -k_{2j} & 0 \\
0 & c_j & 0 & -c_j & 0 & 0 & 1 & 0 & 0
\end{bmatrix}
\right\} \prec 0
\tag{3.31}
$$

where the indexes j indicate that these parameters are uncertain and where

$$
K_1 =
\begin{bmatrix}
0 & 1 & 0 & 0 \\
0 & 0 & 0 & 0 \\
0 & 0 & 0 & 1 \\
0 & 0 & 0 & 0
\end{bmatrix}, \quad
K_2 =
\begin{bmatrix}
0 & 0 \\
1 & 0 \\
0 & 0 \\
0 & 1
\end{bmatrix}.
$$

In this last formula of the SV-LMIs, it can be seen that the 5–7 row/columns are all independent of uncertainties. This feature happens to be exploitable for further reduction of the SV-LMIs.

Lemma 3.2 *If there exists an invertible* T *such that for all* $j = 1, \ldots, L$

$$
N_j = \begin{bmatrix} N_1 & N_{2j} \end{bmatrix} T
\tag{3.32}
$$

with a full column rank matrix N_1 *and assume moreover that*

$$
T^{-T} M_j (P_j) T^{-1} =
\begin{bmatrix}
M_{11} & M_{12} \\
M_{12}^T & M_{22j}(P_j)
\end{bmatrix}
\tag{3.33}
$$

with $M_{11} \succeq 0$, *then* (ii) *implies* (i), *where*

(i) *There exist* (P_j, F) *solution of* (3.27)
(ii) *There exit* $(P_j, \hat{F})$ *solution of*

$$
\hat{M}_j(P_j) + \mathrm{He}\{\hat{F}\hat{N}_j\} \prec 0, \quad j = 1, \ldots, L
\tag{3.34}
$$

where

$$
\hat{M}_j(P_j) = N_{2j}^T N_1^{+T} M_{11} N_1^{+} N_{2j} - \mathrm{He}\{M_{12}^T N_1^{+} N_{2j}\} + M_{22j}(P_j)
$$
$$
\hat{N}_j = N_1^{\perp} N_{2j}
$$

This lemma happens to be a generalization of Lemma 2.5.

Proof Start from (3.34) and by a small perturbation argument note that for all $W \succ 0$ there exists a possibly small positive $\varepsilon > 0$ such that

$$N_{2j}^T N_1^{+T} (M_{11}+\varepsilon W)N_1^+ N_{2j} - \text{He}\{M_{12}^T N_1^+ N_{2j}\}$$
$$+ M_{22j}(P_j) + \text{He}\left\{\hat{F}N_1^\perp N_{2j}\right\} \prec 0$$

for all $j = 1, \ldots, L$. Based on the fact that $M_{11}+\varepsilon W \succ 0$, apply a Schur complement argument to get

$$
\begin{bmatrix}
-(M_{11} + \varepsilon W) & -(M_{11} + \varepsilon W)N_1^+ N_{2j} \\
-((M_{11} + \varepsilon W)N_1^+ N_{2j})^T & M_{22j}(P_j) + \text{He}\left\{(\hat{F}N_1^\perp - M_{12}^T N_1^+)N_{2j}\right\}
\end{bmatrix} \prec 0
$$

It writes also as

$$
\begin{bmatrix}
M_{11} + \varepsilon W & M_{12} \\
M_{12}^T & M_{22j}(P_j)
\end{bmatrix}
$$
$$
+ \text{He}\left\{
\begin{bmatrix}
-(M_{11} + \varepsilon W) & -(M_{11} + \varepsilon W)N_1^+ N_{2j} \\
-M_{12}^T & (\hat{F}N_1^\perp - M_{12}^T N_1^+)N_{2j}
\end{bmatrix}
\right\} \prec 0
$$

and with the choice of $F = \begin{bmatrix} -(M_{11} + \varepsilon W)N_1^+ \\ \hat{F}N_1^\perp - M_{12}^T N_1^+ \end{bmatrix}$ it is exactly (recall that $N_1^+ N_1 = 0$ and since N_1 is full column rank $N_1^+ N_1 = I$)

$$
\begin{bmatrix}
M_{11} + \varepsilon W & M_{12} \\
M_{12}^T & M_{22j}(P_j)
\end{bmatrix}
+ \text{He}\left\{F\begin{bmatrix} N_1 & N_{2j} \end{bmatrix}\right\} \prec 0. \tag{3.35}
$$

Since $\varepsilon W \succeq 0$ one gets (3.27) at the expense of the change of basis defined by T. □

It should be noticed that the condition (ii) of Lemma 3.2 is more conservative than (i). It is a drawback of the numerical burden reduction. To understand the source of conservatism, post and pre-multiply (3.35) by $\begin{bmatrix} N_1^+ N_{2j} \\ -I \end{bmatrix}$ and its transpose respectively. The result is almost alike (3.34) (indeed $N_1^\perp$ can be chosen to be the non-zero rows of $N_1 N_1^+ - I$) except that for the S-Variable $\hat{F}$ replaced by

$$\hat{F}_j = \begin{bmatrix} N_{2j}^T N_1^{+T} & -I \end{bmatrix} F_1.$$

Condition (i) is less conservative since it corresponds to (ii) with a parameter-dependent S-Variable.

We shall now apply this lemma to the SV-LMIs (3.31) of the quarter-car suspension model. The following choice is done for the partition of the N matrix:

$$N_1 = \begin{bmatrix} -1 & 1 & -1 \\ 1 & 0 & 1 \\ 1 & 0 & 0 \\ 0 & 1 & 0 \\ 0 & 0 & 1 \end{bmatrix}, \quad N_{2j} = \left[\begin{array}{cccc|cc} m_j & 0 & 0 & 0 & 0 & 0 \\ 0 & 0 & M_j & 0 & 0 & 0 \\ 0 & 0 & 0 & 0 & k_{1j} & -k_{1j} \\ 0 & 0 & 0 & 0 & -k_{2j} & 0 \\ 0 & c_j & 0 & -c_j & 0 & 0 \end{array}\right].$$

The resulting reduced size SV-LMIs are

$$P_j \succ 0_{4,4}$$

$$\begin{bmatrix} K_1^T P_j + P_j K_1 & P_j K_2 \\ K_2^T P_j & 0_{2,2} \end{bmatrix} \tag{3.36}$$
$$+ \mathrm{He}\left\{ \hat{F} \left[\begin{array}{cccc|cc} -m_j & -c_j & 0 & c_j & -k_{1j} - k_{2j} & k_{1j} \\ 0 & c_j & -M_j & -c_j & k_{1j} & -k_{1j} \end{array}\right] \right\} \prec 0$$

where the S-Variables contains $6 \cdot 2 = 12$ elements. Its more than 84 % reduction compared to the initial 77 elements of the S-Variables. Unfortunately, there may be some conservatism when compared to (3.31). In practice, for this example, we were not able to notice any conservatism in numerical tests.

The size of the constraints has been reduced as well from 11 rows and columns to 6. Since these constraints are repeated L times it is a major reduction. In the considered case when all five parameters are uncertain and independent, the number of vertices of the polytope is $L = 2^5 = 32$. The complete dimension reduction permits to remove $5 \cdot 32 = 160$ rows and columns from the constraints.

3.3 System Augmentation and Conservatism Reduction

3.3.1 Source of Conservatism

All the robust S-Variable LMI results exposed up to this point are sufficient conditions. In order to analyze the sources of conservatism let us reformulate results related to Lemmas 2.2 and 2.4.

Lemma 3.3 *Assume* $M(\theta, P)$ *is a parameter-dependent symmetric-matrix-valued function of one or more variables gathered into a unique notation* P, *where* $M(\theta, P)$ *is affine with respect to* θ *and* P, *respectively. Assume as well a parameter dependent matrix* $N(\theta)$ *be given, where* $N(\theta)$ *is again affine respect to* θ. *With these notations the following four matrix inequality problems are defined.*

(i) *For all uncertainties* $\theta \in \mathbb{E}^L$ *there exists a parameter-dependent* $P(\theta)$ *such that the following parameter-dependent matrix inequalities hold*

$$N^{T\perp}(\theta) M(\theta, P(\theta)) N^{T\perp T}(\theta) \prec 0 \quad \forall \theta \in \mathbb{E}^L. \tag{3.37}$$

(ii) *For all uncertainties $\theta \in \mathbb{E}^L$ there exist parameter-dependent $P(\theta)$ and $F(\theta)$ such that the following parameter-dependent matrix inequalities hold*

$$M(\theta, P(\theta)) + \text{He}\{F(\theta)N(\theta)\} \prec 0 \quad \forall\, \theta \in \mathbb{E}^L. \tag{3.38}$$

(iii) *There exists a matrix F such that for all uncertainties $\theta \in \mathbb{E}^L$ there exist a parameter-dependent $P(\theta)$ such that the following parameter-dependent matrix inequalities hold*

$$M(\theta, P(\theta)) + \text{He}\{FN(\theta)\} \prec 0 \quad \forall\, \theta \in \mathbb{E}^L. \tag{3.39}$$

(iv) *There exist matrices F and P_j $(j \in \mathscr{I}_L)$ such that the following parameter-dependent matrix inequalities hold with $P(\theta) = \sum_{j=1}^{L} \theta_j P_j$*

$$M(\theta, P(\theta)) + \text{He}\{FN(\theta)\} \prec 0 \quad \forall\, \theta \in \mathbb{E}^L. \tag{3.40}$$

Then, among these four conditions, (i) and (ii) are equivalent. (iii) is a sufficient condition for (ii). (iii) and (iv) are equivalent.

Proof Equivalence between (i) and (ii) comes from Lemma 2.2 (or say, Elimination Lemma). It is trivial that (iii) is a sufficient condition for (ii), and (iv) is a sufficient condition for (iii) since (iv) imposes a restriction on $P(\theta)$ which does not exist in (iii). Remains to prove that if (iii) holds then (iv) holds as well.

Assume (iii) holds and define $P_j = P(\theta^{[j]})$ where $\theta^{[j]}$ are the parameters at vertices (*i.e.* $\theta_j^{[j]} = 1, \theta_{i \neq j}^{[j]} = 0$). Since the conditions hold for all $\theta \in \mathbb{E}^L$ these hold at the vertices, that is:

$$M_j(\theta_j, P_j) + \text{He}\{FN_j\} \prec 0$$

Due to the affine property of $M(\theta, P)$ and $N(\theta)$, this implies that (iv) holds. □

The result states the source of conservatism of the robust SV-LMI results: the arbitrary choice of parameter-independent S-Variables (condition (iii)). A by product of this arbitrary choice is that the search of parameter-dependent variables $P(\theta)$ is restricted to the subset of polytopic $P(\theta)$ (condition (iv)). To avoid both these sources of conservatism one strategy is to tackle conditions (i) or (ii) directly with mathematical tools. Most utilized such tools are analyzed in [18]. They include sum-of-squares techniques [19–21], homogeneous polynomials techniques [22–24] and results based on Polyá theorem [18, 25]. All these papers consider the problem assuming polynomially or rationally parameter-dependent variables and data. Historically, they all follow in time results of [26, 27] which, if attentively analyzed, rely on a system augmentation strategy. This strategy is now exposed within our chosen descriptor S-Variables framework.

3.3.2 Preliminary Discussions About Conservatism Reduction

Stability of LTI systems $\dot{x} = A(\theta)x$ is equivalent to proving that poles are located in the left-half of the complex plane. That is, proving regional pole location in

$$\mathbb{C}_- = \mathbb{O}\left[\begin{bmatrix} 0 & 1 \\ 1 & 0 \end{bmatrix}\right].$$

In Chap. 2, Sect. 2.5 it is suggested that one can as well study this problem by augmenting artificially the dimensions of the matrix defining the region. For example as follows:

$$\mathbb{O}\left[\begin{bmatrix} 0 & 1 \\ 1 & 0 \end{bmatrix}\right] = \mathbb{O}\left[\begin{bmatrix} 0 & I_r \\ I_r & 0 \end{bmatrix}\right] \quad \forall\, r \in \mathbb{N}.$$

The resulting SV-LMIs are of increased dimensions:

$$\begin{bmatrix} 0 & I_r \otimes P_j \\ I_r \otimes P_j & 0 \end{bmatrix} + \text{He}\left\{ \begin{bmatrix} F_1 \\ F_2 \end{bmatrix} \begin{bmatrix} I_{rn} & -I_r \otimes A_j \end{bmatrix} \right\} \prec 0 \qquad (3.41)$$

and contain many more degrees of freedom (larger size matrices F_i). However, in Chap. 2, it was proved that these extra degrees of freedom do not bring any improvement in terms of conservatism reduction. Nevertheless, let us start our reasoning from these conditions.

As a first step, notice that the condition implies existence of $\hat{P}_j = I_r \otimes P_j$ such that

$$\begin{bmatrix} 0 & \hat{P}_j \\ \hat{P}_j & 0 \end{bmatrix} + \text{He}\left\{ \begin{bmatrix} F_1 \\ F_2 \end{bmatrix} \begin{bmatrix} I_{rn} & -I_r \otimes A_j \end{bmatrix} \right\} \prec 0. \qquad (3.42)$$

In this new condition $\hat{P}_j$ matrices may as well be full-block (increasing again the degrees of freedom). Written in this way, the condition happens to coincide with SV-LMIs build for the r-times duplicated system $\dot{\hat{x}} = (I_r \otimes A(\theta))\hat{x}$.

These simple manipulations show that there are two means for increasing degrees of freedom of the SV-LMI results. One is the artificial increase of dimensions of the matrix defining the pole location problem. It leads to the LMI conditions (3.41). The second is to apply the SV-LMI result on the replicated system $\dot{\hat{x}} = (I_r \otimes A(\theta))\hat{x}$. It leads to the LMI conditions (3.42). But both these new conditions are no more or less conservative.

Let us go further in the reasoning and choose now $r = 2$ for simplicity. The last inequalities then write as

$$\begin{bmatrix} 0 & \hat{P}_j \\ \hat{P}_j & 0 \end{bmatrix} + \text{He}\left\{ \begin{bmatrix} F_1 \\ F_2 \end{bmatrix} \begin{bmatrix} I & 0 & -A_j & 0 \\ 0 & I & 0 & -A_j \end{bmatrix} \right\} \prec 0.$$

If these inequalities hold it is quite trivial to see that the following also hold (take for example $\tilde{F}_i = 0, i = 1, 2$)

$$
\begin{bmatrix} 0 & \hat{P}_j \\ \hat{P}_j & 0 \end{bmatrix} + \mathrm{He} \left\{ \begin{bmatrix} F_1 \\ F_2 \end{bmatrix} \begin{bmatrix} I & 0 & -A_j & 0 \\ 0 & I & 0 & -A_j \end{bmatrix} \right\}
$$
$$
+ \mathrm{He} \left\{ \begin{bmatrix} \tilde{F}_1 \\ \tilde{F}_2 \end{bmatrix} \begin{bmatrix} 0 & I & -I & 0 \end{bmatrix} \right\} \prec 0.
$$

For $\hat{F}_i = \begin{bmatrix} F_i & \tilde{F}_i \end{bmatrix}, i = 1, 2$ the conditions also read as

$$
\begin{bmatrix} 0 & \hat{P}_j \\ \hat{P}_j & 0 \end{bmatrix} + \mathrm{He} \left\{ \begin{bmatrix} \hat{F}_1 \\ \hat{F}_2 \end{bmatrix} \begin{bmatrix} I & 0 & -A_j & 0 \\ 0 & I & 0 & -A_j \\ 0 & I & -I & 0 \end{bmatrix} \right\} \prec 0 \qquad (3.43)
$$

We have build here some new LMI conditions that are feasible when the classical (or say, original) SV-LMI conditions hold, but from system theoretic viewpoint what do they correspond to? To answer this question, let us rewrite the inequalities (3.43) as

$$
\begin{bmatrix} 0 & \hat{P}_j \\ \hat{P}_j & 0 \end{bmatrix} + \mathrm{He} \left\{ \begin{bmatrix} \hat{F}_1 \\ \hat{F}_2 \end{bmatrix} \begin{bmatrix} \hat{E}_j & -\hat{A}_j \end{bmatrix} \right\} \prec 0.
$$

One recognizes inequalities of Theorem 3.3. These happen to be SV-LMIs proving the robust stability of the descriptor system

$$
\begin{bmatrix} I & 0 \\ 0 & I \\ 0 & I \end{bmatrix} \dot{\check{x}} = \begin{bmatrix} A(\theta) & 0 \\ 0 & A(\theta) \\ I & 0 \end{bmatrix} \check{x}
$$

which is nothing but

$$
\begin{bmatrix} \ddot{x} \\ \dot{x} \end{bmatrix} = \begin{bmatrix} A(\theta) & 0 \\ 0 & A(\theta) \end{bmatrix} \begin{bmatrix} \dot{x} \\ x \end{bmatrix} \qquad (3.44)
$$

Obviously, this is different from the "duplicated system" written by

$$
\begin{bmatrix} \dot{x}_1 \\ \dot{x}_2 \end{bmatrix} = I_2 \otimes A(\theta) \begin{bmatrix} x_1 \\ x_2 \end{bmatrix} = \begin{bmatrix} A(\theta) & 0 \\ 0 & A(\theta) \end{bmatrix} \begin{bmatrix} x_1 \\ x_2 \end{bmatrix}. \qquad (3.45)
$$

The latter is simply a duplication of the system as noted, while the former contains the information that θ is a constant uncertainty. Indeed, if not, one would have

$$
\ddot{x} = A(\theta)\dot{x} + \frac{\partial \theta}{\partial t} \frac{\partial A(\theta)}{\partial \theta} x.
$$

In the following (3.44) is called the first degree augmented system. It is trivially stable if and only if the original system is stable. Thus (3.43) is a sufficient condition for the stability of the system. Moreover, it has been shown to be no more conservative than (3.42) and (3.41). System augmentation by including in the model higher degree derivatives of the state is the strategy that is adopted and analyzed in the following.

Before stating the general theorems, let us understand the reason for conservatism reduction. To this end, apply Lemma 2.3 to transform the condition on vertices to a condition over the polytope

$$
\begin{bmatrix} 0 & \hat{P}(\theta) \\ \hat{P}(\theta) & 0 \end{bmatrix} + \mathrm{He}\left\{ \begin{bmatrix} \hat{F}_1 \\ \hat{F}_2 \end{bmatrix} \begin{bmatrix} I & 0 & -A(\theta) & 0 \\ 0 & I & 0 & -A(\theta) \\ 0 & I & -I & 0 \end{bmatrix} \right\} \prec 0 \qquad (3.46)
$$

where $\hat{P}(\theta) = \sum_{j=1}^{L} \theta_j \hat{P}_j$. Then, pre and post multiply (3.46) by $\begin{bmatrix} \ddot{x}^T & \dot{x}^T & \dot{x}^T & x^T \end{bmatrix}$ and its transpose respectively. It implies that for all $\theta \in \mathbb{E}^L$

$$
2\begin{bmatrix} \ddot{x}^T & \dot{x}^T \end{bmatrix} \hat{P}(\theta) \begin{bmatrix} \dot{x} \\ x \end{bmatrix} < 0, \quad \forall \begin{bmatrix} \ddot{x} \\ \dot{x} \end{bmatrix} = \begin{bmatrix} A(\theta) & 0 \\ 0 & A(\theta) \end{bmatrix} \begin{bmatrix} \dot{x} \\ x \end{bmatrix}.
$$

Equivalently it reads as $\dot{V}(x, \theta) < 0$ negative along the trajectories of the system where

$$
V(x, \theta) = \begin{bmatrix} \dot{x}^T & x^T \end{bmatrix} \hat{P}(\theta) \begin{bmatrix} \dot{x} \\ x \end{bmatrix} = x^T \begin{bmatrix} A^T(\theta) & I_n \end{bmatrix} \hat{P}(\theta) \begin{bmatrix} A(\theta) \\ I_n \end{bmatrix} x.
$$

Stability is proved with a quadratic Lyapunov certificate that depends polynomially of θ as a polynomial of order 3. It is clearly less conservative than when considering an affine dependent certificate. The SV-LMIs (3.43) for the augmented system thus cope with the source of conservatism defined in Lemma 3.3 condition (iv). Yet, as it has been proved in the lemma, the source of conservatism actually comes from choosing parameter-independent S-Variables. The following derivations show that conservatism reduction is indeed related to the usage of a parameter-dependent S-Variable. Indeed, pre and post multiply (3.46) by

$$
\begin{bmatrix} A^T(\theta) I & 0 & 0 \\ 0 & 0 & A^T(\theta) I \end{bmatrix}
$$

and its transpose respectively. The result happens to be exactly

$$
\begin{bmatrix} 0 & P(\theta) \\ P(\theta) & 0 \end{bmatrix} + \mathrm{He}\left\{ \begin{bmatrix} F_1(\theta) \\ F_2(\theta) \end{bmatrix} \begin{bmatrix} I & -A(\theta) \end{bmatrix} \right\} \prec 0
$$

where the parameter dependent matrices are

$$P(\theta) = \begin{bmatrix} A^T(\theta) & I_n \end{bmatrix} \hat{P}(\theta) \begin{bmatrix} A(\theta) \\ I_n \end{bmatrix}, \quad F_i(\theta) = \begin{bmatrix} A^T(\theta) & I_n \end{bmatrix} \hat{F}_i \begin{bmatrix} A(\theta) \\ I_n \\ I_n \end{bmatrix}.$$

This indicates that the SV-LMIs (3.43) for the augmented system do cope with the source of conservatism corresponding to the implication (iii)⇒(ii) in Lemma 3.3. Implicitly, the S-Variables for the augmented system define parameter-dependent S-Variables for the original system. Applying SV-LMI results to the first degree augmented system leads to conditions where the Lyapunov matrix is a polynomial of order three with respect to the parameters and the S-Variables are polynomials of order two. With further degree augmented system representations, the SV-LMIs correspond to Lyapunov matrices and S-Variables of higher order, thus to less and less conservative conditions (at least no more conservative).

3.3.3 Robust Stability

Consider the system (3.15) of order n (i.e., $x \in \mathbb{R}^n$) and let by definition $x^{(0)} = x$, $\pi^{(0)} = \pi$:

$$E_{xx}(\theta)\dot{x}^{(0)} + E_{x\pi}(\theta)\pi^{(0)} = A(\theta)x^{(0)}. \tag{3.47}$$

System augmentation amounts to duplicating this dynamic equation several times

$$E_{xx}(\theta)\dot{x}^{(i)} + E_{x\pi}(\theta)\pi^{(i)} = A(\theta)x^{(i)} \quad i = 0, \ldots, \kappa - 1. \tag{3.48}$$

Here, $\dot{x}^{(i)} := \frac{d^i}{dt^i}\dot{x}(t)$, etc. In all these equations, by definition, $\dot{x}^{(i)}$ is the derivative of $x^{(i)}$. The natural way of linking the time derivatives would be to write that $\dot{x}^{(i)} = x^{(i+1)}$ and $\pi^{(i)} = \dot{\pi}^{(i-1)}$. But for descriptor systems one has to avoid creating a system with impulsive modes when the original one does not contain any. The way to do so is to link the duplicated systems as follows

$$E_{xx}(\theta)\dot{x}^{(i)} + E_{x\pi}(\theta)\pi^{(i)} = E_{xx}(\theta)x^{(i+1)} + E_{x\pi}(\theta)\pi_I^{(i)} \quad i = 0, \ldots, \kappa - 2 \tag{3.49}$$

where $\pi^{(i)}$ $i = 1, \ldots, \kappa - 1$ are considered as derivatives of some implicit states of the system $\dot{\pi}_I^{(i)} = \pi^{(i+1)}$ $i = 0, \ldots, \kappa - 2$. By definition $\pi_I^{(i)}$ is a function of $\pi^{(i+1)}$ and for $i > 1$ (3.48) also reads as

$$E_{xx}(\theta)\dot{x}^{(i)} + E_{x\pi}(\theta)\dot{\pi}_I^{(i-1)} = A(\theta)x^{(i)} \quad i = 1, \ldots, \kappa - 1. \tag{3.50}$$

Notice that any non-dynamic mode of the system indexed by i is such that the left hand side of the linking expression (3.49) is zero. It therefore has no influence on the system indexed by $i + 1$. The system indexed by $i + 1$ contains only the derivatives of the differentiable modes of the one indexed by i.

$\kappa - 1 \geq 0$ is the degree of augmentation: $\kappa = 1$ when there is no augmentation, $\kappa = 2$ when first order derivatives are involved, $\kappa = 3$ when first and second order derivatives are involved, etc.

The following notation is introduced to simplify some formulas:

$$J_{\alpha,\beta,\gamma} = \left[0_{\beta,\alpha} \; I_\beta \; 0_{\beta,\gamma} \right]. \tag{3.51}$$

Some examples of this notation that are used in the next formulas are

$$J_{1,\kappa-1,0} = \left[0_{\kappa-1,1} \; I_{\kappa-1} \right], \quad J_{0,\kappa-1,1} = \left[I_{\kappa-1} \; 0_{\kappa-1,1} \right],$$

$$J_{0,\kappa-1,1}^T = \left[\begin{array}{c} I_{\kappa-1} \\ 0_{1,\kappa-1} \end{array} \right], \quad J_{\kappa-1,1,0}^T = \left[\begin{array}{c} 0_{\kappa-1,1} \\ 1 \end{array} \right].$$

Combining Eqs. (3.47), (3.50) and (3.49), the obtained degree $\kappa - 1$ augmented model writes as

$$\overbrace{\left[E_{xx\kappa}(\theta) \; E_{x\pi\kappa}(\theta) \right]}$$

$$\left[\begin{array}{cc} I_\kappa \otimes E_{xx}(\theta) & I_\kappa \otimes E_{x\pi}(\theta) \\ J_{1,\kappa-1,0} \otimes E_{xx}(\theta) & J_{1,\kappa-1,0} \otimes E_{x\pi}(\theta) \end{array} \right] \left[\begin{array}{c} \dot{x}_\kappa \\ \pi \end{array} \right]$$

$$= \underbrace{\left[\begin{array}{cc} I_\kappa \otimes A(\theta) & 0 \\ J_{0,\kappa-1,1} \otimes E_{xx}(\theta) & I_{\kappa-1} \otimes E_{x\pi}(\theta) \end{array} \right]}_{A_\kappa(\theta)} x_\kappa \tag{3.52}$$

where

$$x_\kappa^T = \left[x^{(\kappa-1)T} \; \cdots \; x^{(1)T} \; x^{(0)T} \; \pi_I^{(\kappa-2)T} \; \cdots \; \pi_I^{(1)T} \; \pi_I^{(0)T} \right].$$

$x^{(0)}$ is the state of the system to be analyzed. It is solely constrained by the system dynamics. $x^{(1)}$ is both constrained by a duplicate of the system dynamics and forced to coincide with the differentiable part of $x^{(0)}$ when evaluated on the descriptor left-hand side linear mapping of the system dynamics. Recursively $x^{(i+1)}$ is constrained in the same way with respect to $x^{(i)}$ for all $i = 0, \ldots, \kappa - 1$. Stability of the augmented system (3.52) implies convergence to zero of $x^{(0)}$ for all initial conditions and thus implies stability of the original system.

Let us apply Theorem 3.3 to the augmented model. The first step is to check that Assumption 3.1 holds for the augmented system. It is indeed the case if Assumption 3.1 holds for the original system then

$$\left[E_{xx\kappa}(\theta) \; E_{x\pi\kappa}(\theta) \right] = \underbrace{\left[\begin{array}{c} I_\kappa \otimes E_{1x}(\theta) \\ J_{1,\kappa-1,0} \otimes E_{1x}(\theta) \end{array} \right]}_{E_{1x\kappa}(\theta)} \underbrace{\left[I_\kappa \otimes E_{2xx} \; I_\kappa \otimes E_{2x\pi} \right]}_{\left[E_{2xx\kappa} \; E_{2x\pi\kappa} \right]}$$

where $E_{1x\kappa}(\theta)$ is full column rank if $E_{1x}(\theta)$ is full column rank. Note that $E_{2x\pi\kappa} = J_{\kappa-1,1,0}^T \otimes E_{2x\pi}$ is the last block column of the $\begin{bmatrix} E_{2xx\kappa} & E_{2x\pi\kappa} \end{bmatrix}$ matrix while $E_{2xx\kappa} = \begin{bmatrix} I_\kappa \otimes E_{2xx} & J_{0,\kappa-1,1}^T \otimes E_{2x\pi} \end{bmatrix}$. With this factorization we have

$$E_{2x\pi\kappa}^\perp = \begin{bmatrix} J_{0,\kappa-1,1} \otimes I_n \\ J_{\kappa-1,1,0} \otimes E_{2x\pi}^\perp \end{bmatrix},$$

$$E_{2\kappa} = \begin{bmatrix} J_{0,\kappa-1,1} \otimes E_{2xx} & I_{\kappa-1} \otimes E_{2x\pi} \\ J_{\kappa-1,1,0} \otimes E_2 & 0 \end{bmatrix}, \qquad (3.53)$$

$$E_{2\kappa}^\perp = \begin{bmatrix} I_{\kappa-1} \otimes \begin{bmatrix} E_{2xx} & E_{2x\pi} \end{bmatrix}^\perp & 0 \\ 0 & E_2^\perp \end{bmatrix}.$$

Corollary 3.1 *Suppose Assumption 3.1 holds and let* $E_2 = E_{2x\pi}^\perp E_{2xx}$. *Take any integer* $\kappa \in \mathbb{N}_+$ *and let the* $E_{2\kappa}$, $E_{2x\pi\kappa}$ *matrices defined above. Then the system* (3.15) *is robustly stable (that is stable for all* $\theta \in \mathbb{E}^L$) *if there exist* $P_{j\kappa} = P_{j\kappa}^T$, $Y_{j\kappa}$ ($j \in \mathscr{I}_L$), $F_{1\kappa}$, $F_{2\kappa}$ *and* $F_{3\kappa}$ *such that for all* $j \in \mathscr{I}_L$:

$$(E_{2\kappa} E_{2\kappa}^\circ)^T P_{j\kappa} (E_{2\kappa} E_{2\kappa}^\circ) \succ 0$$

$$\begin{bmatrix} 0 & P_{ej\kappa}^T \\ P_{ej\kappa} & 0 \end{bmatrix} + \mathrm{He}\left\{ \begin{bmatrix} F_{1\kappa} \\ F_{2\kappa} \\ F_{3\kappa} \end{bmatrix} M_{j\kappa} \right\} \prec 0$$

$$\qquad (3.54)$$

$$P_{ej\kappa} = (E_{2\kappa}^T P_{j\kappa} + Y_{j\kappa}^T E_{2\kappa}^\perp) E_{2x\pi\kappa}^\perp$$
$$M_{j\kappa} = \begin{bmatrix} I_\kappa \otimes E_{1xj} & -I_\kappa \otimes A_j & 0 \\ J_{1,\kappa-1,0} \otimes E_{1xj} & -J_{0,\kappa-1,1} \otimes E_{xxj} & -I_{\kappa-1} \otimes E_{x\pi j} \end{bmatrix}$$

The corollary is a simple application of Theorem 3.3. Following the methodology of the proof of that theorem, the underlying Lyapunov function is the following:

$$V_\kappa(x, \theta) = x_\kappa^T E_{2\kappa}^T P_\kappa(\theta) E_{2\kappa} x_\kappa$$

where $P_\kappa(\theta) = \sum_{j=1}^L \theta_j P_{j\kappa}$ and x_κ is the vector containing the augmented system states. These are not independent. Combining Eqs. (3.48) and (3.49) one notices that the states are cross linked by the parameter-dependent algebraic equations

$$E_{xx}(\theta)x^{(i+1)} + E_{x\pi}(\theta)\pi_I^{(i)} = A(\theta)x^{(i)} \quad i = 0, \dots, \kappa - 2. \qquad (3.55)$$

The Lyapunov function $V_\kappa(x, \theta)$ is thus a parameter-dependent function of $x = x^{(0)}$. To get an explicit formula one has to solve the algebraic equations (3.55).

In the special case when $E_{xx}(\theta)$ is square invertible and $E_{x\pi} = 0$, the algebraic equations can be solved explicitly

$$x^{(i)} = \left[E_{xx}^{-1}(\theta) A(\theta) \right]^i x.$$

In that case one can take $E_2 = E_{2xx} = I$, and the Lyapunov function has the following explicit parameter-dependent expression $V_\kappa(x, \theta) = x^T P(\theta)x$ where

$$
P(\theta) = \begin{bmatrix} (E_{xx}^{-1}(\theta)A(\theta))^{\kappa-1} \\ \vdots \\ E_{xx}^{-1}(\theta)A(\theta) \\ I \end{bmatrix}^T P_\kappa(\theta) \begin{bmatrix} (E_{xx}^{-1}(\theta)A(\theta))^{\kappa-1} \\ \vdots \\ E_{xx}^{-1}(\theta)A(\theta) \\ I \end{bmatrix}
$$

it is polynomial of degree up to $2(\kappa - 1)$ with respect to the rational elements of $E_{xx}^{-1}(\theta)A(\theta)$, times a first order polynomial with respect to θ. When $E_{xx}(\theta) = E_{xx}$ is parameter-independent, $P(\theta)$ is a polynomial of degree up to $2\kappa - 1$ with respect to θ.

Let us now illustrate Corollary 3.1 in the case of non-descriptor models: $E_{xx} = I$, $E_{x\pi} = 0$. In that case the inequalities of Corollary 3.1 write as

$$
\begin{bmatrix} 0 & P_{j\kappa} \\ P_{j\kappa} & 0 \end{bmatrix} + \mathrm{He}\left\{ \begin{bmatrix} F_{1\kappa} \\ F_{2\kappa} \end{bmatrix} \begin{bmatrix} I_{\kappa n_x} & -I_\kappa \otimes A_j \\ J_{1,\kappa-1,0} \otimes I_{n_x} & -J_{0,\kappa-1,1} \otimes I_{n_x} \end{bmatrix} \right\} \prec 0. \tag{3.56}
$$

The dimensions of these inequalities are huge as κ is taken large. Both in terms of size of the constraints and the number of elements in the S-Variables. This issue can be tackled by application of Lemma 3.1. Take

$$
\begin{bmatrix} J_{1,\kappa-1,0} \otimes I_{n_x} & -J_{0,\kappa-1,1} \otimes I_{n_x} \end{bmatrix}^{T\perp T} = \begin{bmatrix} J_{0,\kappa,1} \otimes I_{n_x} \\ J_{1,\kappa,0} \otimes I_{n_x} \end{bmatrix}
$$

the LMI problem reduces without conservatism to the conditions of the next corollary.

Corollary 3.2 *Assume a non-descriptor systems ($E_{xx} = I$, $E_{x\pi} = 0$) and take any integer $\kappa \in \mathbb{N}_+$. The system (3.15) is robustly stable (that is stable for all $\theta \in \mathbb{E}^L$) if there exist $P_{j\kappa} \in \mathbb{S}^{\kappa n_x}$, $Y_{j\kappa}$ ($j \in \mathscr{I}_L$) and $\hat{F}_\kappa$ such that for all $j \in \mathscr{I}_L$:*

$$
P_{j\kappa} \succ 0
$$

$$
\mathrm{He}\left\{ (J_{0,\kappa,1}^T \otimes I_{n_x})P_{j\kappa}(J_{1,\kappa,0} \otimes I_{n_x}) + \hat{F}_\kappa\big[J_{0,\kappa,1} \otimes I_{n_x} - J_{1,\kappa,0} \otimes A_j\big] \right\} \prec 0. \tag{3.57}
$$

The size of the inequalities can be further reduced if some of the rows of $A(\theta)$ are parameter independent.

Corollary 3.1 provides a sequence of LMI conditions as κ is increased. The larger κ the higher order of the polynomial dependence of the Lyapunov function used to assess stability. Conservatism is therefore expected to be reduced when κ grows. It is indeed the case.

Lemma 3.4 *Define $\mathbb{F}_\kappa = \{(P_{j\kappa}, Y_{j\kappa}, F_{1\kappa}, F_{2\kappa}) : (3.54)\}$ the set of solutions to constraints (3.54) for a fixed κ. If $\mathbb{F}_\kappa \neq \emptyset$ then for all $\tilde{\kappa} \geq \kappa$ one has $\mathbb{F}_{\tilde{\kappa}} \neq \emptyset$.*

Lemma 3.4 indicates that the derivative augmentation provides a sequence of LMI conditions of non increasing conservatism. The underlying Lyapunov matrices are of increasing order of dependence with respect to parameters. One can expect that in practice the conservatism may vanish for κ large enough. At this point, there is no clear methodology to guarantee that it is indeed the case whatever considered system. In the following (Sect. 3.4) tests are provided to check on concrete examples if it is actually the case. Before that, the augmented model stability conditions are extended to performance issues.

3.3.4 Regional Finite Pole Location

The upper followed reasoning applies readily to the regional pole location problem. The following is a corollary to Theorem 3.5 when applied to the augmented system.

Corollary 3.3 *Suppose Assumption* 3.1 *holds and let* $E_2 = E_{2x\pi}^{\perp} E_{2xx}$. *Take any integer* $\kappa \in \mathbb{N}_+$ *and let* $E_{2\kappa}$, $E_{2x\pi\kappa}$ *matrices defined in* (3.53). *Then, the system* (3.15) *has finite poles robustly located in* $\mathbb{O}(R)$ *(that is for all* $\theta \in \mathbb{E}^L$ *) if there exist* $P_{j\kappa} = P_{j\kappa}^T$, $Y_{j\kappa}$ *(* $j \in \mathscr{I}_L$ *) and* F_κ *such that for all* $j \in \mathscr{I}_L$:

$$(E_{2\kappa} E_{2\kappa}^{\circ})^T P_{j\kappa} (E_{2\kappa} E_{2\kappa}^{\circ}) \succ 0$$

$$\begin{bmatrix} R_{11} \otimes (E_{2x\pi\kappa}^{\perp T} P_{j\kappa} E_{2x\pi\kappa}^{\perp}) & R_{12} \otimes P_{ej\kappa}^* \\ R_{12}^* \otimes P_{ej\kappa} & R_{22} \otimes (E_{2\kappa}^T P_{j\kappa} E_{2\kappa}) \end{bmatrix} + \mathrm{He}\left\{\mathsf{F}_\kappa M_{j\kappa}\right\} \prec 0$$

$$P_{ej\kappa} = (E_{2\kappa}^T P_{j\kappa} + Y_{j\kappa}^T E_{2\kappa}^{\perp}) E_{2x\pi\kappa}^{\perp}$$

$$M_{j\kappa} = \begin{bmatrix} I_d \otimes I_\kappa \otimes E_{1xj} & -I_d \otimes I_\kappa \otimes A_j & 0 \\ I_d \otimes J_{1,\kappa-1,0} \otimes E_{1xj} & -I_d \otimes J_{0,\kappa-1,1} \otimes E_{xxj} & -I_d \otimes I_{\kappa-1} \otimes E_{x\pi j} \end{bmatrix} \tag{3.58}$$

Moreover, if R_{11} *has one or more non-negative eigenvalue, then the system is robustly impulse free.*

All discussed properties such as conservatism reduction as κ increases and size reduction of LMI conditions when taking into account independent rows of the data apply similarly for stability and pole location. For example, in the special case of non-descriptor systems we have the following reduced size conditions.

Corollary 3.4 *Assume a non-descriptor systems (*$E_{xx} = I$, $E_{x\pi} = 0$*) and take any integer* $\kappa \in \mathbb{N}_+$. *Then, the system* (3.15) *has finite poles robustly in* $\mathbb{O}(R)$ *(that is for all* $\theta \in \mathbb{E}^L$ *) if there exist* $P_{j\kappa} \in \mathbb{S}^{\kappa n_x}$, $Y_{j\kappa}$ *(* $j \in \mathscr{I}_L$ *) and* $\hat{F}_\kappa$ *such that for all* $j \in \mathscr{I}_L$:

$$P_{j\kappa} \succ 0$$

$$
\begin{aligned}
&R_{11} \otimes \left[(J_{0,\kappa,1}^T \otimes I_{n_x}) P_{j\kappa} (J_{0,\kappa,1} \otimes I_{n_x}) \right] \\
&+ \mathrm{He} \left\{ R_{12} \otimes \left[(J_{0,\kappa,1}^T \otimes I_{n_x}) P_{j\kappa} (J_{1,\kappa,0} \otimes I_{n_x}) \right] \right\} \\
&+ R_{22} \otimes \left[(J_{1,\kappa,0}^T \otimes I_{n_x}) P_{j\kappa} (J_{1,\kappa,0} \otimes I_{n_x}) \right] \\
&+ \mathrm{He} \left\{ \hat{F}_\kappa \left[I_d \otimes J_{0,\kappa,1} \otimes I_{n_x} - I_d \otimes J_{1,\kappa,0} \otimes A_j \right] \right\} \prec 0.
\end{aligned}
\tag{3.59}
$$

3.3.5 L_2-induced Norm Performance

The original system with supper-scripts (0), indicating that no derivative augmentation is performed, writes as

$$
E_{xx}(\theta)\dot{x}^{(0)} + E_{x\pi}(\theta)\pi^{(0)} = A(\theta)x^{(0)} + B(\theta)w^{(0)}
\tag{3.60}
$$
$$
E_{zx}(\theta)\dot{x}^{(0)} + E_{z\pi}(\theta)\pi^{(0)} + z = C(\theta)x^{(0)} + D(\theta)w^{(0)}.
$$

The system augmentation amounts to duplicating the input free dynamics

$$
E_{xx}(\theta)\dot{x}^{(i)} + E_{x\pi}(\theta)\dot{\pi}_I^{(i-1)} = A(\theta)x^{(i)} \qquad i = 1, \ldots, \kappa - 1
\tag{3.61}
$$

exactly the same way as previously. To constrain the added dynamics the same linking equations as (3.49) are defined.

$$
E_{xx}(\theta)\dot{x}^{(i)} + E_{x\pi}(\theta)\dot{\pi}_I^{(i-1)} = E_{xx}(\theta)x^{(i+1)} + E_{x\pi}(\theta)\pi_I^{(i)} \quad i = 0, \ldots, \kappa - 2.
\tag{3.62}
$$

The obtained degree $\kappa - 1$ augmented model writes as

$$
E_{xx\kappa}(\theta)\dot{x}_\kappa + E_{x\pi\kappa}(\theta)\pi = A_\kappa(\theta)x_\kappa + B_\kappa(\theta)w
\tag{3.63}
$$
$$
E_{zx\kappa}(\theta)\dot{x}_\kappa + E_{z\pi}(\theta)\pi + z = C_\kappa(\theta)x_\kappa + D(\theta)w
$$

where x_κ, $E_{xx\kappa}(\theta)$, $E_{x\pi\kappa}(\theta)$ and $A_\kappa(\theta)$ are the same as previously, and

$$
E_{zx\kappa}(\theta) = \left[J_{\kappa-1,1,0}^T \otimes E_{zx}(\theta)\ 0 \right],
$$

$$
B_\kappa(\theta) = \begin{bmatrix} J_{\kappa-1,1,0}^T \otimes B(\theta) \\ 0 \end{bmatrix}, \quad C_\kappa(\theta) = \left[J_{\kappa-1,1,0} \otimes C(\theta)\ 0 \right].
$$

Theorem 3.6 can be applied to the augmented system. The result is the following

Corollary 3.5 *Suppose Assumption 3.3 holds and let* $E_2 = E_{2\pi}^\perp E_{2x}$. *Take any integer* $\kappa \in \mathbb{N}_+$ *and let as well*

$$E_{2\pi\kappa}^{\perp} = \begin{bmatrix} J_{0,\kappa-1,1} \otimes I_n \\ J_{\kappa-1,1,0} \otimes E_{2\pi}^{\perp} \end{bmatrix}, \quad E_{2\kappa} = \begin{bmatrix} J_{0,\kappa-1,1} \otimes E_{2x} & I_{\kappa-1} \otimes E_{2\pi} \\ J_{\kappa-1,1,0} \otimes E_2 & 0 \end{bmatrix},$$

$$E_{2\kappa}^{\perp} = \begin{bmatrix} I_{\kappa-1} \otimes \begin{bmatrix} E_{2x} & E_{2\pi} \end{bmatrix} & 0 \\ 0 & E_2^{\perp} \end{bmatrix}.$$

Then, the system (3.23) is robustly stable and satisfies the L_2 induced norm property
$\|z\|_2 \le \gamma_\infty \|w\|_2$ *for all* $\theta \in \mathbb{E}^L$ *if there exist* $P_{j\kappa} = P_{j\kappa}^T$, $Y_{j\kappa}$ $(j \in \mathscr{I}_L)$ *and* F_κ *such that for all* $j \in \mathscr{I}_L$:

$$(E_{2\kappa} E_{2\kappa}^\circ)^T P_{j\kappa} (E_{2\kappa} E_{2\kappa}^\circ) \succ 0$$

$$\begin{bmatrix} 0 & 0 & P_{ej\kappa}^T & 0 \\ 0 & I & 0 & 0 \\ P_{ej\kappa} & 0 & 0 & 0 \\ 0 & 0 & 0 & -\gamma_\infty^2 I \end{bmatrix} + \mathrm{He}\left\{\mathsf{F}_\kappa M_{j\kappa}\right\} \prec 0$$

$$P_{ej\kappa} = (E_{2\kappa}^T P_{j\kappa} + Y_{j\kappa}^T E_{2\kappa}^{\perp}) E_{2\pi\kappa}^{\perp}$$

$$M_{j\kappa} =$$

$$\begin{bmatrix} I_\kappa \otimes E_{1xj} & 0 & -I_\kappa \otimes A_j & 0 & -J_{\kappa-1,1,0}^T \otimes B_j \\ J_{1,\kappa-1,0} \otimes E_{1xj} & 0 & -J_{0,\kappa-1,1} \otimes E_{xxj} & -I_{\kappa-1} \otimes E_{x\pi j} & 0 \\ J_{\kappa-1,1,0} \otimes E_{1zj} & I & -J_{\kappa-1,1,0} \otimes C_j & 0 & -D_j \end{bmatrix}.$$
$$(3.64)$$

Here again the properties such as conservatism reduction as κ increases and size reduction of LMI conditions when taking into account independent rows of the data apply. The special case of non-descriptor systems is considered below as an example of these properties.

Consider $E_{xx} = I$ and $E_{x\pi} = 0$. In that case $M_{j\kappa}$ contains vertex independent rows that satisfy

$$\begin{bmatrix} J_{1,\kappa-1,0} \otimes I_{n_x} & -J_{0,\kappa-1,1} \otimes I_{n_x} & 0 \end{bmatrix}^{T \perp T} = \begin{bmatrix} J_{0,\kappa,1} \otimes I_{n_x} & 0 \\ J_{1,\kappa,0} \otimes I_{n_x} & 0 \\ 0 & I_{n_w} \end{bmatrix}.$$

Moreover, one can notice parameter independents columns (second block columns in $M_{j\kappa}$). Application of both Lemmas 3.1 and 3.2 gives the following corollary.

Corollary 3.6 *Assume a non-descriptor systems ($E_{xx} = I$, $E_{x\pi} = 0$) and take any integer $\kappa \in \mathbb{N}_+$. Then, the system (3.23) is robustly stable and satisfies the L_2 induced norm property $\|z\|_2 \le \gamma_\infty \|w\|_2$ for all $\theta \in \mathbb{E}^L$ if there exist $P_{j\kappa} = P_{j\kappa}^T$, $Y_{j\kappa}$ ($j \in \mathscr{I}_L$) and $\hat{\mathsf{F}}_\kappa$ such that for all $j \in \mathscr{I}_L$:*

$$P_{jκ} \succ 0$$

$$
\left[
\begin{array}{cc}
\text{He}\left\{\left[(J_{0,κ,1}^T \otimes I_{n_x})P_{jκ}(J_{1,κ,0} \otimes I_{n_x})\right]\right\} & 0 \\
0 & 0
\end{array}
\right]
\tag{3.65}
$$
$$
+ \left[
\begin{array}{cc}
J_{κ,1,0} \otimes I_{n_x} & 0 \\
0 & I_{n_w}
\end{array}
\right]^T
\left[
\begin{array}{cc}
\hat{Θ}_{11j} & \hat{Θ}_{12j} \\
\hat{Θ}_{12j}^T & \hat{Θ}_{22j}
\end{array}
\right]
\left[
\begin{array}{cc}
J_{κ,1,0} \otimes I_{n_x} & 0 \\
0 & I_{n_w}
\end{array}
\right]
$$
$$
+ \text{He}\left\{\hat{F}_κ\left[J_{0,κ,1} \otimes I_{n_x} - J_{1,κ,0} \otimes A_j - J_{κ-1,1,0}^T \otimes B_j\right]\right\} \prec 0
$$

and $\hat{Θ}_{11j}$, $\hat{Θ}_{12j}$, $\hat{Θ}_{22j}$ are given by

$$
\left[
\begin{array}{cc}
\hat{Θ}_{11j} & \hat{Θ}_{12j} \\
\hat{Θ}_{12j}^T & \hat{Θ}_{22j}
\end{array}
\right]
=
\left[
\begin{array}{cc}
C_j & D_j \\
0 & I
\end{array}
\right]^T
\left[
\begin{array}{cc}
I & 0 \\
0 & -\overline{γ}_\infty^2 I
\end{array}
\right]
\left[
\begin{array}{cc}
C_j & D_j \\
0 & I
\end{array}
\right].
$$

To further illustrate the result, take $κ = 3$. Applying Lemma 3.1 backwards, we can rewrite the LMI condition (3.65) as

$$
\begin{array}{l}
\left[
\begin{array}{ccc}
0 & P_{j3} & 0 \\
P_{j3}\left[\begin{array}{ccc}0&0&0\\0&0&0\\0&0&\hat{Θ}_{11j}\end{array}\right] & \left[\begin{array}{c}0\\0\\ \hat{Θ}_{12j}\end{array}\right] \\
0 \quad \left[0\ 0\ \hat{Θ}_{12j}^T\right] & \hat{Θ}_{22j}
\end{array}
\right] \\[2em]
+ \text{He}\left\{
\left[
\begin{array}{c}
F_1 \\
F_2 \\
F_3
\end{array}
\right]
\left[
\begin{array}{ccc}
I&0&0 \\
0&I&0 \\
0&0&I \\
0&I&0 \\
0&0&I
\end{array}
\right]
\left[
\begin{array}{ccc}
-A_j & 0 & 0 \\
0 & -A_j & 0 \\
0 & 0 & -A_j \\
-I & 0 & 0 \\
0 & -I & 0
\end{array}
\right]
\left[
\begin{array}{c}
0 \\
0 \\
-B_j \\
0 \\
0
\end{array}
\right]
\right\} \prec 0.
\end{array}
$$

The inequality resembles much to that of (3.43). This is the L_2 performance counterpart and for second degree system augmentation. Similarly to the reasoning that follows (3.43) and with the property that

$$
\left[
\begin{array}{ccc|cc|c}
I&0&0 & -A & 0 & 0 & 0 \\
0&I&0 & 0 & -A & 0 & 0 \\
0&0&I & 0 & 0 & -A & -B \\
0&I&0 & -I & 0 & 0 & 0 \\
0&0&I & 0 & -I & 0 & 0
\end{array}
\right]
\left[
\begin{array}{ccc}
A^2 & 0 & 0 \\
A & 0 & 0 \\
I & 0 & 0 \\
0 & A^2 & AB \\
0 & A & B \\
0 & I & 0 \\
0 & 0 & I
\end{array}
\right]
=
\left[
\begin{array}{c}
A^2 \\
A \\
I \\
A \\
I
\end{array}
\right]
\left[
\begin{array}{ccc}
I & -A & -B
\end{array}
\right]
$$

one obtains that the following parameter-dependent condition holds

$$P(\theta) \succ 0$$

$$\begin{bmatrix} 0 & P(\theta)^T & 0 \\ P(\theta) & \hat{\Theta}_{11}(\theta) & \hat{\Theta}_{12}(\theta) \\ 0 & \hat{\Theta}_{12}^T(\theta) & \hat{\Theta}_{22}(\theta) \end{bmatrix} + \mathrm{He} \left\{ \begin{bmatrix} F_1(\theta) \\ F_2(\theta) \\ F_3(\theta) \end{bmatrix} \begin{bmatrix} I & -A(\theta) & -B(\theta) \end{bmatrix} \right\} \prec 0 \qquad (3.66)$$

$$\text{where} \quad \begin{bmatrix} \hat{\Theta}_{11}(\theta) & \hat{\Theta}_{12}(\theta) \\ \hat{\Theta}_{12}^T(\theta) & \hat{\Theta}_{22}(\theta) \end{bmatrix} = \begin{bmatrix} C(\theta) & D(\theta) \\ 0 & I \end{bmatrix}^T \begin{bmatrix} I & 0 \\ 0 & -\gamma_\infty^2 I \end{bmatrix} \begin{bmatrix} C(\theta) & D(\theta) \\ 0 & I \end{bmatrix},$$

$$P(\theta) = \begin{bmatrix} A^2(\theta) \\ A(\theta) \\ I_n \end{bmatrix}^T \left(\sum_{j=1}^{L} \theta_j P_{j3} \right) \begin{bmatrix} A^2(\theta) \\ A(\theta) \\ I_n \end{bmatrix}, \quad F_i(\theta) = \begin{bmatrix} A^2(\theta) \\ A(\theta) \\ I_n \end{bmatrix}^T \hat{F}_i \begin{bmatrix} A^2(\theta) \\ A(\theta) \\ I_n \\ A(\theta) \\ I_n \end{bmatrix}$$

$$\text{for } i = 1, 2 \text{ and } F_3(\theta) = \begin{bmatrix} \begin{bmatrix} A(\theta) B(\theta) \\ B(\theta) \\ 0 \end{bmatrix}^T \hat{F}_2 + \hat{F}_3 \end{bmatrix} \begin{bmatrix} A^2(\theta) \\ A(\theta) \\ I_n \\ A(\theta) \\ I_n \end{bmatrix}.$$

Compare the last inequality with that of Theorem 2.13. The new condition implies the use of parameter-dependent S-Variables. As discussed in Lemma 3.3 this feature explains the conservatism reduction obtained when applying the system augmentation strategy.

The parameter-dependent S-Variables that are obtained on the example are clearly over parameterized: the matrix "monomials" $A(\theta)$ and I appear twice in the right-hand side factorization above. The reduced size LMI conditions of Corollary 3.6 avoid this over parametrization. This illustrates the reason for which there is no additional conservatism when applying Lemma 3.1: it removes non-useful degrees of freedom for the definition of parameter-dependent S-Variables of given order with respect to $A(\theta)$. On the other hand, conditions of Corollary 3.6 are more cumbersome to manipulate. It is harder to construct the implicit parameter-dependent Lyapunov matrix and S-Variables starting from those conditions.

3.4 Exactness Verification

The previous section has exposed a sequence of LMI conditions for robust analysis with guaranteed non-increasing conservatism. As κ is increased the LMI conditions obtained by applying analysis results to the degree $\kappa - 1$ augmented models are less and less conservative. They are thus expected to become feasible if robustness indeed holds. Unfortunately, none of the conditions invalidates robustness. If one of the LMIs is found feasible, then robustness is guaranteed. But if not, the methodology only suggests to increase κ and test the new LMIs. None of the LMIs of the sequence will be valid if robustness does not hold, that is, if there exists a value of $\theta \in \mathbb{E}^L$ such that the tested criteria (stability, pole location, L_2 induced norm...) is invalidated. If robustness does not hold the methodology suggests to test an infinite sequence of LMI problems none of which being feasible, thus never providing any conclusion.

To avoid this feature, it is needed to look for values $\hat{\theta}$ (if any) that violate the tested criteria. Such value is often called a worst case value in the sense that the criteria may be valid for most values of $\theta \in \mathbb{E}^L$ and fails on possibly few "bad" values $\hat{\theta}$. Looking for worst case values can be done by random testing of values of $\hat{\theta}$ taken from $\mathbb{E}^L$. Finding one value of $\hat{\theta}$ such that the criterion is violated (proof of instability, pole location outside given region, L_2 norm superior to the expected one, etc.) is sufficient to prove that robustness does not hold. Unfortunately, testing randomly values of uncertainties may be costly. This is first because it may be needed to draw many samples for obtaining good probabilistic evaluations of the worst case values [28]. For example, assuming uniform distribution, one needs to test 459 values for getting a 1 %—close estimate of reliability of the worst case performance with violation probability of 1 %. Secondly, the issue of drawing uniformly samples in a polytope is a hard problem. Uniform distribution of $\theta \in \mathbb{E}^L$ do not map to an uniform distribution in a polytope $A(\theta) = \sum_{j=1}^{L} \theta_j A_j$ (except if there are less than $L \leq 3$ vertices).

An alternative which is exposed in details in [18, 29, 30] is to take advantage of duality of the LMI conditions. As is well known, duality arguments can prove that a property described by LMI conditions is invalidated. Those properties are now employed to exhibit worst case values $\hat{\theta}$.

Having in mind this objective, let us define the following notation

$$\mathscr{L}_{1j}(P_j, F) + \mathrm{He}\{\mathscr{Q}(F)\mathscr{L}_{2j}\} \prec 0 \quad \forall\, j \in \mathscr{I}_L \tag{3.67}$$

as a resume of any of the LMI conditions of Corollaries 3.1, 3.3 or 3.5. These notations indicate that all the LMI conditions are gathered into one unique condition to be tested over all vertices $j = 1, \ldots, L$. The decisions variables of these LMIs enter in an affine manner the expressions of $\mathscr{L}_{1j}$ and $\mathscr{Q}$. Some of the decision variables, gathered in the notation P_j, are vertex dependent, the others, gathered in the notation F, are not. The properties of Lemma 2.3 stipulating that if the conditions hold for all vertices $j = 1, \ldots, L$, then they also hold for all values inside the simplex $\theta \in \mathbb{E}^L$:

$$\mathscr{L}_{1\theta}(P(\theta), F) + \mathrm{He}\{\mathscr{Q}(F)\mathscr{L}_{2\theta}\} \prec 0 \quad \forall\, \theta \in \mathbb{E}^L \tag{3.68}$$

where the θ-dependent matrices are affine: $P(\theta) = \sum_{j=1}^{L} \theta_j P_j$, $\mathscr{L}_{1\theta}(P(\theta), F) = \sum_{j=1}^{L} \theta_j \mathscr{L}_{1j}(P_j, F)$ and $\mathscr{L}_{2\theta} = \sum_{j=1}^{L} \theta_j \mathscr{L}_{2j}$. These conditions are relaxations (sufficient conditions) of the original robustness criteria that read as finding for each $\theta \in \mathbb{E}^L$ matrices P_θ and F_θ solution to

$$\mathscr{L}_{1\theta}(P_\theta, F_\theta) + \mathrm{He}\{\mathscr{Q}(F_\theta)\mathscr{L}_{2\theta}\} \prec 0. \tag{3.69}$$

We are interested in finding worst case values $\hat{\theta} \in \mathbb{E}^L$ such that there is no $P_{\hat{\theta}}$ and $F_{\hat{\theta}}$ solution to (3.69).

By convex duality theory [31, 32], and more precisely from the separating hyperplane theorem [32], there is no solution to (3.67) if and only if there exist Lagrange dual variables $\mathcal{H}_j$ ($j \in \mathcal{I}_L$), not all zero, such that

$$\mathcal{H}_j \succeq 0, \quad \sum_{j=1}^{L} \text{trace}\left((\mathcal{L}_{1j}(P_j, F) + \text{He}\{\mathcal{Q}(F)\mathcal{L}_{2j}\})\mathcal{H}_j\right) \geq 0 \quad \forall P_j, F. \quad (3.70)$$

These dual variables can be computed by solving the dual SDP reformulations of (3.67) [18, 32]. Alternatively, these dual variables are also readily available if we solve the SDP (3.67) by means of a primal-dual solver, which is at present the most effective and widely-used method to solve SDPs. A tool such as YALMIP [33], allows to retrieve both primal and dual variables when solving an LMI. If the LMI is infeasible, the dual variables satisfy the constraint (3.70). The next lemma allows to build a worst case value candidate $\hat{\theta}$.

Lemma 3.5 *Let the following notations*

$$\mathcal{H} = \sum_{j=1}^{L} \mathcal{H}_j$$

where $\mathcal{H}_j$ are the Lagrange dual variables proving that (3.67) does not hold whatever P_j and F. Moreover, let us define the following optimization problem

$$\hat{\theta} = \arg \min_{\theta \in \mathbb{E}^L} J(\theta), \quad J(\theta) = \sum_{j=1}^{L} \|\mathcal{H}_j - \theta_j \mathcal{H}\|^2.$$

If $J(\hat{\theta}) = 0$ then $\hat{\theta}$ is a value that invalidates robustness.

Proof Recall

$$\mathcal{H}_j \succeq 0, \quad \sum_{j=1}^{L} \text{trace}\left((\mathcal{L}_{1j}(P_j, F) + \text{He}\{\mathcal{Q}(F)\mathcal{L}_{2j}\})\mathcal{H}_j\right) \geq 0 \quad \forall P_j, F.$$

with $\mathcal{H}_j$ defined as above. If $J(\hat{\theta}) = 0$ then for all $j = 1, \ldots, L$ one has $\mathcal{H}_j = \hat{\theta}_j \mathcal{H}$. In view of the properties of $\mathcal{L}_{ij}$ ($i = 1, 2$), we see that the condition above implies

$$\mathcal{H} \succeq 0, \quad \text{trace}\left((\mathcal{L}_{1\hat{\theta}}(P_{\hat{\theta}}, F_{\hat{\theta}}) + \text{He}\{\mathcal{Q}(F_{\hat{\theta}})\mathcal{L}_{2\hat{\theta}}\})\mathcal{H}\right) \geq 0 \quad \forall P_{\hat{\theta}}, F_{\hat{\theta}}$$

where $P_{\hat{\theta}} = \sum_{j=1}^{L} \hat{\theta}_j P_j$ and $F_{\hat{\theta}} = F$. The matrix $\mathcal{H}$ hence works as a Lagrange dual variable that proves infeasibility of the following LMI

$$\mathscr{L}_{1\hat{\theta}}(P_{\hat{\theta}}, F_{\hat{\theta}}) + \text{He}\{\mathscr{Q}(F_{\hat{\theta}})\mathscr{L}_{2\hat{\theta}}\} \prec 0.$$

Infeasibility of the LMI implies that the criterion is violated for that value of $\hat{\theta}$. □

The minimization problem of Lemma 3.5 is based on the definition of some matrix norm. One choice of such is the Euclidean norm of the vector build out the columns of the matrix. Define $h = \text{vec}(\mathscr{H})$ the column vector build by stacking the columns of the matrix $\mathscr{H}$ in one vector. Define as well $h_j = \text{vec}(\mathscr{H}_j)$. One choice of objective function to minimize can be

$$J(\theta) = \sum_{j=1}^{L} (h_j - \theta_j h)^T (h_j - \theta_j h)$$

along with the constraint $\theta \in \mathbb{E}^L$. It happens to be a quadratic programming problem and can be solved with efficient solvers. Other choices of $J(\theta)$ can be based on the spectral norm of matrices. In that case the optimization problem of Lemma 3.5 can be recast as a semi-definite programming problem. The quadratic programming version is expected to be numerically more efficient.

The exposed method is the one applied in examples of Sect. 2.6.2 (i.e., robust stability analysis of randomly generated uncertain systems). It does allow in many cases to extract worst case values that violate robust stability. In the case of discrete-time systems (Table 2.5) it allows to conclude with either robust stability or existence of destabilizing value of uncertainties for almost all randomly generated examples.

Notice that $\hat{\theta}$ solution to the minimization problem of Lemma 3.5 is guaranteed to be a value that violates the robust criterion only if $J(\hat{\theta}) = 0$. Yet, if $J(\hat{\theta}) \neq 0$ it still may be a value that violates the criterion. To better explain this property, consider the case of L_2 induced norm performance. The results are exposed up-till now assuming one aims at testing if γ_∞ is a robust upper bound on the L_2 induced norm of the uncertain system. Assume for one such value $\gamma_{\infty 1}$ (and a choice of $\kappa \geq 0$) that the LMIs of Corollary 3.5 are found feasible. $\gamma_{\infty 1}$ is hence a guaranteed upper bound. It is guaranteed to be larger than the worst case L_2 induced norms over all admissible values of $\theta \in \mathbb{E}^L$:

$$\sup_{\theta \in \mathbb{E}^L} \|G_\theta\|_\infty = \gamma_{\infty,\text{wc}} \leq \gamma_{\infty 1}.$$

Assume now that for an other value $\gamma_{\infty 2}$ the LMIs of Corollary 3.5 are not feasible. In that case compute the minimizer $\hat{\theta}$ of Lemma 3.5. If $J(\hat{\theta}) = 0$ then one has

$$\gamma_{\infty 2} \leq \|G_{\hat{\theta}}\|_\infty \leq \sup_{\theta \in \mathbb{E}^L} \|G_\theta\|_\infty = \gamma_{\infty,\text{wc}} \leq \gamma_{\infty 1}$$

thus formally indicating that $\gamma_{\infty 2}$ is a lower bound on the worst case performance. But if $J(\hat{\theta}) \neq 0$ the computation still produces a lower bound $\|G_{\hat{\theta}}\|_\infty \leq \gamma_{\infty,\text{wc}}$ that is heuristically expected to be rather close to the worst case value.

Consider now the LMI optimization problem of minimizing γ_∞ under the LMI constraints of Corollary 3.5 for a given $\kappa \geq 0$. The minimum $\gamma_{\infty,\kappa}$ is an upper bound on the worst case value. Moreover, as conservatism is proved to be non-increasing as κ is increased, one has a non-increasing sequence of upper bounds:

$$\sup_{\theta \in \mathbb{E}^L} \|G_\theta\|_\infty = \gamma_{\infty,\mathrm{wc}} \leq \gamma_{\infty,\kappa} \leq \gamma_{\infty,\kappa-1} \leq \cdots \leq \gamma_{\infty,0}.$$

Since $\gamma_{\infty,\kappa}$ are minimizers of the LMI problems, at the optimum the LMIs are marginally feasible as well as their dual. The Lagrange dual variables can hence be utilized to compute minimizers $\hat{\theta}_\kappa$ of Lemma 3.5. Each of these provide lower bounds $\|G_{\hat{\theta}_\kappa}\|_\infty \leq \gamma_{\infty,\mathrm{wc}}$. Moreover, if $J(\hat{\theta}_\kappa) = 0$ then one gets

$$\gamma_{\infty,\kappa} = \|G_{\hat{\theta}_\kappa}\|_\infty \leq \sup_{\theta \in \mathbb{E}^L} \|G_\theta\|_\infty = \gamma_{\infty,\mathrm{wc}} \leq \gamma_{\infty,\kappa}$$

hence guaranteeing that the LMI test is exact (non-conservative).

3.5 Numerical Examples

3.5.1 Quarter-Car Suspension Example

The quarter-car suspension example of Sect. 2.2 is considered once again. This time all five parameters are assumed uncertain in the following intervals

$M \in [\,320,\ 384\,]\,\mathrm{Kg}, \quad m \in [\,38,\ 42\,]\,\mathrm{Kg},$
$k_1 \in [\,171,\ 189\,]\,\mathrm{KN/m}, \quad k_2 \in [\,180,\ 220\,]\,\mathrm{KN/m}, \quad c \in [\,950,\ 1050\,]\,\mathrm{Ns/m}.$

These uncertainties correspond to a possible 20% increase of the mass of the car from the nominal value 320; 5% uncertainty on the parameters m, k_1 and c for example because of variations from one produced suspension system to another; 10% uncertainty on the wheel stiffness k_2 caused by the proper inflation pressure. The affine descriptor modeling is that of (3.3).

Results of Theorem 3.5 are applied to analyze pole location of this suspension system. It is done both in open-loop and in closed-loop with the two following state-feedback controls

$$u = \underbrace{\begin{bmatrix} -46 & -66 & 110 & -4 \end{bmatrix}}_{K_1} \begin{bmatrix} \dot{z}_w \\ z_w \\ \dot{z}_c \\ z_c \end{bmatrix}, \quad u = \underbrace{\begin{bmatrix} 46 & 66 & -110 & 4 \end{bmatrix}}_{K_2=-K_1} \begin{bmatrix} \dot{z}_w \\ z_w \\ \dot{z}_c \\ z_c \end{bmatrix}$$

Two pole location tests are performed. One concerns α-stability (proving that the real parts of all poles are smaller that α), the second concerns damping ratio of all modes (see Sect. 2.5.1 for definition of the pole location regions). Results are given in Table 3.1. The actually tested LMIs are not those of Theorem 3.5 but when applying them we performed the reduction of size by means of Lemmas 3.1 and 3.2. Since the second lemma may bring some conservatism one test is done with only Lemma 3.1 and the second one is combining the two. On the tested examples no conservatism gap is noticed between the two tests. Moreover, there is no conservatism at all since the best tested values for which the SV-LMIs are feasible (fifth column of the table) are equal (with less than 1 digit gap) to the worst values computed on the $2^5 = 32$ extremal values of the uncertainty intervals (fourth column). Notice that compared to the α-stability test, the damping test involves constraints of twice larger dimensions. This is related to the complex-to-real conversion of the LMIs as described in (2.36). Meanwhile, the number of decision variables is unchanged because we performed the tests assuming real-valued decisions variables (which may not be the case). This restriction to real-valued decision variables brings no conservatism for this example.

We now consider the L_2-gain performance analysis for the example. The equations are completed with same inputs/outputs as in Sect. 2.2. The equations in affine descriptor form read as follows.

$$
\begin{bmatrix}
m & 0 & 0 & 0 \\
0 & 1 & 0 & 0 \\
0 & 0 & M & 0 \\
0 & 0 & 0 & 1 \\
0 & 0 & 0 & 0 \\
0 & 0 & 0 & 0 \\
0 & 0 & 0 & 0 \\
\hline
0 & 0 & -1 & 0 \\
0 & 0 & 0 & 0 \\
0 & 0 & 0 & 0
\end{bmatrix}
\dot{x} +
\begin{bmatrix}
-1 & 1 & -1 \\
0 & 0 & 0 \\
1 & 0 & 1 \\
0 & 0 & 0 \\
1 & 0 & 0 \\
0 & 1 & 0 \\
0 & 0 & 1 \\
\hline
0 & 0 & 0 \\
0 & 0 & 0 \\
0 & 0 & 0
\end{bmatrix}
\pi +
\begin{bmatrix}
0 & 0 & 0 \\
0 & 0 & 0 \\
0 & 0 & 0 \\
0 & 0 & 0 \\
0 & 0 & 0 \\
0 & 0 & 0 \\
0 & 0 & 0 \\
\hline
1 & 0 & 0 \\
0 & 1 & 0 \\
0 & 0 & 1
\end{bmatrix}
z
$$

$$
=
\begin{bmatrix}
0 & 0 & 0 & 0 \\
1 & 0 & 0 & 0 \\
0 & 0 & 0 & 0 \\
0 & 0 & 1 & 0 \\
0 & -k_1 & 0 & k_1 \\
0 & k_2 & 0 & 0 \\
-c & 0 & c & 0 \\
\hline
0 & 0 & 0 & 0 \\
0 & 0 & 0 & W_{zc} \\
0 & 0 & 0 & 0
\end{bmatrix}
x +
\begin{bmatrix}
0 \\
0 \\
0 \\
0 \\
0 \\
k_2 \\
0 \\
\hline
0 \\
0 \\
0
\end{bmatrix}
w +
\begin{bmatrix}
-1 \\
0 \\
1 \\
0 \\
0 \\
0 \\
0 \\
\hline
0 \\
0 \\
W_u
\end{bmatrix}
u
$$

The Assumption 3.3 is fulfilled when choosing $\begin{bmatrix} E_{2x} & E_{2\pi} \end{bmatrix} = I$. Theorem 3.6 applies without difficulty. Again we applied Lemmas 3.1 and 3.2 to reduce the size of the

Table 3.1 Pole location of the quarter-car suspension

		Nominal	Worst vertex	SV-LMI (3.21)	nb vars/nb rows/time Lemma 3.1	nb vars/nb rows/time Lemma 3.1 + Lemma 3.2
α	Open-loop	−0.0755	−0.0635	−0.0635	365/292/1.8 s	342/196/1.2 s
	Closed-loop K_1	−0.0421	−0.0318	−0.0317		
	Closed-loop K_2	−0.1051	−0.0920	−0.0919		
ζ	Open-loop	0.1037	0.0921	0.0921	365/580/4.2 s	342/388/1.7 s
	Closed-loop K_1	0.0592	0.0470	0.0469		
	Closed-loop K_2	0.1412	0.1308	0.1307		

Table 3.2 L_2 performance of the quarter-car suspension

		Nominal	Worst vertex	SV-LMI (3.21)	nb vars/nb rows/time Lemma 3.1	nb vars/nb rows/time Lemma 3.1 + Lemma 3.2
$\overline{\gamma}_\infty$	Open-loop	51.292	57.472	57.472	371/324/1.5 s	335/228/0.8 s
	Closed-loop K_1	43.642	47.408	47.408		
	Closed-loop K_2	102.371	124.900	124.900		

LMIs. Numerical results are summarized in Table 3.2. These numerical examples illustrate the efficiency of Lemmas 3.1 and 3.2 to reduce the numerical complexity quite efficiently. In the present case, it is without any conservatism. All SV-LMI results happen to be lossless on the example. It is easily checked by inspection over vertices (the worst cases happen to be on vertices in this example) but it is also the case when applying Lemma 3.5. For all the tests of Tables 3.1 and 3.2 the exactness check by dual LMIs produces the worst case combination $\hat{\theta}$.

3.5.2 Randomly Generated Examples

The following results are based on artificially build random systems. The procedure for generating the random systems is similar to the one described in Sect. 2.6.2. In particular the generated systems are sparse and have few uncertain elements. At the difference with the procedure in Sect. 2.6.2 no optimization of the vertices is made. The reason for it is that computing poles of descriptor systems is slightly more complex and hence optimizing over them is costly. The exact procedure employed here is as follows.

- Generate sparse random matrices E_{1xo}, E_{2xx}, $E_{2x\pi}$, A_o. Compute the minimal singular value of E_{1xo} if it is not greater than 1, generate another E_{1xo}, and continue until it is the case. This requirement on the minimal singular value of E_{1xo} is to avoid as much as possible the uncertain $E_{1x}(\theta)$ matrix defined later to violate Assumption 3.1.

- Generate vertices E_{1xj} and A_j by doing small perturbations on one or two coefficients of the E_{1xo} and A_o matrices.

The systems generated in this way have no particularity in terms of pole location. The goal of the tests is to compute upper bounds on the maximal real part of the finite poles. The worst case maximal real part of the finite poles will be denoted α_{wc}. A test will be said positive if it provides both upper and lower bounds $\underline{\alpha} \le \alpha_{wc} \le \bar{\alpha}$ with a high precision gap (smaller than $\bar{\alpha} - \underline{\alpha} \le 0.001$). Upper bounds are computed using LMI results of Corollary 3.3 while lower bounds are provided applying Lemma 3.5.

For most generated examples (about 90%) the LMI test of Corollary 3.3 with no augmentation ($\kappa = 1$) is valid when fixing $\alpha = \alpha_{vtx} + 0.001$ where α_{vtx} is the maximal real part of finite poles computed on the vertices of the polytope. This fact indicates both that the LMI test is not much conservative and that it is non trivial to generate randomly examples where the worst case is not at vertices. All systems for which this situation occurs are removed from the analysis. They are not helpful to compare the conservatism of the LMI conditions for different values of the augmentation index κ.

All analyzed uncertain systems are hence such that the maximal real part of the poles is not at vertices (or at least this cannot be assessed by the $\kappa = 1$ LMI conditions). For these systems the procedure we adopt is to search for upper and lower bounds by bisection. It is done for three different values of $\kappa = 1, 2, 3$. The bisection is done as follows:

1. Choose two values $0 \le \tilde{\alpha}_\kappa \le \bar{\alpha}_\kappa$. Let $\underline{\alpha}_\kappa = \alpha_{vtx}$ be the worst value known so far.
2. Solve the LMI problem of Corollary 3.3 with R defining the region such that $\mathrm{Re}(\lambda) < \bar{\alpha}_\kappa$. If it is infeasible, increase $\bar{\alpha}_\kappa$ until it is.
3. Take $\alpha = \frac{1}{2}(\tilde{\alpha}_\kappa + \bar{\alpha}_\kappa)$ and solve the LMI problem of Corollary 3.3 with R defining the region such that $\mathrm{Re}(\lambda) < \alpha$. If the LMIs are feasible let $\bar{\alpha}_\kappa = \alpha$, otherwise let $\tilde{\alpha}_\kappa = \alpha$.
4. Compute a worst case estimate $\hat{\theta}$ using Lemma 3.5 applied to the LMI solved at step 3 and compute the largest real part $\hat{\alpha}_\kappa$ of the finite poles of the system evaluated at $\hat{\theta}$. Append the worst known value as $\underline{\alpha}_\kappa := \max(\underline{\alpha}_\kappa, \hat{\alpha}_\kappa)$, and the lower bisection value as $\tilde{\alpha}_\kappa := \max(\tilde{\alpha}_\kappa, \underline{\alpha}_\kappa)$.
5. If $\bar{\alpha}_\kappa - \tilde{\alpha}_\kappa \le 0.001$ stop. Else go to step 3.

In practice the initializations are done with $\tilde{\alpha}_1 = \alpha_{vtx} + 0.001$ and $\bar{\alpha}_1 = \alpha_{vtx} + 10$. This imposes many iterations of the bisection but these are not too costly the LMIs being of rather small dimensions. For the next bisection the number of iterations is drastically reduced by taking $\tilde{\alpha}_2 = \underline{\alpha}_1$ and $\bar{\alpha}_2 = \bar{\alpha}_1$ which are the best lower and upper bounds known at this stage. For the same reason the last bisection is done with $\tilde{\alpha}_3 = \underline{\alpha}_2$ and $\bar{\alpha}_3 = \bar{\alpha}_2$.

The conservatism reduction as κ is increased guarantees that $\bar{\alpha}_3 \le \bar{\alpha}_2 \le \bar{\alpha}_1$ and it is indeed the case on the examples. In practice one also observes that $\underline{\alpha}_1 \le \underline{\alpha}_2 \le \underline{\alpha}_3$.

Table 3.3 Statistics for 100 tests of conservatism based on randomly generated systems

	$\kappa = 1$	$\kappa = 2$	$\kappa = 3$	Total fully nonconservative
rank(E_{1x}) = 3	59 (15)	7 (4)	29	95
rank(E_{1x}) = 4	50 (6)	5 (6)	30	85
rank(E_{1x}) = 5	50 (10)	7 (5)	31	78

The values are the number of systems for which the LMIs are fully nonconservative. In brackets are the number of partially nonconservative LMIs. Tests done for $n = 6, n_x = 4, n_\pi = 2, L = 6$ and different values of rank(E_{1x})

A test for some value of $\kappa = 1, 2, 3$ is said to be fully nonconservative if $\overline{\alpha}_\kappa - \underline{\alpha}_\kappa \leq 0.001$. In such case the primal/dual solutions of the LMIs provide upper and lower bounds with less than 0.001 gap. A test will also be said partially nonconservative if $\overline{\alpha}_\kappa - \underline{\alpha}_3 \leq 0.001$. In this latter case the upper bound happens to be nonconservative (at least with less than 0.001 gap) but the worst case estimate computed for that $\kappa = 1, 2$ does not allow to prove it. Tests for $\kappa + 1$ are done only if the test at κ is not fully nonconservative. The results of the tests are provided in Table 3.3. The average computation time for one of these tests is of one minute and a half. This includes the generation of many systems before getting one that satisfies the upper described criteria, plus all the bisections. As illustrated in Chap. 2 on many examples, the SV LMI result for $\kappa = 1$ is much less conservative than classical "quadratic stability" type results. As shown in Table 3.3 with $\kappa = 1$, even for models selected here that are hard to analyze, the conservatism is shown to be close to zero for more than 50 % of the tests. Conservatism is further reduced when considering the SV-LMIs for the augmented systems $\kappa > 1$. The conservatism reduction surprisingly happens to more important when moving from $\kappa = 2$ to $\kappa = 3$ than when moving from $\kappa = 1$ to $\kappa = 2$. Table 3.3 can give the feeling that the LMIs fail to solve 5, 15 and 22 % of the problems for each row respectively. This is not quite the case. These are the rate of tests where the conservatism gap is greater than 0.001, which is a very small gap. Actually, the gap for all those cases happens to be below 0.01. This would be satisfactory as well in case of a true engineering problem.

3.5.3 Satellite Example

Results are now applied to the robust stability analysis of a satellite. The model is provided by CNES, the french space agency. It corresponds to the DEMETER satellite that has the particularity to have four long appendices wholes first flexible modes (considered identical) are non negligible. The complete model is described in details in [34]. Many publications have used it as a benchmark to illustrate robust analysis tools [35], LPV analysis [36, 37], periodic control features [38], input allocation [39] as well as adaptive control [40, 41]. Here we revisit the robustness analysis based on LFT modeling, but this time taking advantage of polytopic descriptor modeling.

The one axis linearized model of the satellite is as follows:

Table 3.4 Parameters of the DEMETER satellite for each of the 3 axes

Axis	J	l	ω	ζ
x	$31.38 \pm 30\%$	0.7582	$[0.2 , 0.6] \times 2\pi$	$[5 \cdot 10^{-4} , 5 \cdot 10^{-3}]$
y	$21.19 \pm 30\%$	0.7741	$[0.2 , 0.6] \times 2\pi$	$[5 \cdot 10^{-4} , 5 \cdot 10^{-3}]$
z	$35.70 \pm 30\%$	-0.7367	$[0.2 , 0.6] \times 2\pi$	$[5 \cdot 10^{-4} , 5 \cdot 10^{-3}]$

$$J\ddot{\theta} + \sqrt{J}l\ddot{\eta} = T_w$$
$$\sqrt{J}l\ddot{\theta} + \ddot{\eta} + 2\zeta\omega\dot{\eta} + \omega^2\eta = 0 \tag{3.71}$$

where θ is the attitude of the satellite, η is a state that aggregates all contributions of flexible modes, J is the inertia, l is a coefficient modeling the coupling between the attitude and the flexible modes, ω is the natural frequency of the aggregated flexible modes and ζ the damping. The numerical data for the model coefficients are given in Table 3.4. The model (3.71) happens not to be linear in the three uncertain parameters J, ω and ζ. Nevertheless, by introducing two exogenous signals and considering the $\sqrt{J}$ as the actual parameter of inertia, everything becomes affine without any loss of generality:

$$\sqrt{J}\pi_1 + \sqrt{J}l\ddot{\eta} \ = T_w, \quad \pi_1 = \sqrt{J}\ddot{\theta}$$
$$\sqrt{J}l\ddot{\theta} + \ddot{\eta} + \omega\pi_2 = 0, \quad \pi_2 = 2\zeta\dot{\eta} + \omega\eta \tag{3.72}$$

The descriptor state-space representation of the plant with state $x^T = \begin{bmatrix} \theta & \dot{\theta} & \eta & \dot{\eta} \end{bmatrix}$ and exogenous signal $\pi^T = \begin{bmatrix} \pi_1 & \pi_2 \end{bmatrix}$ is as follows:

$$
\begin{bmatrix}
0 & 0 & 0 & \sqrt{J}l \\
0 & \sqrt{J} & 0 & 0 \\
1 & 0 & 0 & 0 \\
0 & \sqrt{J}l & 0 & 1 \\
0 & 0 & 0 & 0 \\
0 & 0 & 1 & 0
\end{bmatrix}
\dot{x} +
\begin{bmatrix}
\sqrt{J} & 0 \\
-1 & 0 \\
0 & 0 \\
0 & \omega \\
0 & 1 \\
0 & 0
\end{bmatrix}
\pi =
\begin{bmatrix}
0 & 0 & 0 & 0 \\
0 & 0 & 0 & 0 \\
0 & 1 & 0 & 0 \\
0 & 0 & 0 & 0 \\
0 & 0 & \omega & 2\zeta \\
0 & 0 & 0 & 1
\end{bmatrix}
x +
\begin{bmatrix}
1 \\
0 \\
0 \\
0 \\
0 \\
0
\end{bmatrix}
T_w \tag{3.73}
$$

The toque T_w is applied to the satellite via reaction wheels. The dynamics of these are modeled by the following state-space equations

$$\dot{x}_w = \begin{bmatrix} -2.4 & 30.5 \\ -0.025 & 0 \end{bmatrix} x_w + \begin{bmatrix} 1.214 \\ 0.025 \end{bmatrix} u_c, \quad T_w = \begin{bmatrix} 1 & 0 \end{bmatrix} x_w \tag{3.74}$$

where u_c is the computed control signal. The control law is decomposed in two terms. One is an LTI filter $u_c = H(s)u_{PD}$ which has different characteristics for each axis (transfer function are given in Table 3.5). These filters have been designed on the nominal model with H_∞/H_2 performance specifications, see [42]. The second term is a filtered PD control with identical coefficients for all axes:

Table 3.5 Filters of the DEMETER satellite control law for each of the 3 axes

Axis	$H(s)$
x	$\dfrac{3.039s^2 + 1.457s + 0.09635}{0.3333s^4 + 1.371s^3 + 1.263s^2 + 0.4489s}$
y	$\dfrac{-0.4457s^3 + 3.309s^2 + 2.032s + 0.1628}{0.25s^4 + 1.361s^3 + 1.636s^2 + 0.7586s}$
z	$\dfrac{-0.3039s^3 + 2.893s^2 + 1.447s + 0.09635}{0.3333s^4 + 1.371s^3 + 1.263s^2 + 0.4489s}$

Table 3.6 Size of the largest cube of stabilizing values of uncertain parameters found by the SV LMI tests, and the associated numerical complexity

	x (%)	y (%)	z (%)	nb. vars.	nb. rows	Time (average) (s)
$\kappa = 1$, $P_j = P$	1.56	2.34	1.56	126	208	2
$\kappa = 1$	71.09	71.88	71.09	588	208	5
$\kappa = 2$	76.51	74.22	79.22	2642	368	60

$$u_{\text{PD}} = -0.1\theta - 2\frac{s}{0.5s + 1}\theta \tag{3.75}$$

The overall feedback loop is trivially of polytopic descriptor form and results of the chapters can apply. The order of the closed-loop plant is $n_x = 4 + 2 + 4 + 1 = 11$ and the number of exogenous signals is $n_\pi = 2$. In this model only four rows out of $n = 13$ contain uncertainties. Lemma 3.1 hence allows to reduce drastically the dimensions of the SV-LMIs.

Robust stability of the closed-loop is not achieved for all uncertainties. This can be seen by computing the closed-loop poles for the minimal value of J, ω and ζ in the defined intervals. Whatever axis, for this minimal value of the uncertain parameters the system is unstable. The robust analysis task is therefore to certify robust stability for the largest set of uncertainties included in the cube corresponding to the intervals on the three parameters. Conservatism of results can then be compared in terms of the size of set for which robust stability is assessed.

A first series of tests is done by searching for the largest hyper cube with one corner at the maximal values of the three intervals (this vertex is stable for all three axes). The search is done by bisection. Results are given in Table 3.6. The LMIs employed here are those of Corollary 3.1 combined with the SV reduction technique of Lemma 3.1. The results given in the first row correspond to "quadratic stability" type conditions where the same Lyapunov matrix P is used for the whole set of uncertainties. The values given in the table are the relative size of the edges compared to the original set. The obtained cubes for axis x are plotted in Fig. 3.1. Figures for the other axes are similar and hence omitted.

Results show that moving from $\kappa = 1$ to $\kappa = 2$ does provide less conservative results. The obtained sets for $\kappa = 2$ are of larger dimensions and include the ones obtained for $\kappa = 1$. This is at the expense of an increased numerical burden. The

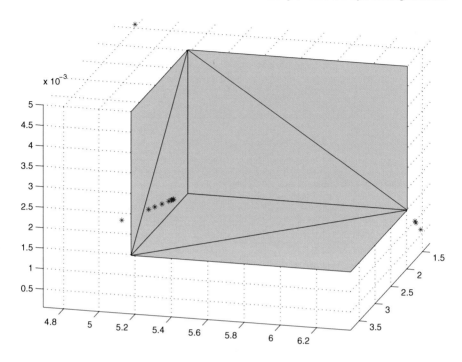

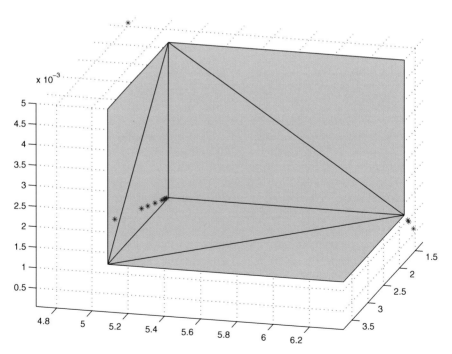

Fig. 3.1 A cube of guaranteed stabilizing values of parameters obtained by the SV LMIs for $\kappa = 1$ (*top*) and $\kappa = 2$ (*bottom*). Results for the x axis

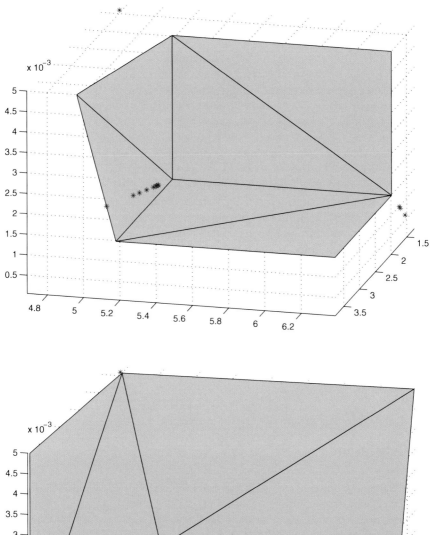

Fig. 3.2 A polytope of guaranteed stabilizing values of parameters obtained by the SV LMIs for $\kappa = 1$ (*top*) and $\kappa = 2$ (*bottom*). Results for the x axis

computation time is 12 times larger. In addition to the cubes of stabilizing values Fig. 3.1 also display ($\star$) values of parameters that are found during the bisection procedure when applying results of Lemma 3.5 to the dual LMIs. These are destabilizing values of the parameters. It can be seen from these that the cube of parameters found for $\kappa = 2$ is non conservative. It cannot be increased in size since there exist destabilizing values of the parameters close to one of its corners (the one corresponding to minimal values of $\sqrt{J}$, ω and ζ). The comparison of the two cubes gives the impression that the conservatism reduction is rather small (only a few percents). As illustrated in the following it happens to be much more significant. Yet at this stage one can state that the SV result with $\kappa = 1$ is already of very low conservatism, at least when compared to the "quadratic stability" conditions.

The actual set of stabilizing parameters is not a cube. While the corner corresponding to the minimal values of the parameters cannot be moved further without incorporating destabilizing values, this does not seem to be the case for the other vertices. The second test we have performed is to increase as much as possible the vertices towards the corners of the original hyper cube. This allows to build a polytope that better approximates the interior the actual set of stabilizing values. The procedure is of bisection type applied to all vertices recursively. The procedure is applied to both SV-LMI tests and obtained polytopes are plotted in Fig. 3.2.

Due to inherent conservatism, the SV-LMI for $\kappa = 1$ do not allow to improve much compared to the cube. Only one vertex can be moved and the improvement is of 12 % for the corresponding edge. In stark contrast, the improvement for $\kappa = 2$ is much more significant. All vertices are pushed either to the limits of the actual set of uncertainties, or stopped close to a value that destabilize the plant. Identical results are obtained for the axes y and z.

3.6 Conclusion

In this chapter, we presented SV-LMI results for the analysis of uncertain LTI systems represented in descriptor form. We then introduced artificial system augmentation procedure to reduce the conservatism of the S-Variable approach. By following this procedure we can build a sequence of SV-LMI problems with reducing conservatism. To conclude that exact analysis is achieved at a certain degree of system augmentation, we showed an exactness verification scheme in terms of dual LMI variables (Lagrange dual variables). We illustrated by numerical examples that the sequence of SV-LMIs via system augmentation and the exactness verification scheme are effective for conservatism reduction, exactness certification, and worst case parameter extraction.

References

1. Hou M, Müller PC (1999) Causal observability of descriptor systems. IEEE Trans Autom Control 44(1):158–163
2. Ishihara JY, Terra MH (2001) Impulse controllability and observability of rectangular descriptor systems. IEEE Trans Autom Control 46(6):991–994
3. Hou M (2004) Controllability and elimination of impulsive modes in descriptor systems. IEEE Trans Autom Control 49(10):1723–1727
4. Verghese G, Levy B, Kailath T (1981) A generalized state-space for singular systems. IEEE Trans Autom Control 26(4):811–831
5. Takaba K, Morihira N, Katayama T (1995) A generalized Lyapunov theorem for descriptor system. Syst Control Lett 24(1):49–51
6. Masubuchi I, Shimemura E (1997) An LMI condition for stability of implicit systems. In: IEEE conference on decision and control, pp 779–780
7. Uezato E, Ikeda M (1999) Strict LMI conditions for stability, robust stabilization, and H_∞ control of descriptor systems. In: IEEE conference on decision and control, pp 4092–4097
8. Ishihara JY, Terra MH (2002) On the lyapunov theorem for singular systems. IEEE Trans Autom Control 47(11):1926–1930
9. Chaabane M, Bachelier O, Souissi M, Mehdi D (2006) Stability and stabilization of continuous descriptor systems: an LMI approach. Math Probl Eng. doi:10.1155/MPE/2006/39367
10. Peaucelle D (2007) Quadratic separation for uncertain descriptor system analysis, strict LMI conditions. In: IEEE conference on decision and control. Section 5 of the manuscript is erroneous. New Orleans. A corrected version is available at http://www.laas.fr/peaucell/papers/cdc07a.pdf
11. Trofino A (2000) Robust stability and domain of attraction of uncertain nonlinear systems. In: American control conference, Chicago, pp 3707–3711
12. Coutinho D, Trofino A, Fu M (2002) Guaranteed cost control of uncertain nonlinear systems via polynomial Lyapunov functions. IEEE Trans Autom Control 47(9):1575–1580
13. Masubuchi I, Akiyama T, Saeki M (2003) Synthesis of output-feedback gain-scheduling controllers based on descriptor LPV system representation. In: IEEE conference on decision and control, pp 6115–6120
14. Doyle J, Packard A, Zhou K (1991) Review of LFTs, LMIs and μ. In: IEEE conference on decision and control, Brignton, England, pp 1227–1232
15. Hecker S, Varga A (2004) Generalized LFT-based representation of parametric uncertain models. Eur J Control 10(4):326–337
16. Peaucelle D, Ebihara Y (2014) Robust stability analysis of discrete-time systems with parametric and switching uncertainties. In: Proceedings of the 19th IFAC world congress, pp 724–729
17. Skelton RE, Iwasaki T, Grigoriadis K (1998) A unified approach to linear control design., Systems and Control. Taylor and Francis, London
18. Scherer CW (2006) LMI relaxations in robust control. Eur J Control 12:3–29
19. Parillo PA (2003) Semidefinite programming relaxations for semialgebraic problems. Math Program 96(2):293–320
20. Henrion D, Garulli A (eds) (2005) Positive polynomials in control. Lecture notes in control and information sciences, vol 312. Springer, Berlin
21. Scherer CW, Hol CWJ (2006) Matrix sum-of-squares relaxations for robust semi-definite programs. Math Program 107(1–2):189–211
22. Chesi G, Garulli A, Tesi A, Vicino A (2003) Robust stability of polytopic systems via polynomially parameter-dependent Lyapunov functions. In: IEEE conference on decision and control, Maui, Hawaii, USA
23. Chesi G, Garulli A, Tesi A, Vicino A (2005) Polynomially parameter-dependent Lyapunov functions for robust stability of polytopic systems: an LMI approach. IEEE Trans Autom Control 50(3):365–370
24. Chesi G (2010) LMI techniques for optimization over polynomials in control: a survey. IEEE Trans Autom Control 55(11):2500–2510

25. Oliveira RCLF, de Oliveira MC, Peres PLD (2008) Convergent LMI relaxations for robust analysis of uncertain linear systems using lifted polynomial parameter-dependent Lyapunov functions. Syst Control Lett 57:680–689
26. Bliman PA (2001) A convex approach to robust stability for linear systems with uncertain scalar parameters. Technical Report RR-4316, INRIA, Rocquencourt, France. http://www.inria.fr/rrrt/rr-4316.html
27. Bliman P-A (2004) A convex approach to robust stability for linear systems with uncertain scalar parameters. SIAM J Control Optim 42:2016–2042
28. Tempo R, Calafiore G, Dabbene F (2005) Randomized algorithms for analysis and control of uncertain systems. Springer, London
29. Scherer CW (2005) Relaxations for robust linear matrix inequality problems with verifications for exactness. SIAM J Matrix Anal Appl 27(2):365–395
30. Ebihara Y, Onishi Y, Hagiwara T (2009) Robust performance analysis of uncertain LTI systems: dual LMI approach and verifications for exactness. IEEE Trans Autom Control 54(5):938–951
31. Balakrishnan V, Vandenberghe L (2003) Semidefinite programming duality and linear time-invariant systems. IEEE Trans Autom Control 48(1):30–41
32. Boyd S, Balakrishnan V (2004) Convex optimization. Cambridge University Press, Cambridge
33. Löfberg J (2004) YALMIP: a toolbox for modeling and optimization in MATLAB. In: Proceedings of the IEEE computer aided control system design, pp 284–289
34. Pittet C, Arzelier D (2006) DEMETER: a benchmark for robust analysis and control of the attitude of flexible microsatellites. In: IFAC symposium on robust control design, Toulouse, France
35. Peaucelle D, Bortott A, Gouaisbaut F, Arzelier D, Pittet C (2010) Robust analysis of DEMETER benchmark via quadratic separation. In: IFAC symposium on automatic control in aerospace, Nara
36. Biannic J-M, Pittet C, Roos C (2010) Lpv analysis of switched controllers in satellite attitude control systems. In: AIAA guidance, navigation, and control conference, Toronto
37. Biannic J-M, Roos C, Pittet C (2011) LPV analysis of switched controllers for attitude control systems. J Guid Control Dyn 34(5):1561–1566
38. Tregouët J-F, Arzelier D, Peaucelle D, Ebihara Y, Pittet C, Falcoz A (2011) Periodic h2 synthesis for spacecraft attitude control with magnetorquers and reaction wheels. In: IEEE conference on decision and control, Orlando
39. Tregouët J-F, Arzelier D, Peaucelle D, Pittet C, Zaccarian L (2014) Reaction wheels desaturation using magnetorquers and static input allocation. IEEE Trans Control Syst Technol. Submitted (LAAS Tech. Report N°13472)
40. Peaucelle D, Drouot A, Pittet C, Mignot J (2011) Simple adaptive control without passivity assumptions and experiments on satellite attitude control DEMETER benchmark. In: IFAC world congress
41. Luzi AR, Peaucelle D, Biannic J-M, Pittet C, Mignot J (2014) Structured adaptive attitude control of a satellite. Int J Adapt Control Signal Process
42. Pittet C, Mignot J, Fallet C (1999) LMI based multi-objective H_∞ control of flexible microsatellites. In: IEEE conference on decision and control, Sydney, Australia

Chapter 4
Robust State-Feedback Synthesis for LTI Systems

4.1 Introduction

In this chapter, we discuss robust state-feedback controller synthesis for LTI systems with polytopic uncertainties using SV-LMIs. In the preceding chapter, we have observed that various robust performance analysis problems can be cast as SV-LMIs (allowing conservatism) where we can introduce S-variables freely without any care. In the state-feedback case, the key issue is how to impose structural constraints on those S-variables, so that we can apply linearizing change of variables and eventually we can obtain SV-LMIs for state-feedback controller synthesis. When imposing such structural constraints, it is of course important to reduce the conservatism of the resulting SV-LMIs as much as possible.

4.2 Preliminaries

In this chapter, usual polytopic uncertain systems are considered. The case corresponds to that of Chap. 2. For such uncertain systems, two types of robust stability conditions are provided in Theorem 2.2:

$$P \succ 0, \quad P A_j + A_j^T P \prec 0 \quad (j \in \mathscr{I}_L). \tag{4.1}$$

and Theorem 2.4:

$$P_j \succ 0, \quad \begin{bmatrix} 0 & P_j \\ P_j & 0 \end{bmatrix} + \mathrm{He} \left\{ \begin{bmatrix} F_1 \\ F_2 \end{bmatrix} \begin{bmatrix} I & -A_j \end{bmatrix} \right\} \prec 0 \quad (j \in \mathscr{I}_L). \tag{4.2}$$

The goal of the chapter is to describe the extensions of these conditions to the state-feedback design problem. The design problem is defined for the LTI uncertain open-loop polytopic system

© Springer-Verlag London 2015
Y. Ebihara et al., *S-Variable Approach to LMI-Based Robust Control*,
Communications and Control Engineering, DOI 10.1007/978-1-4471-6606-1_4

$$\dot{x}(t) = A(\theta)x(t) + B(\theta)u(t), \quad \theta \in \mathbb{E}^L,$$
$$\begin{bmatrix} A(\theta) \ B(\theta) \end{bmatrix} = \sum_{j=1}^{L} \theta_j \begin{bmatrix} A_j \ B_j \end{bmatrix}. \tag{4.3}$$

Here, $\begin{bmatrix} A_j \ B_j \end{bmatrix}$ $(j \in \mathscr{I}_L)$ are known vertices of the polytope and θ is the vector of uncertain parameters. The control to be designed is a static state-feedback $u(t) = Kx(t)$. The closed-loop happens to be polytopic as well:

$$\dot{x}(t) = A(\theta, K)x(t), \quad \theta \in \mathbb{E}^L,$$
$$A(\theta, K) = A(\theta) + B(\theta)K = \sum_{j=1}^{L} \theta_j(A_j + B_j K). \tag{4.4}$$

The classical methodology with respect to the quadratic stability conditions of (4.1) is first to consider the equivalent conditions for the dual system based on Corollary 2.1:

$$X \succ 0, \quad A_j X + X A_j^T \prec 0 \quad (j \in \mathscr{I}_L). \tag{4.5}$$

Second, to apply this condition to the polytopic closed-loop system:

$$X \succ 0, \quad (A_j + B_j K)X + X(A_j^T + K^T B_j^T) \prec 0 \quad (j \in \mathscr{I}_L). \tag{4.6}$$

Finally, to perform the classical linearizing change of variables [1] given by $Y = KX$, which gives, without any conservatism, the following LMI condition:

$$X \succ 0, \quad A_j X + B_j Y + X A_j^T + Y^T B_j^T \prec 0 \quad (j \in \mathscr{I}_L). \tag{4.7}$$

If these LMIs are feasible, then $K = YX^{-1}$ is a robustly stabilizing state feedback. If they are not feasible one cannot conclude anything (i.e., we cannot say anything on the existence of K that robustly stabilizes the uncertain system (4.3)) due to the conservatism of the analysis conditions (4.1) as discussed in Chap. 2.

As the condition (4.2) is proved to be less conservative than (4.1), it is expected that state-feedback conditions build on their basis should help in reducing the conservatism of (4.7). Unfortunately, it is not always the case. To have a quick look on the underlying reasons, let us apply the above mentioned classical methodology of linearization to (4.2).

First, we consider the dual system counterpart of (4.2) that is given in Corollary 2.4:

$$X_j \succ 0, \quad \begin{bmatrix} 0 & X_j \\ X_j & 0 \end{bmatrix} + \mathrm{He}\left\{ \begin{bmatrix} I \\ -A_j \end{bmatrix} \begin{bmatrix} F_1 \ F_2 \end{bmatrix} \right\} \prec 0 \quad (j \in \mathscr{I}_L). \tag{4.8}$$

This condition is not equivalent to (4.2) but still is proved to be less conservative than (4.5). The following step is to apply this condition to the closed-loop system:

$$X_j \succ 0, \quad \begin{bmatrix} 0 & X_j \\ X_j & 0 \end{bmatrix} + \text{He} \left\{ \begin{bmatrix} I \\ -A_j & -B_j K \end{bmatrix} \begin{bmatrix} F_1 & F_2 \end{bmatrix} \right\} \prec 0 \qquad (4.9)$$

which, when developed, reads as

$$X_j \succ 0, \quad \begin{bmatrix} 0 & X_j \\ X_j & 0 \end{bmatrix} + \text{He} \left\{ \begin{bmatrix} F_1 & F_2 \\ -A_j F_1 - B_j K F_1 & -A_j F_2 - B_j K F_2 \end{bmatrix} \right\} \prec 0.$$

$$(4.10)$$

In this case, unfortunately, no linearizing change of variables is possible because of the two products $K F_1$ and $K F_2$. Handling this issue is the central question of this chapter.

4.3 Stabilization of Discrete-Time Systems

For answering the question, let us consider the case of discrete-time systems. The reasons for this choice are: first, the discrete-time case is the one studied in the papers at the origin of the SV approach [2, 3]; second, it gives the opportunity to state that all the results we derived for continuous-time systems up to this point have their equivalent counterparts for discrete-time systems; third, because, as it will be seen, the discrete-time and continuous-time cases have different properties as soon as control design is considered.

4.3.1 Recursive and Variational Representations

Let us consider the uncertain discrete-time system represented by either the recursive equation

$$x_{k+1} = \widehat{A}(\theta)x_k + \widehat{B}(\theta)u_k,$$
$$\theta \in \mathbb{E}^L, \quad [\widehat{A}(\theta) \ \widehat{B}(\theta)] = \sum_{j=1}^{L} \theta_j [\widehat{A}_j \ \widehat{B}_j]. \qquad (4.11)$$

or the variational equation

$$\Delta x_k = A(\theta)x_k + B(\theta)u_k,$$
$$\theta \in \mathbb{E}^L, \quad [A(\theta) \ B(\theta)] = \sum_{j=1}^{L} \theta_j [A_j \ B_j]. \qquad (4.12)$$

Here, T is the sampling time, $\Delta x_k = \frac{1}{T}(x_{k+1} - x_k)$, $A(\theta) = \frac{1}{T}(\widehat{A}(\theta) - I)$, $A_j = \frac{1}{T}(\widehat{A}_j - I)$, $B(\theta) = \frac{1}{T}\widehat{B}(\theta)$, $B_j = \frac{1}{T}\widehat{B}_j$. We emphasize that these are exactly the same system with seemingly different representation. Note that the variational representation is useful to avoid numerical instability especially when T is small.

Note as well that the variational representation has the feature that if T goes to zero, then the sampled system converges to the continuous one.

For the system represented by (4.11) or (4.12), we aim at designing a state-feedback control $u_k = K x_k$. The closed-loop system reads respectively as

$$x_{k+1} = \widehat{A}(\theta, K)x_k ,$$

$$\theta \in \mathbb{E}^L, \quad \widehat{A}(\theta, K) = \widehat{A}(\theta) + \widehat{B}(\theta)K = \sum_{j=1}^{L} \theta_j(\widehat{A}_j + \widehat{B}_j K). \tag{4.13}$$

for the recursive representation and as

$$\Delta x_k = A(\theta, K)x_k ,$$

$$\theta \in \mathbb{E}^L, \quad A(\theta, K) = A(\theta) + B(\theta)K = \sum_{j=1}^{L} \theta_j(A_j + B_j K). \tag{4.14}$$

for the variational representation. Stability of system (4.13) is guaranteed if and only if the eigenvalues of $\widehat{A}(\theta, K)$ are all in the open unit disk $\mathbb{D} = \mathbb{O}\left(\begin{bmatrix} 1 & 0 \\ 0 & -1 \end{bmatrix} \right)$. Based on the linear transformation $A(\theta, K) = \frac{1}{T}(\widehat{A}(\theta, K) - I)$, stability of system (4.14) is equivalent to eigenvalue location of $A(\theta, K)$ in the open disk centered at $-\frac{1}{T}$ with radius $\frac{1}{T}$, that is $\mathbb{O}\left(\begin{bmatrix} 1 & \frac{1}{T} \\ \frac{1}{T} & 0 \end{bmatrix} \right) = \mathbb{O}\left(\begin{bmatrix} T & 1 \\ 1 & 0 \end{bmatrix} \right)$. Applying results of Corollary 2.5 the discrete-time counterparts of the SV result (4.9) are

$$\widehat{X}_j \succ 0, \quad \begin{bmatrix} \widehat{X}_j & 0 \\ 0 & -\widehat{X}_j \end{bmatrix} + \mathrm{He}\left\{ \begin{bmatrix} I \\ -\widehat{A}_j - \widehat{B}_j K \end{bmatrix} \begin{bmatrix} \widehat{F}_1 & \widehat{F}_2 \end{bmatrix} \right\} \prec 0 \; (\forall j \in \mathscr{I}_L) \tag{4.15}$$

for the recursive equation modeling, and

$$X_j \succ 0, \quad \begin{bmatrix} T X_j & X_j \\ X_j & 0 \end{bmatrix} + \mathrm{He}\left\{ \begin{bmatrix} I \\ -A_j - B_j K \end{bmatrix} \begin{bmatrix} F_1 & F_2 \end{bmatrix} \right\} \prec 0 \; (\forall j \in \mathscr{I}_L) \tag{4.16}$$

for the variational representation.

On the other hand, applying Theorem 2.9 the following two discrete-time counterparts of (4.6) are derived:

$$\widehat{\Xi} \succ 0, \quad \begin{bmatrix} (\widehat{A}_j + \widehat{B}_j K) & I \end{bmatrix} \begin{bmatrix} \widehat{\Xi} & 0 \\ 0 & -\widehat{\Xi} \end{bmatrix} \begin{bmatrix} (\widehat{A}_j + \widehat{B}_j K)^T \\ I \end{bmatrix} \prec 0 \; (\forall j \in \mathscr{I}_L) \tag{4.17}$$

for the recursive equation modeling, and

$$\Xi \succ 0, \quad \begin{bmatrix} (A_j + B_j K) & I \end{bmatrix} \begin{bmatrix} T\Xi & \Xi \\ \Xi & 0 \end{bmatrix} \begin{bmatrix} (A_j + B_j K)^T \\ I \end{bmatrix} \prec 0 \; (\forall j \in \mathscr{I}_L) \tag{4.18}$$

for the variational representation. Theorem 2.9 moreover states the following two important properties:

- If (4.17) holds then (4.15) holds as well and a possible choice is $\widehat{X}_j = \widehat{\Xi}$, $\widehat{F}_1 = -\widehat{\Xi}$ and $\widehat{F}_2 = 0$.
- If (4.18) holds then (4.16) holds as well and a possible choice is $X_j = \Xi$, $F_1 = -T\Xi$ and $F_2 = -\Xi$.

4.3.2 LMIs for Stabilization

We now give the central LMI-based results for the state-feedback controller synthesis of discrete-time systems.

Theorem 4.1 *Any of the following LMI conditions, if feasible, provide a robustly stabilizing state-feedback control $u = Kx$ for the uncertain discrete-time LTI system (4.11) [and hence equally (4.12)]:*

(i-r) Exists $\widehat{Y} \in \mathbb{R}^{m \times n}$ and $\widehat{\Xi} \in \mathbb{S}^n$ such that

$$\widehat{\Xi} \succ 0, \quad \begin{bmatrix} -\widehat{\Xi} & \widehat{\Xi} A_j^T + \widehat{Y}^T B_j^T \\ A_j \widehat{\Xi} + B_j \widehat{Y} & -\widehat{\Xi} \end{bmatrix} \prec 0 \quad (\forall j \in \mathcal{I}_L). \qquad (4.19)$$

If these LMI are feasible, then take $K = \widehat{Y}\widehat{\Xi}^{-1}$.

(i-v) Exists $Y \in \mathbb{R}^{m \times n}$ and $\Xi \in \mathbb{S}^n$ such that

$$\Xi \succ 0, \quad \begin{bmatrix} -T\Xi & T\Xi A_j^T + TY^T B_j^T \\ TA_j \Xi + TB_j Y & \mathrm{He}\{A_j \Xi + B_j Y\} \end{bmatrix} \prec 0 \quad (\forall j \in \mathcal{I}_L). \qquad (4.20)$$

If these LMI are feasible, then take $K = Y\Xi^{-1}$.

(ii-r) Exists $\widehat{Y} \in \mathbb{R}^{m \times n}$, $\widehat{X}_j \in \mathbb{S}^n$ and $\widehat{F}_1 \in \mathbb{R}^{n \times n}$ such that

$$\widehat{X}_j \succ 0, \quad \begin{bmatrix} \widehat{X}_j & 0 \\ 0 & -\widehat{X}_j \end{bmatrix} + \mathrm{He}\left\{ \begin{bmatrix} \widehat{F}_1 & 0 \\ -\widehat{A}_j \widehat{F}_1 - \widehat{B}_j \widehat{Y} & 0 \end{bmatrix} \right\} \prec 0 \quad (\forall j \in \mathcal{I}_L). \tag{4.21}$$

If these LMI are feasible, then take $K = \widehat{Y}\widehat{F}_1^{-1}$.

(ii-v) Exists $Y \in \mathbb{R}^{m \times n}$, $X_j \in \mathbb{S}^n$ and $F_2 \in \mathbb{R}^{n \times n}$ such that

$$\begin{aligned} &X_j \succ 0, \\ &\begin{bmatrix} TX_j & X_j \\ X_j & 0 \end{bmatrix} + \mathrm{He}\left\{ \begin{bmatrix} TF_2 & F_2 \\ -TA_j F_2 - TB_j Y & -A_j F_2 - B_j Y \end{bmatrix} \right\} \prec 0 \\ &\hspace{5cm} (\forall j \in \mathcal{I}_L). \end{aligned} \tag{4.22}$$

If these LMI are feasible, then take $K = YF_2^{-1}$.

*Moreover: (i-r) and (i-v) are equivalent; (ii-r) and (ii-v) are equivalent; if (i-r) and
(i-v) are feasible then (ii-r) and (ii-v) are feasible as well.*

Proof First let us prove the validity of (i-r). To this end, substitute the change of
variables $\widehat{Y} = K\widehat{\Xi}$ to (4.19) and further apply a Schur complement argument to the
upper-left block. The result is exactly (4.17).

The same procedure applies to prove (i-v): substitute the change of variables
$Y = K\Xi$ to (4.20) and apply a Schur complement argument to the upper-left block.
The result is exactly (4.18).

The proof of (ii-r) is as follows. First substitute the change of variables $\widehat{Y} = K\widehat{F}_1$
and then notice that the condition is exactly (4.15) for the choice $\widehat{F}_2 = 0$.

(ii-v) is obtained similarly: substitute the change of variables $Y = K F_2$ and then
notice that the condition is exactly (4.16) for the choice $F_1 = T F_2$.

Now let us prove that (i-r) and (i-v) are equivalent. It will be done by proving that
(4.17) and (4.18) are equivalent. Start from (4.18) and replace A_j and B_j by their
respective values $\frac{1}{T}(\widehat{A}_j - I)$ and $\frac{1}{T}\widehat{B}_j$. The matrix inequality then reads as

$$
\left[(\widehat{A}_j - I + \widehat{B}_j K)\ \ I \right]
\begin{bmatrix} \frac{1}{T}\Xi & \frac{1}{T}\Xi \\ \frac{1}{T}\Xi & 0 \end{bmatrix}
\begin{bmatrix} (\widehat{A}_j - I + \widehat{B}_j K)^T \\ I \end{bmatrix} \prec 0.
$$

Multiply the inequality by T and develop the products to get exactly

$$
(\widehat{A}_j + \widehat{B}_j K)\Xi(\widehat{A}_j + \widehat{B}_j K)^T - \Xi \prec 0
$$

which is (4.17) for $\widehat{\Xi} = \Xi$.

Consider now condition (ii-v) and let us prove it is equivalent to (ii-r). To this
end, replace A_j and B_j by their respective values $\frac{1}{T}(\widehat{A}_j - I)$ and $\frac{1}{T}\widehat{B}_j$. The matrix
inequalities in (4.22) then reads as

$$
\begin{bmatrix} T X_j & X_j \\ X_j & 0 \end{bmatrix} + \mathrm{He} \left\{ \begin{bmatrix} T F_2 & F_2 \\ -(\widehat{A}_j - I)F_2 - \widehat{B}_j Y & -\frac{1}{T}(\widehat{A}_j - I)F_2 - \frac{1}{T}\widehat{B}_j Y \end{bmatrix} \right\} \prec 0.
$$

Pre and post multiply by $\begin{bmatrix} T^{-1/2}I & 0 \\ -T^{-1/2}I & T^{1/2}I \end{bmatrix}$ and its transpose respectively to get

$$
\begin{bmatrix} X_j & 0 \\ 0 & -X_j \end{bmatrix} + \mathrm{He} \left\{ \begin{bmatrix} F_2 & 0 \\ -\widehat{A}_j F_2 - \widehat{B}_j Y & 0 \end{bmatrix} \right\} \prec 0
$$

which is exactly the same constraint as (4.21) where $\widehat{X}_j = X_j$, $\widehat{F}_1 = F_2$ and $\widehat{Y} = Y$.

Remains to prove that feasibility of (i-r) and (i-v) imply feasibility of (ii-r) and (ii-
v) respectively. This fact is obtained thanks to the results of Theorem 2.9. As stated
in the preceding subsection, the feasibility of (4.17) implies the feasibility of (4.15)
with $\widehat{F}_2 = 0$, which is indeed the restriction imposed to get result (ii-r). Similarly,
the feasibility of (4.18) implies feasibility of (4.16) with $F_1 = T F_2$ which is indeed
the restriction imposed to get result (ii-v). This completes the proof. $\square$

The important feature of Theorem 4.1 is that it provides two different classes of LMI design conditions with the guarantee that the SV-LMIs are always less (or more precisely, no more) conservative. As shown later, unfortunately, this nice feature cannot be retained for continuous time systems.

4.3.3 More LMIs Parameterized by Schur Stable Matrices

For both conditions (ii-r) and (ii-v) of Theorem 4.1, the SV-LMI conditions impose a structure on the S-variables $\widehat{F}_1$, $\widehat{F}_2$, F_1, F_2 issued from the analysis conditions. The choice of the structure is crucial for proving conservatism reduction. But, it is not the only choice that makes the SV state-feedback problem convex. All other choices can be summarized to $\widehat{F}_2 = -\widehat{F}_1 \widehat{A}_o^T$ or $F_2 = -F_1 A_o^T$ for fixed matrices $\widehat{A}_o$ and A_o. The general result is as follows.

Theorem 4.2 *Let $\widehat{A}_o$ be a Schur stable matrix (all its eigenvalues are in the open unit disk) and let $A_o = \frac{1}{T}(\widehat{A}_o - I)$. Any of the following LMI conditions, if feasible, provide a robustly stabilizing state-feedback control $u = Kx$ for the uncertain discrete-time LTI system (4.11) [and hence equally (4.12)]:*

(iii-r) Exists $\widehat{Y} \in \mathbb{R}^{m \times n}$, $\widehat{X}_j \in \mathbb{S}^n$ and $\widehat{F}_1 \in \mathbb{R}^{n \times n}$ such that

$$\widehat{X}_j \succ 0,$$

$$\begin{bmatrix} \widehat{X}_j & 0 \\ 0 & -\widehat{X}_j \end{bmatrix} + \mathrm{He} \left\{ \begin{bmatrix} \widehat{F}_1 \\ -\widehat{A}_j \widehat{F}_1 - \widehat{B}_j \widehat{Y} \end{bmatrix} \begin{bmatrix} I & -\widehat{A}_o^T \end{bmatrix} \right\} \prec 0 \quad (\forall j \in \mathscr{I}_L). \quad (4.23)$$

If these LMI are feasible, then take $K = \widehat{Y} \widehat{F}_1^{-1}$.

(iii-v) Exists $Y \in \mathbb{R}^{m \times n}$, $X_j \in \mathbb{S}^n$ and $F_2 \in \mathbb{R}^{n \times n}$ such that

$$X_j \succ 0,$$

$$\begin{bmatrix} T X_j & X_j \\ X_j & 0 \end{bmatrix} + \mathrm{He} \left\{ \begin{bmatrix} F_1 \\ -A_j F_1 - B_j Y \end{bmatrix} \begin{bmatrix} I & -A_o^T \end{bmatrix} \right\} \prec 0 \quad (\forall j \in \mathscr{I}_L). \quad (4.24)$$

If these LMI are feasible, then take $K = Y F_1^{-1}$.

Moreover: (iii-r) and (iii-v) are equivalent.

Proof The proof to validate that these conditions are sufficient for the robust stabilization is immediate based on the choices $\widehat{F}_2 = -\widehat{F}_1 \widehat{A}_o^T$ or $F_2 = -F_1 A_o^T$ in (4.15) and (4.16), respectively.

   We next prove the equivalence of these two conditions. The inequality (4.24) with A_o, A_j and B_j replaced by their values gives

$$\begin{bmatrix} T X_j & X_j \\ X_j & 0 \end{bmatrix} + \mathrm{He} \left\{ \begin{bmatrix} F_1 \\ -\frac{1}{T}(\widehat{A}_j - I) F_1 - \frac{1}{T} \widehat{B}_j Y \end{bmatrix} \begin{bmatrix} I & -\frac{1}{T}(\widehat{A}_o - I)^T \end{bmatrix} \right\} \prec 0.$$

Pre and post multiply by $\begin{bmatrix} I & 0 \\ -I & TI \end{bmatrix}$ and its transpose respectively to get exactly

$$\begin{bmatrix} TX_j & 0 \\ 0 & -TX_j \end{bmatrix} + \text{He} \left\{ \begin{bmatrix} F_1 \\ -\widehat{A}_j F_1 - \widehat{B}_j Y \end{bmatrix} \begin{bmatrix} I & -\widehat{A}_o^T \end{bmatrix} \right\} \prec 0$$

which is exactly the same as (4.23) with $\widehat{X}_j = TX_j$, $\widehat{F}_1 = F_1$ and $\widehat{Y} = Y$.   □

An important fact to notice is the requirement that $\widehat{A}_o$ is a Schur stable matrix. If not, the LMIs are infeasible whatever the data. To prove this fact, pre- and post-multiply the constraints in (4.23) by $\begin{bmatrix} \widehat{A}_o & I \end{bmatrix}$ and its transpose, respectively. The result is exactly:

$$\widehat{X}_j \succ 0, \quad \widehat{A}_o \widehat{X}_j \widehat{A}_o^T - \widehat{X}_j \prec 0$$

That is: all matrices $\widehat{X}_j$ need to be Lyapunov certificates proving the Schur stability of $\widehat{A}_o$. This is a strong requirement, except for the trivial case when $\widehat{A}_o = \lambda I$ with $|\lambda| < 1$. Indeed, in this case the constraints $\widehat{X}_j \succ 0$ and $\widehat{A}_o \widehat{X}_j \widehat{A}_o^T - \widehat{X}_j = (\lambda^2 - 1)\widehat{X}_j \prec 0$ are redundant.

4.3.4 Conclusions on the State-Feedback Stabilization of Discrete-Time Systems

Before moving to the continuous-time case, let us summarize the features obtained in the last two theorems.

- Theorem 4.2 provides finitely many LMI conditions for robust state-feedback stabilizing controller synthesis. These conditions are parameterized by the matrix $\widehat{A}_o$ that should be chosen among Schur stable matrices.
- Conditions (ii-r) and (ii-v) of Theorem 4.1 are special cases of conditions (iii-r) and (iii-v) of Theorem 4.2 for the trivial choice of $\widehat{A}_o = 0$ (and $A_o = -\frac{1}{T}I$).
- SV-LMI results of Theorem 4.2 are proved to be no more conservative than quadratic stability type conditions (i-r) and (i-v) of Theorem 4.1, only for the case $\widehat{A}_o = 0$. Any other choice of $\widehat{A}_o$ may provide conditions that are either less or more conservative. This fact is illustrated in examples in the following.
- There is no known systematic methods for choosing $\widehat{A}_o$ that would provide the least conservative conditions. To get around this difficulty, iterative search for decisions variables K and $\widehat{A}_o$ can be considered. One such iterative method is discussed in the following for the pole location issue.

4.4 The Continuous-Time Case

4.4.1 LMIs Parameterized by Hurwitz Stable Matrices

Based on the upper treated discrete-time case, the continuous-time case results are simple to obtain: these correspond to taking $T \rightarrow 0$ in the variational representation (4.12). In that case, Δx_k converges to $\dot{x}(t)$ and the pole location condition in the disc centered at $-\frac{1}{T}$ with radius $\frac{1}{T}$, that is the domain $\mathbb{O}\left(\begin{bmatrix} T & 1 \\ 1 & 0 \end{bmatrix}\right)$, converges to pole location in the left half plane $\mathbb{C}_- = \mathbb{O}\left(\begin{bmatrix} 0 & 1 \\ 1 & 0 \end{bmatrix}\right)$. Hence, one gets without difficulty the following result.

Theorem 4.3 *Let A_o be a Hurwitz stable matrix (all its eigenvalues are in the open left half plane). Any of the following LMI conditions, if feasible, provide a robustly stabilizing state-feedback control $u = Kx$ for the uncertain continuous-time LTI system (4.3):*

(i-c) *Exists $Y \in \mathbb{R}^{m \times n}$ and $\varXi \in \mathbb{S}^n$ such that*

$$\varXi \succ 0, \quad \text{He}\left\{A_j \varXi + B_j Y\right\} \prec 0 \quad (\forall j \in \mathscr{I}_L). \tag{4.25}$$

If these LMI are feasible then take $K = Y \varXi^{-1}$.

(iii-c) *Exists $Y \in \mathbb{R}^{m \times n}$, $X_j \in \mathbb{S}^n$ and $F_1 \in \mathbb{R}^{n \times n}$ such that*

$$X_j \succ 0,$$
$$\begin{bmatrix} 0 & X_j \\ X_j & 0 \end{bmatrix} + \text{He}\left\{\begin{bmatrix} F_1 \\ -A_j F_1 - B_j Y \end{bmatrix}\begin{bmatrix} I & -A_o^T \end{bmatrix}\right\} \prec 0 \quad (\forall j \in \mathscr{I}_L). \tag{4.26}$$

If these LMI are feasible then take $K = Y F_1^{-1}$.

A major difference between the discrete and continuous-time cases is that in the continuous-time case, there is no such result as (ii-v). To understand the reason for it, take $T = 0$ in (4.22) to get

$$\begin{bmatrix} 0 & X_j \\ X_j & 0 \end{bmatrix} + \text{He}\left\{\begin{bmatrix} 0 & F_2 \\ 0 & -A_j F_2 - B_j Y \end{bmatrix}\right\} \prec 0.$$

The upper-left bock of this matrix inequality is zero. The inequality therefore cannot hold in terms of negative definite matrices. For T close to zero, we can nevertheless consider the asymptotic behavior of

$$\begin{bmatrix} T X_j & X_j \\ X_j & 0 \end{bmatrix} + \text{He}\left\{\begin{bmatrix} T F_2 & F_2 \\ -T A_j F_2 - T B_j Y & -A_j F_2 - B_j Y \end{bmatrix}\right\} \prec 0.$$

As T goes to zero, the upper-left block goes to zero. Therefore, the LMI can hold only if the off-diagonal blocks converge to zero as well. Hence, for small T one needs to have $F_2 = -X_j$, for all vertices $j \in \mathscr{I}_L$. The only possible solution to these LMIs as T goes to zero is hence $F_2 = -\Xi = -X_j$ and He $\left\{ A_j \Xi + B_j Y \right\} \prec 0$ which is exactly condition (4.25). That same fact is also stated in Theorem 2.7.

The analysis of the asymptotic behavior of (4.22) as T goes to zero allows to conclude the following facts:

- SV-LMI results for the state-feedback design in the continuous-time case need to choose a priori some Hurwitz stable matrix A_o. A natural choice of such matrix is to take $A_o = -\lambda I$ for some positive $\lambda > 0$.
- There exists no a priori choice of λ that would guarantee condition (iii-c) to be no more conservative than (i-c).
- The choice $\lambda = \frac{1}{T}$ for some small T makes conditions (iii-c) and (i-c) almost equivalent and should hence not be considered since it contains many unnecessary decision variables compared to (i-c).
- If (i-c) holds there exists a matrix A_o such that (iii-c) holds as well. Finding such A_o for fixed K obtained when solving (i-c) needs only to solve the following LMI problem

$$X_j \succ 0, \quad \begin{bmatrix} 0 & X_j \\ X_j & 0 \end{bmatrix} + \text{He} \left\{ \begin{bmatrix} I \\ -A_j - B_j K \end{bmatrix} \begin{bmatrix} F_1 & F_2 \end{bmatrix} \right\} \prec 0$$

and take $A_o = -(F_1^{-1} F_2)^T$. Existence of a solution is guaranteed by Corollary 2.3.

4.4.2 An Unsolved Issue

Before moving further to more complex situations, let us relate a discussion with P. Apkarian. The authors remember having discussed with him the issue of choosing A_o in the continuous-time case and he suggested to convert the continuous-time problem into a discrete-time problem using a bijective map:

$$s = \frac{z-1}{z+1} \in \mathbb{C}_- \Leftrightarrow z = \frac{1+s}{1-s} \in \mathbb{D}$$

Applying this transformation, Hurwitz stability of $\dot{x} = A(\theta, K)x$ becomes a problem of Schur stability of $(I - A(\theta, K))x_{k+1} = (I + A(\theta, K))x_k$. Based on results of Theorem 3.5 one gets the following condition for the stabilizing controller synthesis of this descriptor system:

$$X_j \succ 0, \quad \begin{bmatrix} X_j & 0 \\ 0 & -X_j \end{bmatrix} + \text{He} \left\{ \begin{bmatrix} I - A_j - B_j K \\ I + A_j + B_j K \end{bmatrix} \begin{bmatrix} F_1 & F_2 \end{bmatrix} \right\} \prec 0.$$

Based on the results for discrete-time systems state above, the S-variables can be chosen as $F_2 = -F_1 A_o$ and an intuitive choice of A_o is the zero matrix. That choice gives the following LMI condition

$$X_j \succ 0, \quad \begin{bmatrix} X_j & 0 \\ 0 & -X_j \end{bmatrix} + \mathrm{He}\left\{ \begin{bmatrix} F_1 - A_j F_1 - B_j Y & 0 \\ F_1 + A_j F_1 + B_j Y & 0 \end{bmatrix} \right\} \prec 0.$$

Although seemingly new, this condition is just a special case of those in Theorem 4.3. To confirm this fact, pre- and post-multiply the last inequality by $\frac{1}{\sqrt{2}} \begin{bmatrix} I & I \\ -I & I \end{bmatrix}$ and its transpose respectively. The obtained formula is exactly that of (4.26) for $A_o = -I$. The result is not surprising since $s = -1$ is the image of $z = 0$ by the bijective map.

4.5 Pole Location in Subregions of the Complex Plane

4.5.1 The Multiperformance Pole Location Problem

In practical engineering problems, finding a stabilizing state feedback is often not enough. A situation where all closed-loop poses are close to the imaginary axis is usually to be avoided in order to prevent slowly converging and oscillatory time responses. The classical state-feedback problem is hence to perform some pole place-ment in predefined places of the complex plane. In case of systems with uncertainties, precise pole placement is not achievable and the problem is relaxed into pole location in a region of the complex plane. The simplest type of regions that are usually con-sidered are LMI or QMI regions and intersections of such, see [4, 5]. See also [6–8] for more complex regions described by unions and intersections of QMI regions.

Typical pole location design problems are to find a state-feedback gain such that the closed-loop poles have simultaneously: real parts smaller than some $\alpha = -\frac{1}{\tau}$ to ensure exponential decay has a time constant faster than τ; be in a cone defined by a line that passes trough the origin of the complex plane with and angle $\phi = \arcsin \zeta$ to guarantee a minimal damping of ζ for all poles; be in a disc centered at zero and with radius ω to guarantee all modes have undamped natural frequency smaller than ω. As discussed in Sect. 2.5.1, all such regions are QMI regions $\mathbb{O}(R_l)$ defined by matrices $R_l = \begin{bmatrix} R_{11l} & R_{12l} \\ R_{12l}^* & R_{22l} \end{bmatrix}$ with $R_{11l} \in \mathbb{H}^{d_l}$, $R_{12l} \in \mathbb{C}^{d_l \times d_l}$, $R_{22l} \in \mathbb{H}^{d_l}$.

Regions delimited by circles or lines, which are the most common cases (if not the only relevant cases), are all such that $d_l = 1$, or eventually $d_l = 2$ when considering their real-valued version. See Sect. 2.5.1 for precise description of these. Regions that describe interior of circles are such that $R_{11l} \succeq 0$ and exteriors of circles are such that $R_{11l} \preceq 0$. Half planes such as $\mathbb{C}_-$ are defined by matrices satisfying $R_{11l} = 0$. These are hence at the border between these two convex and concave cases.

Our goal is to provide LMI conditions for robust pole location state-feedback design in intersections of QMI regions. Such intersections happen to be QMI regions $\mathbb{O}(R)$ defined by taking

$$
R = \begin{bmatrix} R_{11} & R_{12} \\ R_{12}^* & R_{22} \end{bmatrix} = \left[\begin{array}{cccc|cccc}
R_{111} & 0 & \cdots & R_{121} & 0 & \cdots \\
0 & R_{112} & & 0 & R_{122} & \\
\vdots & & \ddots & \vdots & & \ddots \\
\hline
R_{121}^* & 0 & \cdots & R_{221} & 0 & \cdots \\
0 & R_{122}^* & & 0 & R_{222} & \\
\vdots & & \ddots & \vdots & & \ddots
\end{array} \right]. \tag{4.27}
$$

If the intersection combines concave and convex regions, the matrix R_{11} will not be sign definite and would render the derivation of LMI conditions more complex. As seen further, this is one of the reasons for which we do not recommend to treat intersections of QMI regions as a unique QMI region $\mathbb{O}(R)$ defined by a matrix R of larger dimensions.

4.5.2 LMIs Parameterized by a Matrix Pencil

Based on Corollary 2.5 and the reasonings developed for discrete and continuous-time stabilizing state-feedback controller synthesis problems, we can obtain the following result.

Theorem 4.4 *Let (F_{o1l}, F_{o2l}) be a couple of matrices of appropriate dimensions. The following LMI conditions, if feasible, provide a state-feedback control $u = Kx$ that locates all poles of the uncertain continuous-time LTI system (4.3) in $\mathbb{O}(R) = \bigcap_{l=1}^{M} \mathbb{O}(R_l)$:*

(iii-p) Exists $Y \in \mathbb{R}^{m \times n}$, $X_{jl} \in \mathbb{H}^n$ and $\tilde{F}_1 \in \mathbb{R}^{n \times n}$ such that

$$
X_{jl} \succ 0,
$$
$$
\begin{bmatrix} R_{11l} \otimes X_{jl} & R_{12l}^* \otimes X_{jl} \\ R_{12l} \otimes X_{jl} & R_{22l} \otimes X_{jl} \end{bmatrix}
$$
$$
+ \mathrm{He} \left\{ \begin{bmatrix} I_{d_l} \otimes \tilde{F}_1 \\ -I_{d_l} \otimes (A_j \tilde{F}_1 + B_j Y) \end{bmatrix} \begin{bmatrix} F_{o1l} & F_{o2l} \end{bmatrix} \right\} \prec 0 \tag{4.28}
$$
$$
(\forall j \in \mathscr{I}_L, \ \forall l \in \mathscr{I}_M).
$$

If these LMI are feasible then take $K = Y \tilde{F}_1^{-1}$.

Proof The result readily follows from the following choice of SVs $F_1 = (I_{d_l} \otimes \tilde{F}_1) F_{o1l}$ and $F_2 = (I_{d_l} \otimes \tilde{F}_1) F_{o2l}$ in the conditions of Corollary 2.5. Indeed, for

$Y = K\tilde{F}_1$ and that choice of SVs one has

$$
\begin{bmatrix} I_{d_l n} \\ -I_{d_l} \otimes (A_j + B_j K) \end{bmatrix} \begin{bmatrix} F_1 & F_2 \end{bmatrix} = \begin{bmatrix} I_{d_l} \otimes \tilde{F}_1 \\ -I_{d_l} \otimes (A_j \tilde{F}_1 + B_j Y) \end{bmatrix} \begin{bmatrix} F_{o1l} & F_{o2l} \end{bmatrix} \quad \square
$$

As in the discrete-time and continuous-time stabilization cases treated previously, the result is based on a conservative choice of structure on the S-variables. This structure is here characterized by two matrices F_{o1l} and F_{o2l} and there is a need for choosing appropriately these matrices a priori. There is unfortunately no ideal known choice, but the following suggestions have interesting rationalities:

[Ia] Take $F_{o1l} = I_{d_l n}$ and $F_{o2l} = -I_{d_l} \otimes A_{ol}^*$ where A_{ol} is a matrix with all eigenvalues located in $\mathbb{O}(R_l)$.

[Ib] Take $F_{o1l} = I_{d_l} \otimes E_{ol}^*$ and $F_{o2l} = -I_{d_l n}$ where E_{ol} is a matrix with all eigenvalues located in $\mathbb{O}\left(\begin{bmatrix} R_{22l} & R_{12l} \\ R_{12l}^* & R_{11l} \end{bmatrix} \right)$.

[Ic] If $R_{11l} \succeq 0$ take $F_{o1l} = (R_{11l} + \varepsilon_l I) \otimes I_n$ and $F_{o2l} = R_{12l}^* \otimes I_n$ for some small positive scalar $\varepsilon_l > 0$.

[Id] If $R_{22l} \succeq 0$ take $F_{o1l} = R_{12l} \otimes I_n$ and $F_{o2l} = (R_{22l} + \varepsilon_l I) \otimes I_n$ for some small positive scalar $\varepsilon_l > 0$.

[Ie] Take $F_{o1l} = \check{F}_{o1l} \otimes E_{ol}^*$ and $F_{o2l} = -\check{F}_{o2l} \otimes A_{ol}^*$ where $\lambda E_{ol} - A_{ol}$ is a matrix pencil with all eigenvalues located in the modified region $\mathbb{O}(\check{R})$ where:

$$
\begin{bmatrix} \check{F}_{o1l}^* \\ \check{F}_{o2l}^* \end{bmatrix}^{\perp *} = \begin{bmatrix} \check{R}_{1l} \\ \check{R}_{2l} \end{bmatrix}, \quad \check{R} = \begin{bmatrix} \check{R}_{1l}^* R_{11l} \check{R}_{1l} & \check{R}_{2l}^* R_{12l} \check{R}_{1l} \\ \check{R}_{1l}^* R_{12l}^* \check{R}_{2l} & \check{R}_{2l}^* R_{22l} \check{R}_{2l} \end{bmatrix}.
$$

The rationale of these suggestions follow the reasoning adopted to derive result in the continuous and discrete-time stabilization cases. The features of these choices are that when pre- and post-multiplying (4.28) by $N = \begin{bmatrix} F_{o1l}^* \\ F_{o2l}^* \end{bmatrix}^{\perp}$ and its conjugate transpose respectively one should get a feasible LMI constraint.

To see this more concretely, consider the instance [Ia]. The operation gives:

$$
N(4.28)N^* = \begin{bmatrix} I_{d_l} \otimes A_{ol} & I_{d_l n} \end{bmatrix} \begin{bmatrix} R_{11l} \otimes X_{jl} & R_{12l}^* \otimes X_{jl} \\ R_{12l} \otimes X_{jl} & R_{22l} \otimes X_{jl} \end{bmatrix} \begin{bmatrix} I_{d_l} \otimes A_{ol}^* \\ I_{d_l n} \end{bmatrix} \prec 0
$$

which, based on Lemma 2.6, can hold only if A_{ol} has all its poles located in $\mathbb{O}(R_l)$. For the instance [Ib], the operation produces

$$
\begin{aligned}
N(4.28)N^* &= \begin{bmatrix} I_{d_l n} & I_d \otimes E_{ol} \end{bmatrix} \begin{bmatrix} R_{11l} \otimes X_{jl} & R_{12l}^* \otimes X_{jl} \\ R_{12l} \otimes X_{jl} & R_{22l} \otimes X_{jl} \end{bmatrix} \begin{bmatrix} I_{d_l n} \\ I_d \otimes E_{ol}^* \end{bmatrix} \\
&= \begin{bmatrix} I_d \otimes E_{ol} & I_{d_l n} \end{bmatrix} \begin{bmatrix} R_{22l} \otimes X_{jl} & R_{12l} \otimes X_{jl} \\ R_{12l}^* \otimes X_{jl} & R_{11l} \otimes X_{jl} \end{bmatrix} \begin{bmatrix} I_d \otimes E_{ol}^* \\ I_{d_l n} \end{bmatrix} \prec 0
\end{aligned}
$$

which, based on Lemma 2.6, can hold only if E_{ol} has all its poles located in the region specified in [Ib]. One can by the way notice at this point that the inverse of the eigenvalues of E_{ol} are then in $\mathbb{O}(R_l)$. For a region defined as the exterior of a disc, the instance allows hence to specify some E_{ol} chosen in the interior of an other disc which is a simpler problem to solve.

For the instance [Ic], the operation produces

$$N = \left[(R_{12l}(R_{11l} + \varepsilon_l I)^{-1}) \otimes I_n \; -I_{d_l n} \right],$$
$$N(4.28)N^* = \left(-\varepsilon R_{12l}(R_{11l} + \varepsilon I)^{-2} R^*_{12l} \right.$$
$$\left. - R_{12l}(R_{11l} + \varepsilon I)^{-1} R^*_{12l} + R_{22l} \right) \otimes X_{jl} \prec 0.$$

Since X_{jl} is positive definite, the result indeed holds if the left-hand side term of the Kronecker product is negative definite. The term $-\varepsilon R_{12l}(R_{11l} + \varepsilon I)^{-2} R^*_{12l}$ is trivially negative semidefinite. The fact that the second term $-R_{12l}(R_{11l} + \varepsilon I)^{-1} R^*_{12l} + R_{22l}$ is negative definite as well comes from the assumption that the region is nonempty. Indeed, assuming the region is nonempty, it contains a value $\lambda \in \mathbb{C}$ such that

$$\lambda^* \lambda R_{11l} + \lambda R^*_{12l} + \lambda^* R_{12l} + R_{22l} \prec 0.$$

A small perturbation argument implies that there exists $\varepsilon_l > 0$ such that

$$\lambda^* \lambda (R_{11l} + \varepsilon_l I) + \lambda R^*_{12l} + \lambda^* R_{12l} + R_{22l} \prec 0.$$

Since $R_{11l} \succeq 0$ one gets that $R_{11l} + \varepsilon_l I \succ 0$ is invertible, and the last inequality also reads as

$$(\lambda I + R_{12l}(R_{11l} + \varepsilon_l I))(R_{11l} + \varepsilon_l I)^{-1} (\lambda^* I + (R_{11l} + \varepsilon_l I) R^*_{12l})$$
$$\prec R_{12l}(R_{11l} + \varepsilon_l I)^{-1} R^*_{12l} - R_{22l}$$

and the matrix on the right-hand side is hence positive definite.

The rationale of [Id] follows the same lines as that of [Ic] but when assuming the region $\mathbb{O} \left(\begin{bmatrix} R_{22} & R_{12} \\ R^*_{12} & R_{11} \end{bmatrix} \right)$, region in which lie the inverse conjugate of the eigenvalues, is nonempty.

The complicated instance [Ie] is the combination of the previous ones. Its justification is as follows. By definition, one has $\check{F}_{1ol} \check{R}_{1l} + \check{F}_{2ol} \check{R}_{2l} = 0$. Assume λ is an eigenvalue of the matrix pencil and ξ is the associated eigenvector for the conjugate transpose pencil, that is $\lambda^* E^*_{ol} \xi = A^*_{ol} \xi$. It is easy to see that the following result holds

$$\left[F_{1ol} \; F_{2ol} \right] \begin{bmatrix} \check{R}_{1l} \otimes (\lambda^* \xi) \\ \check{R}_{2l} \otimes \xi. \end{bmatrix} = 0$$

Hence when performing the following congruence operation, only one term remains which is

$$\begin{bmatrix} \check{R}_{1l} \otimes (\lambda^*\xi) \\ \check{R}_{2l} \otimes \xi \end{bmatrix}^* (4.28) \begin{bmatrix} \check{R}_{1l} \otimes (\lambda^*\xi) \\ \check{R}_{2l} \otimes \xi \end{bmatrix}$$

$$= (\xi^* X_{jl}\xi) \begin{bmatrix} \lambda I_d & I_d \end{bmatrix} \begin{bmatrix} \check{R}_{1l}^* R_{11l} \check{R}_{1l} & \check{R}_{1l}^* R_{12l}^* \check{R}_{2l} \\ \check{R}_{2l}^* R_{12l} \check{R}_{1l} & \check{R}_{2l}^* R_{22l} \check{R}_{2l} \end{bmatrix} \begin{bmatrix} \lambda^* I_d \\ I_d \end{bmatrix}$$

$$= (\xi^* X_{jl}\xi) \begin{bmatrix} \lambda^* I_d & I_d \end{bmatrix} \begin{bmatrix} \check{R}_{1l}^* R_{11l} \check{R}_{1l} & \check{R}_{2l}^* R_{12l} \check{R}_{1l} \\ \check{R}_{1l}^* R_{12l}^* \check{R}_{2l} & \check{R}_{2l}^* R_{22l} \check{R}_{2l} \end{bmatrix} \begin{bmatrix} \lambda I_d \\ I_d \end{bmatrix} \prec 0.$$

It is negative definite only if λ is in the interior of the specified region $\mathbb{O}(\check{R})$.

4.5.3 LMIs for Pole Location in Convex Regions

Before moving on to state-feedback design in intersections of regions, let us notice that instance [Ic] with $\varepsilon_l = 0$ coincides with conditions (ii-r) and (ii-v) of Theorem 4.1. That instance allows to relate the SV design results with more classical "quadratic stability" type results of Theorem 2.9. If limiting the regions to convex ones with $R_{11} \succeq 0$, Theorem 2.9 allows to build the following additional result.

Theorem 4.5 *Suppose $R_{11l} = \bar{R}_l^* \bar{R}_l \succeq 0$ and $\varepsilon_l \geq 0$. Then, any of the following LMI conditions, if feasible, provides a state-feedback control $u = Kx$ that locates all poles of the uncertain continuous-time LTI system (4.3) in $\mathbb{O}(R_l)$:*

(i-p)   *There exists $Y \in \mathbb{R}^{m \times n}$ and $\varXi \in \mathbb{H}^n$ such that*

$$\varXi \succ 0,$$
$$\begin{bmatrix} I \otimes \varXi & \bar{R}_l \otimes (\varXi A_j^* + Y^* B_j^*) \\ \bar{R}_l^* \otimes (A_j \varXi + B_j Y) & \mathrm{He}\left\{R_{12l}^* \otimes (A_j \varXi + B_j Y)\right\} + R_{22l} \otimes \varXi \end{bmatrix} \prec 0$$
$$(\forall j \in \mathscr{I}_L).$$
(4.29)

If these LMI are feasible then take $K = Y\varXi^{-1}$.

(ii-p)   *There exists $Y \in \mathbb{R}^{m \times n}$, $X_{jl} \in \mathbb{H}^n$ and $\tilde{F}_1 \in \mathbb{C}^{n \times n}$ such that*

$$X_{jl} \succ 0,$$
$$\begin{bmatrix} R_{11l} \otimes X_{jl} & R_{12l}^* \otimes X_{jl} \\ R_{12l} \otimes X_{jl} & R_{22l} \otimes X_{jl} \end{bmatrix}$$
$$+ \mathrm{He}\left\{\begin{bmatrix} I_d \otimes \tilde{F}_1 \\ -I_d \otimes (A_j \tilde{F}_1 + B_j Y) \end{bmatrix} \begin{bmatrix} -(R_{11l} + \varepsilon_l I) \otimes I_n & R_{12l}^* \otimes I_n \end{bmatrix}\right\} \prec 0$$
$$(\forall j \in \mathscr{I}_L).$$
(4.30)

If these LMI are feasible, then take $K = Y\tilde{F}_1^{-1}$.

Moreover, if (4.29) holds then (4.30) holds as well for some small enough $\varepsilon_l > 0$.

Proof To prove (i-p) make the change of variables $Y = K \Xi$ and perform a Schur complement on the upper-left block. The result is

$$\left[I_d \otimes (A_j + B_j K) \ I \right] \begin{bmatrix} R_{11} \otimes \Xi & R_{12}^* \otimes \Xi \\ R_{12} \otimes \Xi & R_{22} \otimes \Xi \end{bmatrix} \begin{bmatrix} I_d \otimes (A_j + B_j K)^* \\ I \end{bmatrix} \prec 0$$

which is the dual version of (2.39) for the closed-loop system.

(ii-p) follows from the fact that it is exactly (iii-p) of Theorem 4.4 for the instance [Ic] of matrices F_{1ol} and F_{2ol}.

The implication (i-p) $\Rightarrow$ (ii-p) is a direct consequence of Theorem 2.9.   □

Recall that regional pole location includes as special cases the discrete and continuous-time stabilization problems. Theorem 4.5 hence allows to explain the paradoxical difference between these two stabilization problems. In the discrete-time case, there exists an a priori choice of structure on the SVs that makes the SV-LMI result no more conservative than the usual "quadratic stability" result, while in the continuous-time case, there is no such a priori choice. What Theorem 4.5 indicates is that the SV-LMI result applies for both convex regions for which $R_{11l} \succeq 0$ and for nonconvex regions of the complex plane. Meanwhile, the "quadratic stability" type result applies only to convex regions. The case of regions defined by $R_{11l} = 0$, such as the left-hand side half of the complex plane $\mathbb{C}_-$ used to assess stability of continuous-time system, is clearly at the border between these two situations. It is not surprising that some properties are lost when considering such borderline instance.

4.5.4 Pole Location in Intersections of Regions

Theorems 4.4 and 4.5 cope with any pole location problem in QMI regions, including the case of regions defined as intersections of QMI regions as in (4.27). For that case of intersections of QMI regions an alternative to Theorem 4.4 is, rather than considering the intersection as a unique QMI region $\mathbb{O}(R)$, to solve a multiperformance problem, each performance specification being the pole location in one region $\mathbb{O}(R_l)$. Let us study these two approaches.

When considering the intersection of QMI regions as a unique QMI region $\mathbb{O}(R)$ with R defined by (4.27), the state-feedback design conditions impose to chose a priori some matrices F_{1o} and F_{2o} then to find a feasible solution to the LMIs

$$X_j \succ 0,$$
$$\begin{bmatrix} R_{11} \otimes X_j & R_{12}^* \otimes X_j \\ R_{12} \otimes X_j & R_{22} \otimes X_j \end{bmatrix} + \text{He} \left\{ \begin{bmatrix} I_d \otimes \tilde{F}_1 \\ -I_d \otimes (A_j \tilde{F}_1 + B_j Y) \end{bmatrix} \left[F_{o1} \ F_{o2} \right] \right\} \prec 0$$
$$(\forall j \in \mathscr{I}_L).$$

$$(4.31)$$

The alternative, multiperformance type approach, is to select for each individual region $\mathbb{O}(R_l)$ a pair of matrices (F_{1ol}, F_{2ol}) and to solve simultaneously for all l the LMIs

$$
X_{jl} \succ 0,
$$
$$
\begin{bmatrix} R_{11l} \otimes X_{jl} & R_{12l} \otimes X_{jl} \\ R_{12l} \otimes X_{jl} & R_{22l} \otimes X_{jl} \end{bmatrix} + \mathrm{He} \left\{ \begin{bmatrix} I_{d_l} \otimes \tilde{F}_1 \\ -I_{d_l} \otimes (A_j \tilde{F}_1 + B_j Y) \end{bmatrix} \begin{bmatrix} F_{o1l} & F_{o2l} \end{bmatrix} \right\} \prec 0
$$
$$
(\forall j \in \mathscr{I}_L).
$$

$$(4.32)$$

For each l, the LMIs impose the pole location in the region $\mathbb{O}(R_l)$. If the problem is feasible, the state-feedback gain locates the poles in all regions simultaneously, that is, in their intersection.

The immediate to notice advantage of the latter approach is that there are available heuristics for choosing the matrices (F_{o1l}, F_{o2l}) (see instances [Ia], [Ib], [Ic] and [Id]), while it is more complex to choose the huge (F_{o1}, F_{o2}) matrices. The second advantage is that (4.32) is a collection of LMIs of low dimensions while (4.31) involves matrix constraints of large size. The overall size when concatenating all matrices is identical, yet the multiperformance approach is expected to be more reliable numerically. The last difference between the two approaches is that in the former one the same Lyapunov certificate X_j is used to asses pole location in all regions $\mathbb{O}(R_l)$, while in the multiperformance approach extra degrees of freedom are introduced by allowing the Lyapunov certificates X_{jl} to differ from one region to another. The question that immediately arises is whether these degrees of freedom allow conservatism reduction. The answer is yes, as stated in the following theorem.

Theorem 4.6 *If there exists a solution X_j, Y, $\tilde{F}_1$ to the LMI conditions (4.31), then there also exists a solution to (4.32) for appropriately chosen matrices (F_{o1l}, F_{o2l}).*

Proof Let $T_l = \begin{bmatrix} 0 & I_{d_l} & 0 \end{bmatrix}^T$ be the matrix such that

$$
\begin{bmatrix} T_l^T & T_l^T \end{bmatrix} R \begin{bmatrix} T_l \\ T_l \end{bmatrix} = R_l.
$$

Pre- and post-multiply (4.31) by $\begin{bmatrix} T_l^T \otimes I_n & T_l^T \otimes I_n \end{bmatrix}$ and its transpose respectively. The result is exactly (4.32) with $X_{jl} = X_j$, $F_{o1l} = (T_l^T \otimes I_n) F_{o1}(T_l \otimes I_n)$ and $F_{o2l} = (T_l^T \otimes I_n) F_{o2}(T_l \otimes I_n)$. □

Theorem 4.6 indicates that SV-LMIs allow having different Lyapunov certificates X_{jl} for each region, and states that it might on examples produce results with reduced conservatism. Such possibility does not hold for "quadratic stability" type results. Not only, as noticed in Chap. 2, S-variable approach allows to search for parameter-dependent Lyapunov certificates, but it also permits to relax the so-called "Lyapunov shaping paradigm" [9] that imposes to search for common Lyapunov matrices in case of multiperformance design. This advantage is unfortunately associated to a drawback: the need for choosing a priori the matrices F_{o1l} and F_{o2l} that structure the SVs. The chapter that follows is dedicated to multiperformance design including H_∞, H_2 and impulse-to-peak performances. It will fully exploit the decoupling between Lyapunov matrices and controller gains that allows the S-variable approach.

4.5.5 Heuristic Algorithms

The priori choice of matrices F_{o1l} and F_{o2l} strongly affects the success of the design procedure. If the LMIs reveal infeasible, one may aim at searching for other values and solve the LMIs again. Some strategies for doing so are exposed in the following.

4.5.5.1 Randomized Strategy

Since there is no indication on how to select the matrices F_{o1l} and F_{o2l} a priori except for the suggestions given in Sect. 4.5.2, the randomized strategy can be employed which generates F_{o1l} and F_{o2l} randomly that satisfy conditions of [Ie]. These are the most general conditions that include the instances [Ia]–[Id]. The difficulty is of course to generate uniformly the samples.

The simplest case is $\mathbb{O}(R_l) = \mathbb{D}$, where $\mathbb{D}$ is the interior of the unit disc. Because $d_l = 1$ the choice $\check{F}_{o1l} = 1$ and $\check{F}_{o2l} = 1$ brings no conservatism. Moreover, it is rather simple to show that choosing $E_{ol} = I$ is not restrictive either. Hence, the only difficulty is to generate a Schur stable A_{ol}. To do so, methods exposed in [10] are applicable.

Another rather simple case is when $R_l = \begin{bmatrix} -1 & 0 \\ 0 & 1 \end{bmatrix}$ defines the exterior of the unit disc. In that case, the inverse of the eigenvalues belong to the interior of the unit disc and following the strategy of instance [Ib] the randomization is possible choosing $A_{ol} = I$ and E_{ol} a Schur stable matrix.

For all other regions defined as interiors or exteriors of discs, randomization methods can be derived from the two upper defined cases by translation and homothecy.

More complicated case is half planes since these are unbounded. As suggested in [10], a possibility is then to close the half planes. For example, by approximating these as discs of large radius and tangent to the line defying the half plane, that is, replacing $\mathbb{O}\left(\begin{bmatrix} 0 & R_{12} \\ R_{12}^* & R_{22} \end{bmatrix} \right)$ by $\mathbb{O}\left(\begin{bmatrix} \varepsilon I & R_{12} \\ R_{12}^* & R_{22} \end{bmatrix} \right)$ for some small $\varepsilon > 0$. Doing so eliminates the choice of matrix pencils $\lambda E_{ol} - A_{ol}$ with poles going to infinity. That happens not to be an issue. Indeed, as discussed in the continuous-time stabilization case, such test converge asymptotically to the "quadratic stability" type test (4.29) which can be solved independently.

4.5.5.2 Iterative Analysis/Synthesis Strategy

An alternative strategy that follows the classical D-K iterative algorithm is to search alternatively for K the control gain and for matrices F_{o1l} and F_{o2l}. To do so, one needs to render the LMI problems feasible, for example, by replacing the negative definite constraints $\prec 0$ by $\prec \alpha_l t_l I$ where t_l is some positive scalar to minimize and $\alpha_l = 0$ or $= 1$ depending on whether that specification is satisfied or not. The iterative algorithm amounts to iterating between the following two sets of constraints. The first one is the set of design constraints and assumes the matrices F_{o1l} and F_{o2l} are fixed:

$$X_{jl} \succ 0,$$
$$\begin{bmatrix} R_{11l} \otimes X_{jl} & R_{12l} \otimes X_{jl} \\ R_{12l} \otimes X_{jl} & R_{22l} \otimes X_{jl} \end{bmatrix} \tag{4.33}$$
$$+\mathrm{He}\left\{ \begin{bmatrix} I_{d_l} \otimes \tilde{F}_1 \\ -I_{d_l} \otimes (A_j \tilde{F}_1 + B_j Y) \end{bmatrix} \begin{bmatrix} F_{o1l} & F_{o2l} \end{bmatrix} \right\} \prec \alpha_l t_l I \ (\forall j \in \mathscr{I}_L).$$

The second one is the set of analysis constraints and assumes the gain K is fixed:

$$X_{jl} \succ 0,$$
$$\begin{bmatrix} R_{11l} \otimes X_{jl} & R_{12l} \otimes X_{jl} \\ R_{12l} \otimes X_{jl} & R_{22l} \otimes X_{jl} \end{bmatrix} \tag{4.34}$$
$$+\mathrm{He}\left\{ \begin{bmatrix} I_{d_l n} \\ -I_{d_l} \otimes (A_j + B_j K) \end{bmatrix} \begin{bmatrix} F_{1l} & F_{2l} \end{bmatrix} \right\} \prec \alpha_l t_l I \ (\forall j \in \mathscr{I}_L).$$

The algorithm is as follows:

[i0] For all l take $\alpha_l = 1$ and select some initial values of F_{1ol} and F_{2ol}, for example following the rules in instances [Ic] and [Id].

[iD] For fixed F_{o1l} and F_{o2l} minimize the sum $t_1 + t_2 + \cdots$, with respect to variables X_{jl}, Y, $\tilde{F}_1$ and t_l, subject to the LMI constraints (4.33). At the optimum take $K = Y \tilde{F}_1^{-1}$ and for all l such that $t_l \leq 0$ set $\alpha_l = 0$.

[iA] For fixed K minimize the sum $t_1 + t_2 + \cdots$, with respect to variables X_{jl}, F_{1l}, F_{2l} and t_l, subject to the LMI constraints (4.34). At the optimum take $F_{o1l} = F_{1l}$ and $F_{o2l} = F_{1l}$ and for all l such that $t_l \leq 0$ set $\alpha_l = 0$.

[iS] As soon as all the α_l are zero, stop, the control gain K is proved to locate the poles in the intersection of all regions $\mathbb{O}(R_l)$. If any of the α_l is non zero, start again at the design step [iD]. Continue until a control gain is found or some maximal number of steps is reached.

Such algorithm is not guaranteed to converge to a feasible solution, yet, each step is guaranteed to improve the previous one. Indeed, if (4.33) is feasible, then (4.34) is also feasible with the same X_{jl} and t_l variables and for the choice $F_{1l} = (1_{d_l} \otimes \tilde{F}_1) F_{o1l}$, $F_{2l} = (1_{d_l} \otimes \tilde{F}_1) F_{o2l}$. Conversely, if (4.34) is feasible, then (4.33) is feasible as well, with same values of X_{jl}, t_l and for $\tilde{F}_1 = I_n$, $Y = K$, $F_{o1l} = F_{1l}$, $F_{o2l} = F_{2l}$.

Unfortunately, in practice we have observed on examples that the algorithm is very much dependent on the initialization scheme. If the initialization [i0] is good enough, then at the first step [iD], the problem is solved. If not, the algorithm most often iterates without converging to a feasible solution and encounters numerical problems.

4.6 Numerical Examples on Quarter-Car Suspension Model

The same quarter-car suspension model as in Sect. 2.6.1 is considered. The state-feedback SV-LMIs are tested for various pole location requirements.

4.6.1 Intersection of Interiors of Two Discs

In this subsection, the aim is to locate the closed-loop poles in a region defined as the intersection of the two following discs:

- the interior of the disc centered at -2 with radius 1.7
- the interior of the disc centered at 0 with radius ρ.

LMI state-feedback conditions are tested while searching by bisection the minimal ρ. The tolerance on the bisection is chosen to be 10^{-4}.

Since the pole location region of interest is convex, the LMI conditions (i-p) of Theorem 4.5 are applicable. These correspond to searching a solution with a single Lyapunov certificate for all uncertainties and common to both regions. The bisection stops with the feasible region of radius $\rho_{QS} = 1.7380$. The obtained state feedback is

$$K_{QS} = \begin{bmatrix} -151.2427 & -852.3131 & 228.8416 & 898.0351 \end{bmatrix}.$$

Figure 4.1 illustrates the closed-loop poles with that state-feedback gain. The bullets are the closed-loop poles computed at the four vertices of the polytope. The crosses are poles computed on random samples of the uncertain system. The left and right curves are the limits of the discs. The figure illustrates that the poles are located in the intersection of the two discs. One can notice that the curve on the left-hand side is distant from the actual location of the poles. This illustrates the conservatism of the LMI conditions (i-p) of Theorem 4.5. One can prove that the poles are actually in a smaller disc than the one with radius $\rho_{QS} = 1.7380$ attested by the LMIs.

To further illustrate this fact, the closed-loop can be analyzed with SV-LMIs (4.34) with all $\alpha_l = 0$. Doing so not only allows to prove that the poles are in a smaller disc but it also provides an initialization for matrices F_{o1l} and F_{o2l} of (4.33) with all $\alpha_l = 0$. These LMIs are guaranteed by construction to be feasible for values of ρ smaller or equal than ρ_{QS}. When performing the bisection for these LMIs one obtains a new pole location with $\rho_{SV/QS} = 1.6459$. The corresponding state-feedback gain

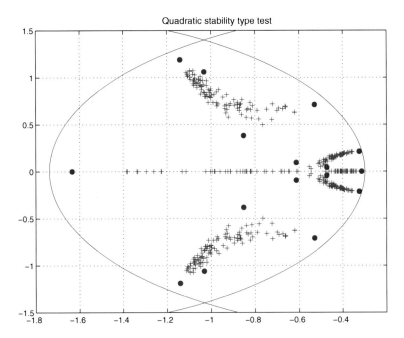

Fig. 4.1 Closed-loop pole location with K_{QS} as state feedback

is close to the previous one

$$K_{SV/QS} = \begin{bmatrix} -151.4491 & -853.3730 & 230.7741 & 898.7668 \end{bmatrix}$$

yet, the pole location is not exactly identical (slightly improved). Poles location is illustrated in Fig. 4.2. The eigenvalues of the matrices $-F_{o1l}^{-1}F_{o2l}$ for $l = 1, 2$ for the first and second disc are also plotted. Based on the discussion that follows Theorem 4.4 the eigenvalues of these matrices should lie in the specified regions. As illustrated by the figure, it is indeed the case. The values represented by the sign ◁ are in the disc centered at -2 with radius 1.7 and the values represented by the sign ▷ are in the disc centered as 0 with radius $\rho_{SV/QS} = 1.6459$.

To illustrate the influence of the choice of the F_{o1l} and F_{o2l} matrices, some other possibilities are tested. The first tested possibility is to choose for both regions $F_{o1l} = -I$ and $F_{o2l} = -I$. It corresponds to placing all eigenvalues of $-F_{o1l}^{-1}F_{o2l}$ at -1 which is in the intersection of the discs. This choice is coherent with suggestion [Ia] of the discussion that follows Theorem 4.4. Performing the bisection on the SV-LMIs for this choice of matrices F_{o1l}, F_{o2l} one obtains the following minimal radius $\rho_{SV/-1} = 1.7834$ associated to the state-feedback gain:

$$K_{SV/-1} = \begin{bmatrix} -151.8576 & -851.5639 & 225.8756 & 893.5785 \end{bmatrix}.$$

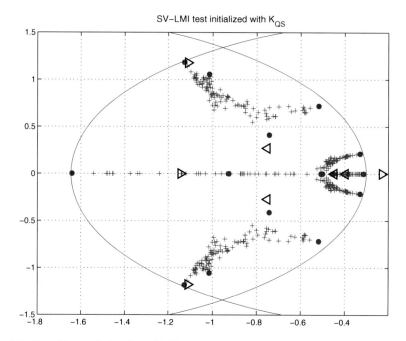

Fig. 4.2 Closed-loop pole location with $K_{SV/QS}$ as state feedback

Clearly, compared to the pole location obtained by $K_{SV/QS}$, the result is worse (larger disc) (Figs. 4.2 and 4.3).

The second tested possibility is to choose $F_{o11} = -I$, $F_{o21} = -2I$ and $F_{o12} = -I$, $F_{o22} = 0$. This corresponds to the suggestion [Ic] of the discussion that follows Theorem 4.4 or equivalently to the LMIs (ii-p) of Theorem 4.5. It also corresponds to locating the eigenvalues of the $-F_{o11}^{-1} F_{o21}$ matrices at the centers of the two discs respectively. As stated in Theorem 4.5, these SV-LMIs are guaranteed to improve the disc radius compared to ρ_{QS}. The bisection indeed stops with an improved value $\rho_{SV/c} = 1.5980$ and the associated state-feedback gain is

$$K_{SV/c} = \begin{bmatrix} -154.6944 & -868.5590 & 243.4151 & 904.2677 \end{bmatrix}.$$

From all tests up to this point, this choice of matrices F_{o11} and F_{o21} may seem the best one (see Fig. 4.4). Yet it can be further improved. To do so, one can analyze the closed-loop obtained with the gain $K_{SV/c}$. This amounts to solving the analysis SV-LMI of (4.34) with all $\alpha_l = 0$. It provides a new initialization for matrices F_{o11} and F_{o21}. For this new choice, the design SV-LMIs are solved again and the bisection gives an improved $\rho_{SV/SV/c} = 1.5846$ with the associated gain

$$K_{SV/SV/c} = \begin{bmatrix} -154.7339 & -868.7642 & 244.0250 & 904.4172 \end{bmatrix}.$$

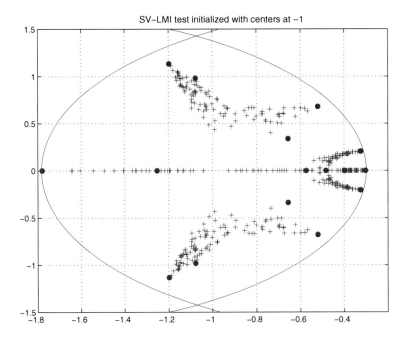

Fig. 4.3 Closed-loop pole location with $K_{SV/-1}$ as state feedback

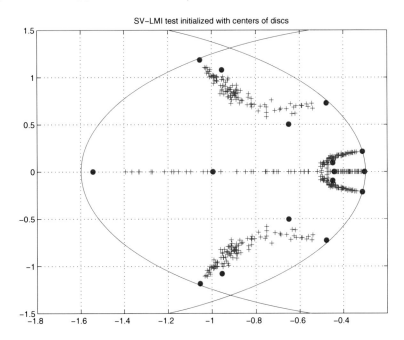

Fig. 4.4 Closed-loop pole location with $K_{SV/c}$ as state feedback

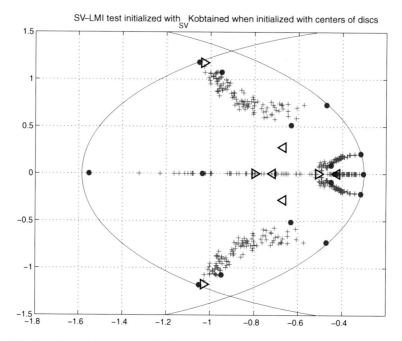

Fig. 4.5   Closed-loop pole location with $K_{SV/SV/c}$ as state feedback

The obtained pole location is illustrated in Fig. 4.5. This time, the eigenvalues of the two matrices $-F_{o1l}^{-1}F_{o2l}$ happen to be in the intersection of the two discs. They moreover are close to being among possible realization of the closed-loop poles. This is not a required feature and, as illustrated in the examples that follow, is not the case in general.

4.6.2 Intersection of the Interior of a Disc and Exterior of Another

In this subsection, the aim is to locate the closed-loop poles in a region defined as the intersection of the two following discs:

- the interior of the disc centered at 0 with radius 2
- the exterior of the disc centered at 0 with radius ρ.

LMI state-feedback conditions are tested while searching by bisection the maximal ρ. The tolerance on the bisection is chosen to be 10^{-4}.

One of the two regions being nonconvex (exterior of a disc), the 'quadratic stability' type conditions are not applicable. The only manner to solve the state-feedback problem is to apply SV-LMI results of Theorem 4.4, but this imposes to choose a priori the F_{o1l} and F_{o2l} matrices.

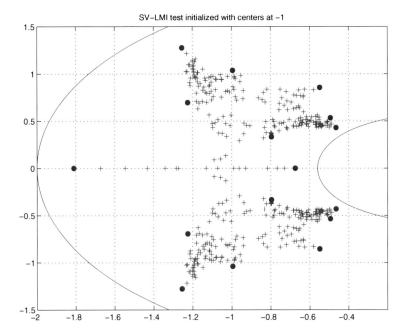

Fig. 4.6 Closed-loop pole location with $K_{SV/-1}$ as state feedback

A first considered choice is following suggestion [Ia] to take $F_{o1l} = I$ and $F_{o2l} = I$. It corresponds to choosing some A_{ol} matrix common to the two regions and with poles in the expected intersection of the two regions (all eigenvalues located at -1). The bisection stops with a value $\rho_{SV/-1} = 0.5599$ and a state-feedback gain

$$K_{SV/-1} = \begin{bmatrix} -90.6096 & -722.8869 & 172.3557 & 891.5088 \end{bmatrix}.$$

The closed-loop pole location is illustrated on Fig. 4.6. Closed-loop poles are indeed in the intersection of the exterior of one disc and the interior of the second.

By construction, the value of $\rho_{SV/-1}$ could not exceed 1 (the eigenvalues of $-F_{o1l}^{-1}F_{o2l}$ should satisfy the pole location constraints). Because of that one could expect that better solutions are achievable when taking, for example, $F_{o1l} = I$ and $F_{o2l} = 1.5I$ (all eigenvalues located at -1.5). Actually it is not the case and one gets $\rho_{SV/-1.5} = 0.5026$ illustrating that the choice of the F_{o1l}, F_{o2l} matrices is not simple.

Another choice of matrices F_{o1l} and F_{o2l} is now made applying the suggestions [Ic] for the interior of the first disc and [Id] for the exterior of the second disc. For that choice the bisection ends with $\rho_{SV/c} = 1.0189$ and a state-feedback gain

$$K_{SV/c} = \begin{bmatrix} 123.0328 & -524.9451 & 156.8014 & 923.6124 \end{bmatrix}.$$

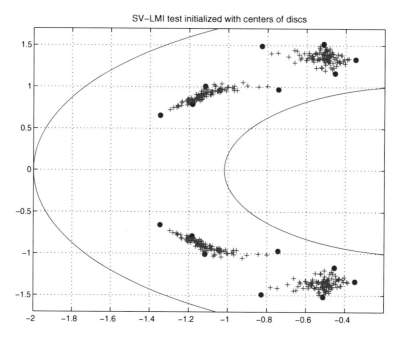

Fig. 4.7   Closed-loop pole location with $K_{SV/c}$ as state feedback

The closed-loop pole location is illustrated in Fig. 4.7.

The space between the plotted poles and the curves suggests that the LMI tests are conservative. A closed-loop analysis allows to confirm this fact and also to design new values for the F_{o1l}, F_{o2l} matrices. Based on these a new state-feedback gain can be computed and it gives $\rho_{SV/SV/c} = 1.2398$ and a state-feedback gain

$$K_{SV/SV/c} = \left[\, 124.4211 \;\; {-}522.9432 \;\; 162.1160 \;\; 923.2392 \,\right].$$

The closed-loop pole location and the eigenvalue location of the $-F_{o1l}^{-1} F_{o2l}$ matrices are illustrated in Fig. 4.8. The poles represented by the sign $\triangleleft$ are as expected in the interior of the disc centered at 0 with radius 2. The poles represented by the sign $\triangleright$ are in the exterior of the disc centered at 0 with radius $\rho_{SV/SV/c}$.

4.6.3 Intersection of a Disc and of a Half Plane

In this subsection the aim is to locate the closed-loop poles in a region defined as the intersection of the three following half planes such that:

- the complex modulus of the poles is less than 2
- the real part of the poles is less than α.

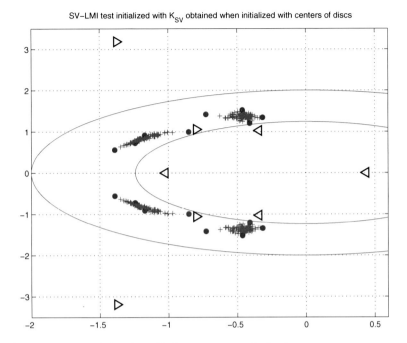

Fig. 4.8 Closed-loop pole location with $K_{SV/SV/c}$ as state feedback

LMI state-feedback conditions are tested while searching by bisection the minimal α. The tolerance on the bisection is chosen to be 10^{-4}.

The LMI conditions (i-p) of Theorem 4.5 are applicable. These correspond to searching a solution with a single Lyapunov certificate for all uncertainties and common to both regions. The bisection stops with the feasible region of radius $\alpha_{QS} = -0.6153$ and the state-feedback gain

$$K_{QS} = \begin{bmatrix} -4.5406 & -516.6913 & 108.0114 & 900.6544 \end{bmatrix}.$$

Figure 4.9 illustrates the closed-loop poles with that state-feedback gain. The bullets are the closed-loop poles computed at the four vertices of the polytope. The crosses are poles computed on random samples of the uncertain system. The left and right curves are the limits of the disc and the half plane. The figure illustrates that the poles are located in the intersection of the two regions. One can notice that the line on the right-hand side is distant from the actual location of the poles. This illustrates the conservatism of the LMI conditions (i-p) of Theorem 4.5. One can prove that the poles are actually with lower real parts than the value of $\alpha_{QS} = -0.6153$ attested by the LMIs.

This fact is confirmed when performing the closed-loop analysis with SV-LMIs (4.34). Doing so not only allows to prove that the poles have lower real parts but it also provides an initialization for matrices F_{o1l} and F_{o2l} of (4.33). These LMIs

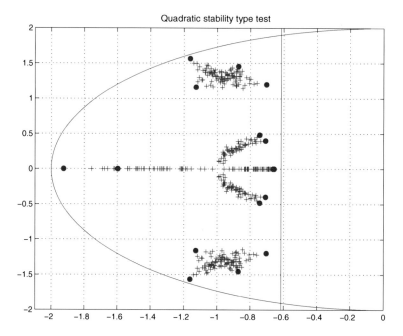

Fig. 4.9 Closed-loop pole location with K_{QS} as state feedback

are guaranteed by construction to be feasible for values of α smaller or equal than α_{QS}. When performing the bisection for these LMIs, one obtains a new pole location with $\alpha_{SV/QS} = -0.6562$. The corresponding state-feedback gain is close to the previous one

$$K_{SV/QS} = \begin{bmatrix} -9.0270 & -534.2979 & 118.7075 & 904.8231 \end{bmatrix}$$

yet, the pole location is not exactly identical (slightly improved). Poles location is illustrated in Fig. 4.10. The eigenvalues of the matrices $-F_{o1l}^{-1}F_{o2l}$ for $l = 1, 2$ for the two regions are also plotted. Based on the discussion that follows Theorem 4.4 the eigenvalues of these matrices should lie in the specified regions. As illustrated by the figure it is indeed the case. The values represented by the sign ◁ have a complex modulus smaller than 2 and the values represented by the sign ▷ have real parts below $\rho_{SV/QS} = -0.6562$.

The considered pole location problem includes the interior of a disc centered at zero. For that region, the a priori choice of $F_{o11} = -I$, $F_{o21} = 0$ guarantees to build SV-LMI conditions (ii-p) of Theorem 4.5 with reduced conservatism compared to the 'quadratic stability' type test. Considering now the half plane, the a priori choice for these matrices having the same nonincreasing conservatism property is $F_{o12} = -\varepsilon I$, $F_{o22} = -I$ with small enough value of ε. The following experiments are done for different values of ε.

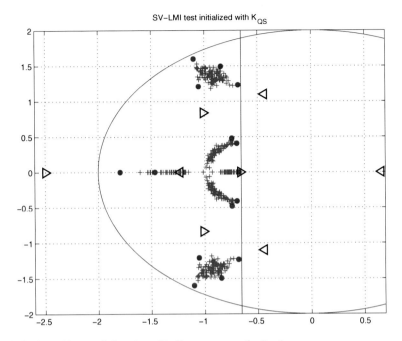

Fig. 4.10 Closed-loop pole location with $K_{SV/QS}$ as state feedback

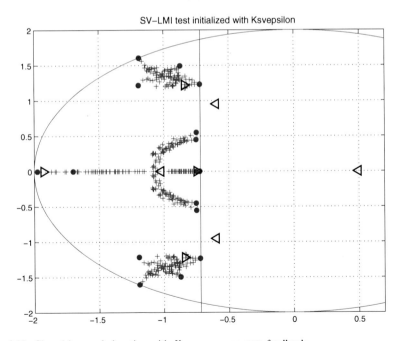

Fig. 4.11 Closed-loop pole location with $K_{SV/SV/10^{-2}}$ as state feedback

For $\varepsilon = 10^{-4}$ the bisection stops with value $\alpha_{SV/10^{-4}} = -0.6406$ which is indeed lower than α_{QS}. The analysis of the obtained closed-loop provides an other set of matrices F_{o1l}, F_{o2l} for which an additional design step gives an improved valued of $\alpha_{SV/SV/10^{-4}} = -0.6797$ with the state-feedback gain of

$$K_{SV/SV/10^{-4}} = \begin{bmatrix} 8.9656 & -502.2417 & 107.3028 & 904.1516 \end{bmatrix}.$$

For $\varepsilon = 10^{-2}$ and following the same procedure, we obtain $\alpha_{SV/10^{-2}} = -0.6719$, $\alpha_{SV/SV/10^{-2}} = -0.7188$ and

$$K_{SV/10^{-2}} = \begin{bmatrix} 30.3822 & -462.6287 & 94.4810 & 903.2435 \end{bmatrix}$$
$$K_{SV/SV/10^{-2}} = \begin{bmatrix} 33.7855 & -453.9355 & 89.0452 & 901.6782 \end{bmatrix}.$$

For $\varepsilon = 1$ and following the same procedure, we obtain $\alpha_{SV/1} = -0.5156$, $\alpha_{SV/SV/1} = -0.5312$ and

$$K_{SV/1} = \begin{bmatrix} -75.7504 & -652.0572 & 141.3068 & 890.4552 \end{bmatrix}$$
$$K_{SV/SV/1} = \begin{bmatrix} -79.0913 & -672.5785 & 160.0374 & 899.1795 \end{bmatrix}.$$

All these results make it clear that the initial choice of the initial F_{o1l}, F_{o2l} matrices is crucial. The suggestion [Ic] of the discussion that follows Theorem 4.4 is quite efficient on the treated examples and can be completed by a line search over the values ε. Among those that were tested the value $\varepsilon = 10^{-2}$ is the one that provides the best results (see Fig. 4.11).

The other conclusion of these tests is that it is always valuable, based on a first state-feedback design, to perform the analysis/design procedure above. It provides an improved state-feedback gain together with a less conservative evaluation of the pole location. Yet, it can be seen that this analysis/design procedure does not chance significantly the gains of the state feedback. This explains the reason for which the iterative algorithm of Sect. 4.5.5.2 fails in general when not initialized properly.

4.7 Conclusion

In this chapter, we presented SV-LMI-based results on robust state-feedback controller synthesis for uncertain LTI systems. We have shown that SV-LMIs are again useful for conservatism reduction, yet, there are significant difference between discrete-time systems and continuous-time systems. More precisely, in the discrete-time system synthesis, it is proved that we can impose structural constraints on S-variables without being more conservative than the "quadratic stability" conditions, but this is not possible for continuous-time systems. We finally note that, even though we concentrated on stability and pole location performance in this chapter, almost parallel arguments are possible for other performance criteria such as H_2 and H_∞ norms.

References

1. Bernussou J, Geromel JC, Peres PLD (1989) A linear programing oriented procedure for quadratic stabilization of uncertain systems. Syst Control Lett 13:65–72
2. de Oliveira MC, Bernussou J, Geromel JC (1999) A new discrete-time stability condition. Syst Control Lett 37(4):261–265
3. de Oliveira MC, Geromel JC, Hsu L (1999) LMI characterization of structural and robust stability: the discrete-time case. Linear Algebra Appl 296(1–3):27–38
4. Chilali M, Gahinet P (1996) H_∞ design with pole placement constraints: an LMI approach. IEEE Trans Automat Control 41:358–367
5. Chilali M, Gahinet P, Apkarian P (December 1999) Robust pole placement in LMI regions. IEEE Trans Automat Control 44(12):2257–2270
6. Ashokkumar CR, Yedavalli RK (1997) Eigenstructure preturbation analysis in disjointed domains for linear uncertain systems. Int J Control 67:887–899
7. Bachelier O, Henrion D, Pradin B, Mehdi D (2004) Root-clustering of a matrix in intersections or unions of regions. Siam J Control Optimization 43(3):1078–1093
8. Rejichi O, Bachelier O, Chaabane M, Mehdi D (2008) Robust root clustering analysis in a union of subregions for descriptor systems. In: IEE Proceedings of Control Theory and applications
9. Scherer CW, Gahinet P, Chilali M (1997) Multiobjective output-feedback control via LMI optimization. IEEE Trans Autom Control 42(7):896–911
10. Shcherbakov P, Dabbene F (2011) On the generation of random stable polynomials. European Journal of Control 17(2):145–159

Chapter 5
Multiobjective Controller Synthesis for LTI Systems

5.1 Introduction

In the preceding chapters, we have illustrated that SV-LMIs are effective for robust performance analysis problems of LTI systems affected by polytopic-type uncertainties. When dealing with those problems, the problem amounts to solving simultaneous LMIs defined on vertices of the polytope. In that situation, SV-LMIs allow us to employ distinct Lyapunov variables for each vertex, so that we can achieve less (no more) conservative results than the single Lyapunov-based, or say, quadratic stability-based approach. Qualitatively, the former approach employs parameter-dependent Lyapunov functions to assess the robust performance, whereas the latter employs parameter-independent Lyapunov functions and hence inherently conservative. This advantage comes from the "decoupling" feature of SV-LMIs, where in the SV-LMIs Lyapunov variables and system matrices are decoupled, whereas system matrices have multiplications with S-variables.

Such decoupling feature of SV-LMIs is advantageous also when dealing with multiobjective controller synthesis problems. The objective is to design a single controller that satisfy multiple performance specifications. The problem again amounts to solving simultaneous LMIs defined on each design specification, and the standard analysis LMIs that have coupling between Lyapunov variables and controller variables enforce us to employ a single (common) Lyapunov variable for linearization. In stark contrast, thanks to the "decoupling" feature, the SV-LMIs allow us to employ distinct (noncommon) Lyapunov variables for each design specification, so that less conservative results can be achieved. As previously noted, in the SV-LMI-based controller synthesis, there is a strong gap between discrete- and continuous-time systems settings. Namely, in the discrete-time system setting, the effectiveness of SV-LMIs can be shown clearly over the standard LMI-based approach. This is true also when dealing with multiobjective controller synthesis problems. Due to this reason, we concentrate our attention on the treatment of discrete-time multiobjective controller synthesis problems in this chapter. We finally note that the contents of this chapter

© Springer-Verlag London 2015
Y. Ebihara et al., *S-Variable Approach to LMI-Based Robust Control*,
Communications and Control Engineering, DOI 10.1007/978-1-4471-6606-1_5

is heavily relies on the pioneering work of Oliveira et al. [1]. For partial extension of discrete-time system results to continuous-time systems, see [2, 3].

5.2 Multiobjective Controller Synthesis for Discrete-Time LTI Systems

Consider the discrete-time LTI system P described by

$$
P : \begin{cases}
x(k+1) = Ax(k) & + B_{w_2} w_2(k) & + B_{w_\infty} w_\infty(k) & + B_u u(k), \\
z_2(k) = C_{z_2} x(k) & + D_{z_2,w_2} w_2(k) + & & + D_{z_2,u} u(k), \\
z_\infty(k) = C_{z_\infty} x(k) & & + D_{z_\infty,w_\infty} w_\infty(k) & + D_{z_\infty,u} u(k), \\
y(k) = C_y x(k) & + D_{y,w_2} w_2(k) & + D_{y,w_\infty} w_\infty(k)
\end{cases}
\tag{5.1}
$$

where $x \in \mathbb{R}^n$ is the state, $w_j \in \mathbb{R}^{n_{w_j}}$ ($j = 2, \infty$) the disturbance inputs, $z_j \in \mathbb{R}^{n_{z_j}}$ ($j = 2, \infty$) the performance outputs, $u \in \mathbb{R}^{n_u}$ the control input, and $y \in \mathbb{R}^{n_y}$ the measured output, respectively. For this LTI system, we first design static state-feedback controller K of the form

$$
K : u(k) = Kx(k), \quad K \in \mathbb{R}^{n_u \times n}
\tag{5.2}
$$

assuming that $C_y = I$, $D_{y,w_2} = 0$ and $D_{y,w_\infty} = 0$. The resulting closed-loop system can be written as

$$
P_{j,K} : \begin{cases}
x(k+1) = \mathscr{A} x(k) + \mathscr{B}_j w_j(k), \\
z_j(k) = \mathscr{C}_j x(k) + \mathscr{D}_j w_j(k)
\end{cases}
\quad (j = 2, \infty)
\tag{5.3}
$$

where

$$
\left[\begin{array}{c|c} \mathscr{A} & \mathscr{B}_j \\ \hline \mathscr{C}_j & \mathscr{D}_j \end{array} \right] :=
\left[\begin{array}{c|c} A + B_u K & B_{w_j} \\ \hline C_{z_j} + D_{z_j u} K & D_{z_j w_j} \end{array} \right]
\quad (j = 2, \infty).
\tag{5.4}
$$

We also consider the synthesis of full-order dynamic controller K given by

$$
K : \begin{cases}
x_K(k+1) = A_K x_K(k) + B_K y(k), \\
u(k) = C_K x_K(k) + D_K y(k)
\end{cases}
\tag{5.5}
$$

where $A_K \in \mathbb{R}^{n \times n}$. In this case the close-loop system becomes

$$
P_{j,K} : \begin{cases}
x_{cl}(k+1) = \mathscr{A} x_{cl}(k) + \mathscr{B}_j w_j(k), \\
z_j(k) = \mathscr{C}_j x_{cl}(k) + \mathscr{D}_j w_j(k)
\end{cases}
\quad (j = 2, \infty)
\tag{5.6}
$$

where $x_{cl} := [\, x^T \; x_K^T \,]^T$ and

$$
\left[\begin{array}{c|c} \mathscr{A} & \mathscr{B}_j \\ \hline \mathscr{C}_j & \mathscr{D}_j \end{array} \right]
$$

$$
:= \left[\begin{array}{cc|c} A + B_u D_K C_y & B_u C_K & B_{w_j} + B_u D_K D_{yw_j} \\ B_K C_y & A_K & B_K D_{yw_j} \\ \hline C_{z_j} + D_{z_ju} D_K C_y & D_{z_ju} C_K & D_{z_jw_j} + D_{z_ju} D_K D_{yw_j} \end{array} \right] \quad (j = 2, \infty).
\tag{5.7}
$$

Our goal in this section is to provide LMI conditions for the synthesis of K, static state-feedback (5.2) or dynamic output-feedback (5.5), such that multiple design specifications such as pole location, H_2, and H_∞ specifications on the closed-loop system P_K is satisfied. Such controller synthesis problems are often called multiobjective controller synthesis problems [4, 5]. One of the typical examples of multiobjective controller synthesis problems is the so-called multiobjective H_2/H_∞ controller synthesis problem, which is formulated in terms of (5.1), by

$$
\inf_K \| P_{2,K} \|_2 \quad \text{subject to} \quad \| P_{\infty,K} \|_\infty < \gamma_\infty.
\tag{5.8}
$$

Here, γ_∞ is a prescribed H_∞ performance level to be satisfied.

To show the effectiveness of the SV-LMIs for multiobjective controller synthesis problems with minimum complexity of descriptions, in this chapter, we concentrate our attention on the multiobjective H_2/H_∞ controller synthesis problem for discrete-time LTI systems described by (5.8).

5.3 Basic Results for Discrete-Time System Analysis

Consider the discrete-time LTI system P_{cl} described by

$$
P_{\text{cl}} : \begin{cases} \xi(k+1) = \mathscr{A}\xi(k) + \mathscr{B}w(k), \\ z(k) = \mathscr{C}\xi(k) + \mathscr{D}w(k). \end{cases}
\tag{5.9}
$$

The following lemmas are important to show the effectiveness of the SV-LMIs in dealing with multiobjective H_2/H_∞ controller synthesis problem. In these lemmas, we show standard and SV-LMIs for H_2 and H_∞ performance analysis of the discrete-time system (5.9). These are counterpart results of Chap. 2 already given for continuous-time systems.

Lemma 5.1 [1] *For given $\gamma_2 > 0$ and the LTI system P_{cl} represented by (5.9), the following statements are equivalent:*

(i) *The system P_{cl} is stable (i.e., the matrix $\mathscr{A}$ is Schur stable) and $\| P_{\text{cl}} \|_2 < \gamma_2$.*
(ii) *There exist $\mathscr{X}_2$ and $\mathscr{Z}$ such that*

$$\begin{bmatrix} -\mathcal{X}_2 + \mathcal{B}\mathcal{B}^\mathsf{T} & \mathcal{A}\mathcal{X}_2 \\ (\mathcal{A}\mathcal{X}_2)^\mathsf{T} & -\mathcal{X}_2 \end{bmatrix} \prec 0,$$

$$\begin{bmatrix} -\mathcal{Z} + \mathcal{D}\mathcal{D}^\mathsf{T} & \mathcal{C}\mathcal{X}_2 \\ (\mathcal{C}\mathcal{X}_2)^\mathsf{T} & -\mathcal{X}_2 \end{bmatrix} \prec 0,$$

$$\text{trace}\,(\mathcal{Z}) < \gamma_2^2 . \tag{5.10}$$

(iii) *There exist* $\mathcal{X}_2$, $\mathcal{Z}$ *and* $\mathcal{G}_2$ *such that*

$$\begin{bmatrix} -\mathcal{X}_2 + \mathcal{B}\mathcal{B}^\mathsf{T} & 0 \\ 0 & \mathcal{X}_2 \end{bmatrix} + \text{He}\left\{ \begin{bmatrix} \mathcal{A}\mathcal{G}_2 \\ -\mathcal{G}_2 \end{bmatrix} \begin{bmatrix} 0 & I \end{bmatrix} \right\} \prec 0,$$

$$\begin{bmatrix} -\mathcal{Z} + \mathcal{D}\mathcal{D}^\mathsf{T} & 0 \\ 0 & \mathcal{X}_2 \end{bmatrix} + \text{He}\left\{ \begin{bmatrix} \mathcal{C}\mathcal{G}_2 \\ -\mathcal{G}_2 \end{bmatrix} \begin{bmatrix} 0 & I \end{bmatrix} \right\} \prec 0,$$

$$\text{trace}\,(\mathcal{Z}) < \gamma_2^2 . \tag{5.11}$$

Moreover, if the LMI (5.10) *holds with* $\mathcal{X}_2 = \mathcal{X}_2^\star$ *and* $\mathcal{Z} = \mathcal{Z}^\star$, *then the LMI* (5.11) *holds with* $\mathcal{X}_2 = \mathcal{X}_2^\star$, $\mathcal{G}_2 = \mathcal{X}_2^\star$ *and* $\mathcal{Z} = \mathcal{Z}^\star$.

Lemma 5.2 [1] *For given* $\gamma_\infty > 0$ *and the LTI system* P_{cl} *represented by* (5.9), *the following statements are equivalent:*

(i) *The system* P_{cl} *is stable (i.e., the matrix* $\mathcal{A}$ *is Schur stable) and* $\| P_{cl} \|_\infty < \gamma_\infty$.

(ii) *There exists* $\mathcal{X}_\infty$ *such that*

$$\begin{bmatrix} -\mathcal{X}_\infty + \mathcal{B}\mathcal{B}^\mathsf{T} & \mathcal{A}\mathcal{X}_\infty & \mathcal{B}\mathcal{D}^\mathsf{T} \\ (\mathcal{A}\mathcal{X}_\infty)^\mathsf{T} & -\mathcal{X}_\infty & (\mathcal{C}\mathcal{X}_\infty)^\mathsf{T} \\ \mathcal{D}\mathcal{B}^\mathsf{T} & \mathcal{C}\mathcal{X}_\infty & \mathcal{D}\mathcal{D}^\mathsf{T} - \gamma_\infty^2 I \end{bmatrix} \prec 0. \tag{5.12}$$

(iii) *There exist* $\mathcal{X}_\infty$ *and* $\mathcal{G}_\infty$ *such that*

$$\begin{bmatrix} -\mathcal{X}_\infty + \mathcal{B}\mathcal{B}^\mathsf{T} & 0 & \mathcal{B}\mathcal{D}^\mathsf{T} \\ 0 & \mathcal{X}_\infty & 0 \\ \mathcal{D}\mathcal{B}^\mathsf{T} & 0 & \mathcal{D}\mathcal{D}^\mathsf{T} - \gamma_\infty^2 I \end{bmatrix} + \text{He}\left\{ \begin{bmatrix} \mathcal{A}\mathcal{G}_\infty \\ -\mathcal{G}_\infty \\ \mathcal{C}\mathcal{G}_\infty \end{bmatrix} \begin{bmatrix} 0 & I & 0 \end{bmatrix} \right\} \prec 0. \tag{5.13}$$

Moreover, if the LMI (5.12) *holds with* $\mathcal{X}_\infty = \mathcal{X}_\infty^\star$, *then the LMI* (5.13) *holds with* $\mathcal{X}_\infty = \mathcal{X}_\infty^\star$ *and* $\mathcal{G}_\infty = \mathcal{X}_\infty^\star$.

The SV-LMI (5.11) follows immediately from (5.10) by Lemma 2.2. Similarly, the SV-LMI (5.13) follows immediately from (5.12) by Lemma 2.2. Let us illustrate the derivation of the SV-LMI (5.13). To this end, we first rewrite (5.12) by permutations of rows and columns and by Schur complement as

$$\begin{bmatrix} -\mathcal{X}_\infty + \mathcal{B}\mathcal{B}^\mathsf{T} & \mathcal{B}\mathcal{D}^\mathsf{T} \\ \mathcal{D}\mathcal{B}^\mathsf{T} & \mathcal{D}\mathcal{D}^\mathsf{T} - \gamma_\infty^2 I \end{bmatrix} + \begin{bmatrix} \mathcal{A} \\ \mathcal{C} \end{bmatrix} \mathcal{X}_\infty \begin{bmatrix} \mathcal{A}^\mathsf{T} & \mathcal{C}^\mathsf{T} \end{bmatrix} \prec 0.$$

This can be rewritten as

$$\begin{bmatrix} \mathscr{A}^T & \mathscr{C}^T \\ I & 0 \\ 0 & I \end{bmatrix}^T \begin{bmatrix} \mathscr{X}_\infty & 0 & 0 \\ 0 & -\mathscr{X}_\infty + \mathscr{B}\mathscr{B}^T & \mathscr{B}\mathscr{D}^T \\ 0 & \mathscr{D}\mathscr{B}^T & \mathscr{D}\mathscr{D}^T - \gamma_\infty^2 I \end{bmatrix} \begin{bmatrix} \mathscr{A}^T & \mathscr{C}^T \\ I & 0 \\ 0 & I \end{bmatrix} \prec 0$$

which is conformable with (2.18) with $E = I$, $\mathsf{A} = [\mathscr{A}^T \ \mathscr{C}^T]$, $\mathsf{M}_{11} = \mathscr{X}_\infty$, $\mathsf{M}_{12} = 0$, and $\mathsf{M}_{22} = \begin{bmatrix} -\mathscr{X}_\infty + \mathscr{B}\mathscr{B}^T & \mathscr{B}\mathscr{D}^T \\ \mathscr{D}\mathscr{B}^T & \mathscr{D}\mathscr{D}^T - \gamma_\infty^2 I \end{bmatrix}$. Therefore, by (c) of Lemma 2.2, we see that (5.12) holds if and only if

$$\begin{bmatrix} \mathscr{X}_\infty & 0 & 0 \\ 0 & -\mathscr{X}_\infty + \mathscr{B}\mathscr{B}^T & \mathscr{B}\mathscr{D}^T \\ 0 & \mathscr{D}\mathscr{B}^T & \mathscr{D}\mathscr{D}^T - \gamma_\infty^2 I \end{bmatrix} + \mathrm{He}\left\{ \begin{bmatrix} \mathsf{F}_1 \\ \mathsf{F}_2 \end{bmatrix} [I \ -\mathscr{A}^T \ -\mathscr{C}^T] \right\} \prec 0.$$

holds, and we can let without loss of generality that $\mathsf{F}_1 = -\mathscr{X}_\infty$ and $\mathsf{F}_2 = 0$. By permutations of rows and columns, we can rewrite the above LMI as

$$\begin{bmatrix} -\mathscr{X}_\infty + \mathscr{B}\mathscr{B}^T & 0 & \mathscr{B}\mathscr{D}^T \\ 0 & \mathscr{X}_\infty & 0 \\ \mathscr{D}\mathscr{B}^T & 0 & \mathscr{D}\mathscr{D}^T - \gamma_\infty^2 I \end{bmatrix} + \mathrm{He}\left\{ \begin{bmatrix} \mathscr{A} \\ -I \\ \mathscr{C} \end{bmatrix} [0 \ \mathscr{X}_\infty \ 0] \right\} \prec 0.$$

This is nothing but (5.13) with $\mathscr{G}_\infty = \mathscr{X}_\infty$.

We note that the SV-LMI (5.11) *recovers* the standard LMI (5.10) by letting $\mathscr{G}_2 = \mathscr{X}_2$ as stated in (iii) of Lemma 5.1. Similarly, the S-LMI (5.13) recovers the standard LMI (5.12) by letting $\mathscr{G}_\infty = \mathscr{X}_\infty$. This recovery property of the SV-LMIs plays an important role in theoretically ensuring the advantage of SV-LMIs in dealing with multiobjective H_2/H_∞ controller synthesis problem.

5.4 State-Feedback Multiobjective H_2/H_∞ Controller Synthesis

Consider the multiobjective H_2/H_∞ control problem formulated by (5.8) where the controller to be designed is a state-feedback controller given by (5.2). In the state-feedback case, the effectiveness of the SV-LMIs can be readily seen from the following two theorems that follow immediately from Lemmas 5.1, and 5.2 as well as simple linearizing change of variables.

Theorem 5.1 *For given $\gamma_2 > 0$ and the LTI system P represented by (5.1), the following statements are equivalent:*

(i) *There exists a static state-feedback controller K of the form (5.2) such that the closed-loop system $P_{2,K}$ given by (5.3) becomes stable and $\|P_{2,K}\|_2 < \gamma_2$.*

(ii) *There exist X_2, Z, and W such that*

$$\begin{bmatrix} -X_2 + B_{w_2} B_{w_2}^{\mathrm{T}} & AX_2 + B_u W \\ (AX_2 + B_u W)^{\mathrm{T}} & -X_2 \end{bmatrix} \prec 0,$$

$$\begin{bmatrix} -Z + D_{z_2,w_2} D_{z_2,w_2}^{\mathrm{T}} & C_{z_2} X_2 + D_{z_2,u} W \\ (C_{z_2} X_2 + D_{z_2,u} W)^{\mathrm{T}} & -X_2 \end{bmatrix} \prec 0, \qquad (5.14)$$

$$\mathrm{trace}\,(Z) < \gamma_2^2.$$

If the LMI (5.14) is feasible, then a stabilizing controller K of the form (5.2) satisfying $\| P_{2,K} \|_2 < \gamma_2$ *is given by*

$$K = W X_2^{-1}. \qquad (5.15)$$

(iii) *There exist* X_2, Z, W *and* G_2 *such that*

$$\begin{bmatrix} -X_2 + B_{w_2} B_{w_2}^{\mathrm{T}} & 0 \\ 0 & X_2 \end{bmatrix} + \mathrm{He}\left\{ \begin{bmatrix} AG_2 + B_u W \\ -G_2 \end{bmatrix} \begin{bmatrix} 0 & I \end{bmatrix} \right\} \prec 0,$$

$$\begin{bmatrix} -Z + D_{z_2,w_2} D_{z_2,w_2}^{\mathrm{T}} & 0 \\ 0 & X_2 \end{bmatrix} + \mathrm{He}\left\{ \begin{bmatrix} C_{z_2} G_2 + D_{z_2,u} W \\ -G_2 \end{bmatrix} \begin{bmatrix} 0 & I \end{bmatrix} \right\} \prec 0, \qquad (5.16)$$

$$\mathrm{trace}\,(Z) < \gamma_2^2.$$

If the LMI (5.16) is feasible, then a stabilizing controller K of the form (5.2) satisfying $\| P_{2,K} \|_2 < \gamma_2$ *is given by*

$$K = W G_2^{-1}. \qquad (5.17)$$

Moreover, if the LMI (5.14) holds with $X_2 = X_2^\star$, $Z = Z^\star$ *and* $W = W^\star$, *then the LMI (5.20) holds with* $X_2 = X_2^\star$, $G_2 = X_2^\star$, $Z = Z^\star$ *and* $W = W^\star$.

Theorem 5.2 *For given* $\gamma_\infty > 0$ *and the LTI system P represented by (5.1), the following statements are equivalent:*

(i) *There exists a static state-feedback controller K of the form (5.2) such that the closed-loop system* $P_{\infty,K}$ *given by (5.3) becomes stable and* $\| P_{\infty,K} \|_\infty < \gamma_\infty$.

(ii) *There exist* X_∞ *and* W *such that*

$$\begin{bmatrix} -X_\infty + B_{w_\infty} B_{w_\infty}^{\mathrm{T}} & AX_\infty + B_u W & B_{w_\infty} D_{z_\infty,w_\infty}^{\mathrm{T}} \\ (AX_\infty + B_u W)^{\mathrm{T}} & -X_\infty & (C_{z_\infty} X_\infty + D_{z_\infty,u} W)^{\mathrm{T}} \\ D_{z_\infty,w_\infty} B_{w_\infty}^{\mathrm{T}} & C_{z_\infty} X_\infty + D_{z_\infty,u} W & D_{z_\infty,w_\infty} D_{z_\infty,w_\infty}^{\mathrm{T}} - \gamma_\infty^2 I \end{bmatrix} \prec 0. \qquad (5.18)$$

If the LMI (5.18) is feasible, then a stabilizing controller K of the form (5.2) satisfying $\| P_{\infty,K} \|_\infty < \gamma_\infty$ *is given by*

$$K = W X_\infty^{-1}. \qquad (5.19)$$

(iii) *There exist X_∞, W and G_∞ such that*

$$
\begin{bmatrix}
-X_\infty + B_{w\infty}B_{w\infty}^{\mathrm{T}} & 0 & B_{w\infty}D_{z\infty,w\infty}^{\mathrm{T}} \\
0 & X_\infty & 0 \\
D_{z\infty,w\infty}B_{w\infty}^{\mathrm{T}} & 0 & D_{z\infty,w\infty}D_{z\infty,w\infty}^{\mathrm{T}} - \gamma_\infty^2 I
\end{bmatrix}
$$

$$
+ \mathrm{He}\left\{
\begin{bmatrix}
AG_\infty + B_u W \\
-G_\infty \\
C_{z\infty}G_\infty + D_{z\infty,u}W
\end{bmatrix}
\begin{bmatrix} 0 & I & 0 \end{bmatrix}
\right\} \prec 0.
$$

(5.20)

*If the LMI (5.20) is feasible, then a stabilizing controller K of the form (5.2)
satisfying $\|P_{\infty,K}\|_\infty < \gamma_\infty$ is given by*

$$
K = WG_\infty^{-1}.
$$

(5.21)

*Moreover, if the LMI (5.18) holds with $X_\infty = X_\infty^\star$ and $W = W^\star$, then the LMI
(5.20) holds with $X_\infty = X_\infty^\star$, $G_\infty = X_\infty^\star$ and $W = W^\star$.*

The validity of these theorems readily follows from Lemmas 5.1 and 5.2. Indeed,
(ii) of Theorem 5.1 follows from (ii) of Lemma 5.1 by letting $\mathscr{X}_2 = X_2$, $\mathscr{Z} = Z$
and applying linearizing change of variables $W := K X_2$. On the other hand, (iii) of
Theorem 5.1 follows from (iii) of Lemma 5.1 by letting $\mathscr{X}_2 = X_2$, $\mathscr{Z} = Z$, $\mathscr{G}_2 = G_2$
and applying linearizing change of variables $W := K G_2$. Similarly, the results in
Theorem 5.2 follows directly from Lemma 5.2.

Now, we are ready to show explicitly the advantage of SV-LMIs in dealing with
multiobjective H_2/H_∞ controller synthesis problem. We see from the standard LMI
given in Theorem 5.1 that a controller K satisfying $\|P_{2,K}\|_2 < \gamma_2$ can be char-
acterized by (5.14) and (5.15). Similarly, from (ii) of Theorem 5.2, a controller K
satisfying $\|P_{\infty,K}\|_\infty < \gamma_\infty$ can be characterized by (5.18) and (5.19). Of course
the controllers generated by (5.15) and (5.19) are different in general unless we
enforce $X_2 = X_\infty$. Since we are actually required to satisfy both the H_2 and H_∞
specifications, it is natural to consider the following SDP by means of the standard
LMIs:

$$
\gamma_X^\star := \inf_{X_2, X_\infty, W, Z} \gamma_2 \text{ subject to (5.14), (5.18) and } X_2 = X_\infty.
$$

(5.22)

Note that this formulation using standard LMIs is conservative due to the restriction
of a common Lyapunov variable $X_2 = X_\infty$.

On the other hand, if we employ the SV-LMIs given by (5.16) and (5.20), we see
that these generate a same controller if $G_2 = G_\infty$, where the Lyapunov variables
X_2 and X_∞ are not necessarily to be the same. We thus obtain the following SDP by
means of the SV-LMIs:

$$
\gamma_G^\star := \inf_{X_2, X_\infty, W, Z, G_2, G_\infty} \gamma_2 \text{ subject to (5.16), (5.20) and } G_2 = G_\infty.
$$

(5.23)

Note again that this formulation is conservative due to the restriction of a common S-variable $G_2 = G_\infty$. However, the advantage of the SV-LMI formulation (5.23) over the standard LMI formulation (5.22) is clear from the recovery property stated in (iii) of Theorems 5.1 and 5.2. Indeed, if the LMIs (5.14) and (5.18) hold with $X_2 = X_\infty = X^\star$, $Z = Z^\star$ and $W = W^\star$ for $\gamma_2 = \gamma_2^\star$, we see that the LMIs (5.16) and (5.20) hold with $G_2 = G_\infty = X^\star$, $X_2 = X_\infty = X^\star$, $Z = Z^\star$ and $W = W^\star$ for $\gamma_2 = \gamma_2^\star$. It follows that $\gamma_G^\star \le \gamma_X^\star$ holds in terms of the SDPs (5.22) and (5.23). Namely, we can obtain better (no worse) results by the SV-LMI formulation in comparison with the standard LMI formulation.

5.5 Output-Feedback Multiobjective H_2/H_∞ Controller Synthesis

We next consider output-feedback multiobjective H_2/H_∞ controller synthesis. Even though the linearizing change of variables become much more complicated in output-feedback controller synthesis, we eventually show the advantage of SV-LMIs over the standard LMIs exactly the same way as the state-feedback case.

5.5.1 Linearizing Change of Variables for Standard LMIs

To facilitate the description of linearizing change of variables for the output-feedback controller synthesis, consider the discrete-time LTI system P_0 described by

$$
P_0 : \begin{cases} x(k+1) = Ax(k) \; + B_w w(k) \; + B_u u(k), \\ \;\; z(k) \;\;\; = C_z x(k) + D_{zw} w(k) + D_{zu} u(k), \\ \;\; y(k) \;\;\; = C_y x(k) + D_{yw} w(k). \end{cases} \tag{5.24}
$$

The closed-loop system constructed by P_0 and K given by (5.5) is given by

$$
P_{0,K} : \begin{cases} x_{\mathrm{cl}}(k+1) = \mathscr{A} x_{\mathrm{cl}}(k) + \mathscr{B} w(k), \\ \;\; z(k) \;\;\;\;\; = \mathscr{C} x_{\mathrm{cl}}(k) + \mathscr{D} w(k) \end{cases} \tag{5.25}
$$

where $x_{\mathrm{cl}} := [\, x^\mathrm{T} \; x_K^\mathrm{T} \,]^\mathrm{T}$ and

$$
\left[\begin{array}{c|c} \mathscr{A} & \mathscr{B} \\ \hline \mathscr{C} & \mathscr{D} \end{array} \right] := \left[\begin{array}{cc|c} A + B_u D_K C_y & B_u C_K & B_w + B_u D_K D_{yw} \\ B_K C_y & A_K & B_K D_{yw} \\ \hline C_z + D_{zu} D_K C_y & D_{zu} C_K & D_{zw} + D_{zu} D_K D_{yw} \end{array} \right] . \tag{5.26}
$$

If we consider the standard LMI-based H_2 and H_∞ conditions (5.10) and (5.12) for the closed-loop system $P_{0,K}$, we find that these matrix inequalities include bilinear terms $\mathscr{A}\mathscr{X}$ and $\mathscr{C}\mathscr{X}$ as well as linear terms $\mathscr{X}$, $\mathscr{B}$ and $\mathscr{D}$ (here we identify $\mathscr{X}_j$ ($j = 2, \infty$) with $\mathscr{X}$). In the output-feedback case, these terms can be linearized by change of variables with a similarity transformation [4, 5]. To see this, let us write

$$\mathscr{X} = \begin{bmatrix} X & M \\ M^{\mathsf{T}} & W \end{bmatrix}, \quad X \in \mathbb{S}^n_{++}, \quad W \in \mathbb{S}^n_{++} \tag{5.27}$$

and denote

$$\mathscr{X}^{-1} =: \begin{bmatrix} Y & N \\ N^{\mathsf{T}} & V \end{bmatrix}, \quad Y \in \mathbb{S}^n_{++}, \quad V \in \mathbb{S}^n_{++}. \tag{5.28}$$

We further define

$$\Pi_1 := \begin{bmatrix} I & Y \\ 0 & N^{\mathsf{T}} \end{bmatrix}, \quad \Pi_2 := \begin{bmatrix} X & I \\ M^{\mathsf{T}} & 0 \end{bmatrix}, \tag{5.29}$$

Note that we can assume without loss of generality that N is nonsingular since LMIs of interest are all strict. We further define

$$\begin{aligned}
\widehat{A}_K &:= \begin{bmatrix} Y & N \end{bmatrix} \begin{bmatrix} A + B_u D_K C_y & B_u C_K \\ B_K C_y & A_K \end{bmatrix} \begin{bmatrix} X \\ M^{\mathsf{T}} \end{bmatrix}, \\
\widehat{B}_K &:= \begin{bmatrix} Y & N \end{bmatrix} \begin{bmatrix} B_u D_K \\ B_K \end{bmatrix}, \\
\widehat{C}_K &:= \begin{bmatrix} D_K C_y & C_K \end{bmatrix} \begin{bmatrix} X \\ M^{\mathsf{T}} \end{bmatrix}, \\
\widehat{D}_K &:= D_K
\end{aligned} \tag{5.30}$$

These correspond to the change of variables for the output-feedback controller synthesis. Indeed, those terms $\mathscr{X}$, $\mathscr{A}\mathscr{X}$, $\mathscr{B}$, $\mathscr{C}\mathscr{X}$ can be linearized in terms of the variables X, Y, $\widehat{A}_K$, $\widehat{B}_K$, $\widehat{C}_K$ and $\widehat{D}_K$ in the following way:

$$\Pi_1^{\mathsf{T}} \mathscr{X} \Pi_1 = \Pi_1^{\mathsf{T}} \Pi_2 = \begin{bmatrix} X & I \\ I & Y \end{bmatrix}, \tag{5.31a}$$

$$\Pi_1^{\mathsf{T}} \mathscr{A} \mathscr{X} \Pi_1 = \Pi_1^{\mathsf{T}} \mathscr{A} \Pi_2 = \begin{bmatrix} AX + B_u \widehat{C}_K & A + B_u \widehat{D}_K C_y \\ \widehat{A}_K & YA + \widehat{B}_K C_y \end{bmatrix}, \tag{5.31b}$$

$$\Pi_1^{\mathsf{T}} \mathscr{B} = \begin{bmatrix} B_w + B_u \widehat{D}_K D_{yw} \\ Y B_w + \widehat{B}_K D_{yw} \end{bmatrix}, \tag{5.31c}$$

$$\mathscr{C} \mathscr{X} \Pi_1 = \mathscr{C} \Pi_2 = \begin{bmatrix} C_z X + D_{zu} \widehat{C}_K & C_z + D_{zu} \widehat{D}_K C_y \end{bmatrix}. \tag{5.31d}$$

From X, Y, $\widehat{A}_K$, $\widehat{B}_K$, $\widehat{C}_K$ and $\widehat{D}_K$, the controller variables A_K, B_K, C_K and D_K can be obtained by the reverse transformation of (5.30) given by

$$D_K = \widehat{D}_K,$$
$$C_K = (\widehat{C}_K - D_K C_y X) M^{-T},$$
$$B_K = N^{-1}(\widehat{B}_K - Y B_u D_K),$$
$$A_K = N^{-1}\left(\widehat{A}_K - N B_K C_y X - Y B_u C_K M^T - Y(A + B_u D_K C_y)X\right) M^{-T}$$
$$(5.32)$$

where M and N are arbitrary nonsingular matrix satisfying $N M^T = I - Y X$.

5.5.2 Linearizing Change of Variables for SV-LMIs

As shown in [1, 2], we can conceive similar similarity transformation and linearizing change of variables for the output-feedback controller synthesis using SV-LMIs (5.11) and (5.13). Note that these LMIs include bilinear terms $\mathscr{A}\mathscr{G}$ and $\mathscr{C}\mathscr{G}$ as well as linear terms $\mathscr{X}$, $\mathscr{G}$, $\mathscr{B}$ and $\mathscr{D}$ with respect to the decision variables (here we identify $\mathscr{X}_j$ ($j = 2, \infty$) with $\mathscr{X}$ and $\mathscr{G}_j$ ($j = 2, \infty$) with $\mathscr{G}$). To see the basic idea for the linearization, we partition $\mathscr{G}$ as in

$$\mathscr{G} = \begin{bmatrix} X & J \\ M^T & W \end{bmatrix}, \quad X \in \mathbb{R}^{n \times n}, \quad W \in \mathbb{R}^{n \times n} \tag{5.33}$$

and define

$$\mathscr{G}^{-1} =: \begin{bmatrix} Y^T & H \\ N^T & V \end{bmatrix}, \quad Y \in \mathbb{R}^{n \times n}, \quad V \in \mathbb{R}^{n \times n}. \tag{5.34}$$

Note that X and Y are no longer symmetric in the SV-LMI cases. We then define

$$\Pi_1 := \begin{bmatrix} I & Y^T \\ 0 & N^T \end{bmatrix}, \quad \Pi_2 := \begin{bmatrix} X & I \\ M^T & 0 \end{bmatrix}. \tag{5.35}$$

Note again that we can assume N is nonsingular since SV-LMIs of interest are all strict. With the variables X, Y, M, N and controller variables A_K, B_K, C_K, D_K, we further define $\widehat{A}_K, \widehat{B}_K, \widehat{C}_K, \widehat{D}_K$ exactly the same way as (5.30). Then, we see that those terms $\mathscr{G}, \mathscr{A}\mathscr{G}, \mathscr{B}, \mathscr{C}\mathscr{G}$ can be linearized in terms of variables $X, Y, S, \widehat{A}_K, \widehat{B}_K, \widehat{C}_K$ and $\widehat{D}_K$ in the following way:

$$\Pi_1^T \mathscr{G} \Pi_1 = \Pi_1^T \Pi_2 = \begin{bmatrix} X & I \\ S & Y \end{bmatrix}, \quad S := YX + NM^T, \tag{5.36a}$$

$$\Pi_1^T \mathscr{A}\mathscr{G} \Pi_1 = \Pi_1^T \mathscr{A} \Pi_2 = \begin{bmatrix} AX + B_u \widehat{C}_K & A + B_u \widehat{D}_K C_y \\ \widehat{A}_K & YA + \widehat{B}_K C_y \end{bmatrix}, \tag{5.36b}$$

$$\Pi_1^T \mathscr{B} = \begin{bmatrix} B_w + B_u \widehat{D}_K D_{yw} \\ Y B_w + \widehat{B}_K D_{yw} \end{bmatrix}, \tag{5.36c}$$

$$\mathscr{C}\mathscr{G}\Pi_1 = \mathscr{C}\Pi_2 = \begin{bmatrix} C_z X + D_{zu}\widehat{C}_K & C_z + D_{zu}\widehat{D}_K C_y \end{bmatrix}. \tag{5.36d}$$

We also represent $\Pi_1^{\mathrm{T}}\mathscr{X}\Pi_1$ by

$$\Pi_1^{\mathrm{T}}\mathscr{X}\Pi_1 = \begin{bmatrix} P & R \\ R^{\mathrm{T}} & Q \end{bmatrix}, \quad P \in \mathbb{S}_{++}^n, \quad Q \in \mathbb{S}_{++}^n.$$

From X, Y, $\widehat{A}_K$, $\widehat{B}_K$, $\widehat{C}_K$ and $\widehat{D}_K$, the controller variables A_K, B_K, C_K and D_K can be obtained by the reverse transformation of (5.30) given by (5.32) where M and N are arbitrary nonsingular matrix satisfying $NM^{\mathrm{T}} = S - YX$.

5.5.3 LMIs for Output-Feedback H_2 and H_∞ Controller Synthesis

We next give LMI conditions for the output-feedback H_2 and H_∞ controller synthesis by means of the standard and SV-LMIs.

Theorem 5.3 *For given $\gamma_2 > 0$ and the LTI system P represented by (5.1), the following statements are equivalent:*

(i) *There exists a full-order output-feedback controller K of the form (5.5) such that the closed-loop system $P_{2,K}$ becomes stable and $\|P_{2,K}\|_2 < \gamma_2$.*

(ii) *There exist X_2, Y_2, Z, $\widehat{A}_K$, $\widehat{B}_K$, $\widehat{C}_K$, and $\widehat{D}_K$ such that*

$$\begin{bmatrix} -X_2 & -I & AX_2 + B_u\widehat{C}_K & A + B_u\widehat{D}_K C_y & B_{w_2} + B_u\widehat{D}_K D_{y,w_2} \\ * & -Y_2 & \widehat{A}_K & Y_2 A + \widehat{B}_K C_y & Y_2 B_{w_2} + \widehat{B}_K D_{y,w_2} \\ * & * & -X_2 & -I & 0 \\ * & * & * & -Y_2 & 0 \\ * & * & * & * & -I_{n_{w_2}} \end{bmatrix} \prec 0, \tag{5.37a}$$

$$\begin{bmatrix} -Z & C_{z_2}X_2 + D_{z_2,u}\widehat{C}_K & C_{z_2} + D_{z_2,u}\widehat{D}_K C_y & D_{z_2,w_2} + D_{z_2,u}\widehat{D}_K D_{y,w_2} \\ * & -X_2 & -I & 0 \\ * & * & -Y_2 & 0 \\ * & * & * & -I_{n_{w_2}} \end{bmatrix} \prec 0, \tag{5.37b}$$

$$\mathrm{trace}(Z) < \gamma_2^2. \tag{5.37c}$$

If the system of LMIs (5.37a)–(5.37c) is feasible, then a stabilizing controller K of the form (5.5) satisfying $\|P_{2,K}\|_2 < \gamma_2$ is given by

$$\begin{aligned}
D_K &:= \widehat{D}_K, \\
C_K &:= (\widehat{C}_K - D_K C_y X_2) M_2^{-T}, \\
B_K &:= N_2^{-1}(\widehat{B}_K - Y_2 B_u D_K), \\
A_K &:= N_2^{-1}\left(\widehat{A}_K - N_2 B_K C_y X_2 - Y_2 B_u C_K M_2^{T} \right. \\
&\quad \left. - Y_2(A + B_u D_K C_y)X_2\right) M_2^{-T}
\end{aligned} \tag{5.38}$$

where M_2 and N_2 are arbitrary matrices satisfying $N_2 M_2^T = I - Y_2 X_2$.

(iii) *There exist X_2, Y_2, S_2, Z, P_2, Q_2, R_2, $\widehat{A}_K$, $\widehat{B}_K$, $\widehat{C}_K$, and $\widehat{D}_K$ such that*

$$\begin{bmatrix}
-P_2 & -R_2 & AX_2 + B_u\widehat{C}_K & A + B_u\widehat{D}_K C_y & B_{w_2} + B_u\widehat{D}_K D_{y,w_2} \\
* & -Q_2 & \widehat{A}_K & Y_2 A + \widehat{B}_K C_y & Y_2 B_{w_2} + \widehat{B}_K D_{y,w_2} \\
* & * & P_2 - \mathrm{He}\{X_2\} & R_2 - I - S_2^T & 0 \\
* & * & * & Q_2 - \mathrm{He}\{Y_2\} & 0 \\
* & * & * & * & -I_{n_{w_2}}
\end{bmatrix} \prec 0, \tag{5.39a}$$

$$\begin{bmatrix}
-Z & C_{z_2} X_2 + D_{z_2,u}\widehat{C}_K & C_{z_2} + D_{z_2,u}\widehat{D}_K C_y & D_{z_2,w_2} + D_{z_2,u}\widehat{D}_K D_{y,w_2} \\
* & P_2 - \mathrm{He}\{X_2\} & R_2 - I - S_2^T & 0 \\
* & * & Q_2 - \mathrm{He}\{Y_2\} & 0 \\
* & * & * & -I_{n_{w_2}}
\end{bmatrix} \prec 0, \tag{5.39b}$$

$$\mathrm{trace}(Z) < \gamma_2^2. \tag{5.39c}$$

If the system of LMIs (5.39a)–(5.39c) is feasible, then a stabilizing controller K of the form (5.5) satisfying $\| P_{2,K} \|_2 < \gamma_2$ is given by (5.38) where M_2 and N_2 are arbitrary matrices satisfying $N_2 M_2^T = S_2 - Y_2 X_2$. Moreover, if the LMI (5.37a)–(5.37c) holds with

$$\begin{bmatrix} X_2 & I \\ I & Y_2 \end{bmatrix} = \begin{bmatrix} X_2^\star & I \\ I & Y_2^\star \end{bmatrix}, \quad Z = Z^\star, \quad \begin{bmatrix} \widehat{A}_K & \widehat{B}_K \\ \widehat{C}_K & \widehat{D}_K \end{bmatrix} = \begin{bmatrix} \widehat{A}_K^\star & \widehat{B}_K^\star \\ \widehat{C}_K^\star & \widehat{D}_K^\star \end{bmatrix},$$

then the LMI (5.39a)–(5.39c) holds with

$$\begin{bmatrix} P_2 & R_2 \\ R_2^T & Q_2 \end{bmatrix} = \begin{bmatrix} X_2^\star & I \\ I & Y_2^\star \end{bmatrix}, \quad Z = Z^\star, \quad \begin{bmatrix} \widehat{A}_K & \widehat{B}_K \\ \widehat{C}_K & \widehat{D}_K \end{bmatrix} = \begin{bmatrix} \widehat{A}_K^\star & \widehat{B}_K^\star \\ \widehat{C}_K^\star & \widehat{D}_K^\star \end{bmatrix},$$

$$\begin{bmatrix} X_2 & I \\ S_2 & Y_2 \end{bmatrix} = \begin{bmatrix} X_2^\star & I \\ I & Y_2^\star \end{bmatrix}.$$

Theorem 5.4 *For given $\gamma_\infty > 0$ and the LTI system P represented by (5.1), the following statements are equivalent:*

(i) *There exists a full-order output-feedback controller K of the form (5.5) such that the closed-loop system $P_{\infty,K}$ becomes stable and $\| P_{\infty,K} \|_\infty < \gamma_\infty$.*

(ii) *There exist* X_∞, Y_∞, $\widehat{A}_K$, $\widehat{B}_K$, $\widehat{C}_K$, *and* $\widehat{D}_K$ *such that*

$$
\begin{bmatrix}
-X_\infty & -I & AX_\infty + B_u\widehat{C}_K & A + B_u\widehat{D}_K C_y & 0 & B_{w\infty} + B_u\widehat{D}_K D_{y,w\infty} \\
* & -Y_\infty & \widehat{A}_K & Y_\infty A + \widehat{B}_K C_y & 0 & Y_\infty B_{w\infty} + \widehat{B}_K D_{y,w\infty} \\
* & * & -X_\infty & -I & (C_{z\infty}X_\infty + D_{z\infty,u}\widehat{C}_K)^T & 0 \\
* & * & * & -Y_\infty & (C_{z\infty} + D_{z\infty,u}\widehat{D}_K C_y)^T & 0 \\
* & * & * & * & -\gamma_\infty^2 I_{n_{z\infty}} & D_{z\infty,w\infty} + D_{z\infty,u}\widehat{D}_K D_{y,w\infty} \\
* & * & * & * & * & -I_{n_{w\infty}}
\end{bmatrix} < 0,
$$
(5.40)

If the LMI (5.40) *is feasible, then a stabilizing controller K of the form* (5.5) *satisfying* $\|P_{\infty,K}\|_\infty < \gamma_\infty$ *is given by*

$$
\begin{aligned}
D_K &:= \widehat{D}_K, \\
C_K &:= (\widehat{C}_K - D_K C_y X_\infty) M_\infty^{-T}, \\
B_K &:= N_\infty^{-1}(\widehat{B}_K - Y_\infty B_u D_K), \\
A_K &:= N_\infty^{-1}\left(\widehat{A}_K - N_\infty B_K C_y X_\infty - Y_\infty B_u C_K M_\infty^T \right. \\
&\qquad \left. - Y_\infty(A + B_u D_K C_y)X_\infty\right) M_\infty^{-T}
\end{aligned}
$$
(5.41)

where M_∞ *and* N_∞ *are arbitrary matrices satisfying* $N_\infty M_\infty^T = I - Y_\infty X_\infty$.

(iii) *There exist* X_∞, Y_∞, S_∞, P_∞, Q_∞, R_∞, $\widehat{A}_K$, $\widehat{B}_K$, $\widehat{C}_K$, *and* $\widehat{D}_K$ *such that*

$$
\begin{bmatrix}
-P_\infty & -R_\infty & AX_\infty + B_u\widehat{C}_K & A + B_u\widehat{D}_K C_y & 0 & B_{w\infty} + B_u\widehat{D}_K D_{y,w\infty} \\
* & -Q_\infty & \widehat{A}_K & Y_\infty A + \widehat{B}_K C_y & 0 & Y_\infty B_{w\infty} + \widehat{B}_K D_{y,w\infty} \\
* & * & P_\infty - \text{He}\{X_\infty\} & R_\infty - I - S_\infty^T & (C_{z\infty}X_\infty + D_{z\infty,u}\widehat{C}_K)^T & 0 \\
* & * & * & Q_\infty - \text{He}\{Y_\infty\} & (C_{z\infty} + D_{z\infty,u}\widehat{D}_K C_y)^T & 0 \\
* & * & * & * & -\gamma_\infty^2 I_{n_{z\infty}} & D_{z\infty,w\infty} + D_{z\infty,u}\widehat{D}_K D_{y,w\infty} \\
* & * & * & * & * & -I_{n_{w\infty}}
\end{bmatrix} < 0,
$$
(5.42)

If the LMI (5.42) *is feasible, then a stabilizing controller K of the form* (5.5) *satisfying* $\|P_{\infty,K}\|_\infty < \gamma_\infty$ *is given by* (5.41) *where* M_∞ *and* N_∞ *are arbitrary matrices satisfying* $N_\infty M_\infty^T = S_\infty - Y_\infty X_\infty$. *Moreover, if the LMI* (5.40) *holds with*

$$
\begin{bmatrix} X_\infty & I \\ I & Y_\infty \end{bmatrix} = \begin{bmatrix} X_\infty^\star & I \\ I & Y_\infty^\star \end{bmatrix}, \quad \begin{bmatrix} \widehat{A}_K & \widehat{B}_K \\ \widehat{C}_K & \widehat{D}_K \end{bmatrix} = \begin{bmatrix} \widehat{A}_K^\star & \widehat{B}_K^\star \\ \widehat{C}_K^\star & \widehat{D}_K^\star \end{bmatrix},
$$

then the LMI (5.42) *holds with*

$$
\begin{bmatrix} P_\infty & R_\infty \\ R_\infty^T & Q_\infty \end{bmatrix} = \begin{bmatrix} X_\infty^\star & I \\ I & Y_\infty^\star \end{bmatrix}, \quad \begin{bmatrix} \widehat{A}_K & \widehat{B}_K \\ \widehat{C}_K & \widehat{D}_K \end{bmatrix} = \begin{bmatrix} \widehat{A}_K^\star & \widehat{B}_K^\star \\ \widehat{C}_K^\star & \widehat{D}_K^\star \end{bmatrix}, \quad \begin{bmatrix} X_\infty & I \\ S_\infty & Y_\infty \end{bmatrix} = \begin{bmatrix} X_\infty^\star & I \\ I & Y_\infty^\star \end{bmatrix}.
$$

Let us confirm how the standard LMIs and SV-LMIs in Theorems 5.3 and 5.4 can be derived by means of the linearizing change of variables stated in Sects. 5.5.1 and 5.5.2. From (5.6) and (ii) of Lemma 5.1, we see that $\|P_{2,K}\| < \gamma_2$ holds if and only if there exist $\mathscr{X}$, $\mathscr{Z}$, A_K, B_K, C_K and D_K such that

$$\begin{bmatrix} -\mathscr{X} + \mathscr{B}_2\mathscr{B}_2^{\mathrm{T}} & \mathscr{A}\,\mathscr{X} \\ (\mathscr{A}\,\mathscr{X})^{\mathrm{T}} & -\mathscr{X} \end{bmatrix} \prec 0,$$

$$\begin{bmatrix} -\mathscr{L} + \mathscr{D}_2\mathscr{D}_2^{\mathrm{T}} & \mathscr{C}_2\mathscr{X} \\ (\mathscr{C}_2\mathscr{X})^{\mathrm{T}} & -\mathscr{X} \end{bmatrix} \prec 0, \tag{5.43}$$

$$\mathrm{trace}\,(\mathscr{L}) < \gamma_2^2.$$

If we define Π_1 by (5.29) which can be assumed to be nonsingular and apply similarity transformations by means of Π_1, we see that the above system of LMIs holds if and only if

$$\begin{bmatrix} -\Pi_1^{\mathrm{T}}\mathscr{X}\Pi_1 + \Pi_1\mathscr{B}_2\mathscr{B}_2^{\mathrm{T}}\Pi_1^{\mathrm{T}} & \Pi_1^{\mathrm{T}}\mathscr{A}\,\mathscr{X}\Pi_1 \\ (\Pi_1^{\mathrm{T}}\mathscr{A}\,\mathscr{X}\Pi_1)^{\mathrm{T}} & -\Pi_1^{\mathrm{T}}\mathscr{X}\Pi_1 \end{bmatrix} \prec 0,$$

$$\begin{bmatrix} -\mathscr{L} + \mathscr{D}_2\mathscr{D}_2^{\mathrm{T}} & \mathscr{C}_2\mathscr{X}\Pi_1 \\ (\mathscr{C}_2\mathscr{X}\Pi_1)^{\mathrm{T}} & -\Pi_1^{\mathrm{T}}\mathscr{X}\Pi_1 \end{bmatrix} \prec 0, \tag{5.44}$$

$$\mathrm{trace}\,(\mathscr{L}) < \gamma_2^2.$$

By applying linearizing change of variables shown in (5.30) and (5.31a)–(5.31d), expanding matrices via Schur complement, and putting subscript "2" to the variables appropriately, we see that (5.44) can be written in the form of (5.37a)–(5.37c).

On the other hand, from (5.6) and (iii) of Lemma 5.1, we see that $\|P_{2,K}\| < \gamma_2$ holds if and only if there exist $\mathscr{X}$, $\mathscr{L}$, $\mathscr{G}$, A_K, B_K, C_K and D_K such that

$$\begin{bmatrix} -\mathscr{X} + \mathscr{B}_2\mathscr{B}_2^{\mathrm{T}} & 0 \\ 0 & \mathscr{X} \end{bmatrix} + \mathrm{He}\left\{ \begin{bmatrix} \mathscr{A}\,\mathscr{G} \\ -\mathscr{G} \end{bmatrix} \begin{bmatrix} 0 & I \end{bmatrix} \right\} \prec 0,$$

$$\begin{bmatrix} -\mathscr{L} + \mathscr{D}_2\mathscr{D}_2^{\mathrm{T}} & 0 \\ 0 & \mathscr{X} \end{bmatrix} + \mathrm{He}\left\{ \begin{bmatrix} \mathscr{C}_2\,\mathscr{G} \\ -\mathscr{G} \end{bmatrix} \begin{bmatrix} 0 & I \end{bmatrix} \right\} \prec 0, \tag{5.45}$$

$$\mathrm{trace}\,(\mathscr{L}) < \gamma_2^2.$$

If we define Π_1 by (5.35) that can be assumed to be nonsingular and apply similarity transformations with Π_1, we see that the above system of LMIs holds if and only if

$$\begin{bmatrix} -\Pi_1^{\mathrm{T}}\mathscr{X}\Pi_1 + \Pi_1^{\mathrm{T}}\mathscr{B}_2\mathscr{B}_2^{\mathrm{T}}\Pi_1 & 0 \\ 0 & \Pi_1^{\mathrm{T}}\mathscr{X}\Pi_1 \end{bmatrix} + \mathrm{He}\left\{ \begin{bmatrix} \Pi_1^{\mathrm{T}}\mathscr{A}\,\mathscr{G}\Pi_1 \\ -\Pi_1^{\mathrm{T}}\mathscr{G}\Pi_1 \end{bmatrix} \begin{bmatrix} 0 & I \end{bmatrix} \right\} \prec 0,$$

$$\begin{bmatrix} -\mathscr{L} + \mathscr{D}_2\mathscr{D}_2^{\mathrm{T}} & 0 \\ 0 & \Pi_1^{\mathrm{T}}\mathscr{X}\Pi_1 \end{bmatrix} + \mathrm{He}\left\{ \begin{bmatrix} \mathscr{C}_2\,\mathscr{G}\Pi_1 \\ -\Pi_1^{\mathrm{T}}\mathscr{G}\Pi_1 \end{bmatrix} \begin{bmatrix} 0 & I \end{bmatrix} \right\} \prec 0,$$

$$\mathrm{trace}\,(\mathscr{L}) < \gamma_2^2.$$

$$\tag{5.46}$$

By applying linearizing change of variables shown in (5.30) and (5.36a)–(5.36d) we see that (5.46) can be written in the form of (5.39a)–(5.39c) where we again apply Schur complement and put subscript "2" to the variables appropriately. We finally note that the LMIs in Theorem 5.4 for the output H_∞ controller synthesis can be derived by following similar procedures.

5.5.4 Advantages of SV-LMIs for Output-Feedback Multiobjective H_2/H_∞ Controller Synthesis

Now, we go back to the multiobjective H_2/H_∞ controller synthesis problem by output-feedback. We first consider the standard LMI-based formulation. From (ii) of Theorem 5.3, a controller K satisfying $\|P_{2,K}\|_2 < \gamma_2$ is given by (5.37a)–(5.37c) and (5.38). Similarly, from (ii) of Theorem 5.4, a controller K satisfying $\|P_{\infty,K}\|_2 < \gamma_\infty$ is given by (5.40) and (5.41). In view of the controller parametrizations given in (5.38) and (5.41), it is inevitable to enforce $X_2 = X_\infty$ and $Y_2 = Y_\infty$ to generate a single controller satisfying both the H_2 and H_∞ specifications. We thus obtain the following SDP formulation based on the standard LMIs:

$$\gamma_X^\star := \inf_{\substack{X_2, Y_2, Z, X_\infty, Y_\infty, \\ \widehat{A}_K, \widehat{B}_K, \widehat{C}_K, \widehat{D}_K}} \gamma_2 \text{ subject to (5.37), (5.40) and } X_2 = X_\infty, Y_2 = Y_\infty.$$

$$(5.47)$$

Note that in this formulation we enforced $X_2 = X_\infty$ and $Y_2 = Y_\infty$ which is essentially enforcing a common Lyapunov matrix variable for both the H_2 and H_∞ specifications (after linearizing change of variables). Due to this restriction, the standard LMI formulation is conservative.

On the other hand, as for the SV-LMIs and corresponding controller parametrizations given by (5.39a)–(5.39c), (5.38) and (5.42), (5.41), we see that these generate a same controller if $X_2 = X_\infty$, $Y_2 = Y_\infty$, and $S_2 = S_\infty$, i.e., the same SV-variable. It is not necessary to enforce (P_2, Q_2, R_2) and $(P_\infty, Q_\infty, R_\infty)$ being the same where these are submatrices of the transformed Lyapunov matrices. We thus obtain the following SDP by means of SV-LMIs:

$$\gamma_G^\star := \inf_{\substack{P_2, Q_2, R_2, X_2, Y_2, S_2, Z, \\ P_\infty, Q_\infty, R_\infty, X_\infty, Y_\infty, S_\infty \\ \widehat{A}_K, \widehat{B}_K, \widehat{C}_K, \widehat{D}_K}} \gamma_2 \text{ subject to (5.39), (5.42) and} \qquad (5.48)$$
$$X_2 = X_\infty, \ Y_2 = Y_\infty, \ S_2 = S_\infty.$$

This formulation is again conservative due to the restriction of common SV-variables $X_2 = X_\infty$, $Y_2 = Y_\infty$ and $S_2 = S_\infty$. However, from the recovery property stated in (iii) of Theorems 5.3 and 5.4, we see that if the LMIs (5.37a)–(5.37c) and (5.40) hold with

$$\begin{bmatrix} X_2 & I \\ I & Y_2 \end{bmatrix} = \begin{bmatrix} X_\infty & I \\ I & Y_\infty \end{bmatrix} = \begin{bmatrix} X^\star & I \\ I & Y^\star \end{bmatrix}, \quad Z = Z^\star,$$
$$\begin{bmatrix} \widehat{A}_K & \widehat{B}_K \\ \widehat{C}_K & \widehat{D}_K \end{bmatrix} = \begin{bmatrix} \widehat{A}_K^\star & \widehat{B}_K^\star \\ \widehat{C}_K^\star & \widehat{D}_K^\star \end{bmatrix},$$

we see that the LMIs (5.39a)–(5.39c) and (5.42) hold with

$$\begin{bmatrix} P_2 & R_2 \\ R_2^T & Q_2 \end{bmatrix} = \begin{bmatrix} P_\infty & R_\infty \\ R_\infty^T & Q_\infty \end{bmatrix} = \begin{bmatrix} X^\star & I \\ I & Y^\star \end{bmatrix}, \ Z = Z^\star,$$

$$\begin{bmatrix} X_2 & I \\ S_2 & Y_2 \end{bmatrix} = \begin{bmatrix} X_\infty & I \\ S_\infty & Y_\infty \end{bmatrix} = \begin{bmatrix} X^\star & I \\ I & Y^\star \end{bmatrix},$$

$$\begin{bmatrix} \widehat{A}_K & \widehat{B}_K \\ \widehat{C}_K & \widehat{D}_K \end{bmatrix} = \begin{bmatrix} \widehat{A}_K^\star & \widehat{B}_K^\star \\ \widehat{C}_K^\star & \widehat{D}_K^\star \end{bmatrix}.$$

Therefore, again we can ensure that $\gamma_G^\star \le \gamma_X^\star$ holds in terms of the SDPs (5.47) and (5.48). Namely, the SV-LMI formulation is better (no worse) than the standard LMI formulation.

5.6 Numerical Examples

Consider the cart-inverted-pendulum schematically shown in Fig. 5.1.

The cart moves back and forward by rotating a DC motor equipped with the backward two tires (of course the cart has four tires, two at front and two at back). A pendulum is implemented with the pivot point without actuator mounted on a cart as shown in the figure. The control objective is to keep the pendulum upright by moving the cart horizontally back and forward.

We first derive the mathematical model of the cart-inverted-pendulum. Denote by z [m] the position of the cart, θ [rad] the angle of the pendulum, and i [A] the current of the DC motor. We assume that these are measurable. Even though the dynamics of the cart-inverted-pendulum is nonlinear, we can linearize it at the equilibrium with $\theta = 0, \dot{\theta} = 0$ and obtain the state-space model as follows.

Fig. 5.1 Cart-inverted pendulum

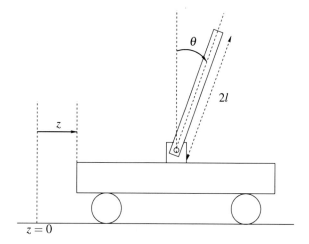

Table 5.1 Parameters of the cart-inverted-pendulum

Parameter	Physical meaning	Value	Unit
l	Half length of the pendulum	2.00×10^{-1}	m
m	Weight of the pendulum	5.36×10^{-2}	kg
M	Weight of the cart	6.86×10^{-1}	kg
a	Gain from i to the force applied to the cart	$[1.90, 2.80]$	N/A
J_c	Moment of inertia of tires and armature	1.34×10^{-4}	$\text{kg} \cdot \text{m}^2$
J_p	Moment of inertia of the pendulum	7.15×10^{-4}	$\text{kg} \cdot \text{m}^2$
r	Radius of the tires	2.49×10^{-2}	m
F	Viscous friction coefficient of the cart	3.60	$\text{kg} \cdot \text{s}^{-1}$
c	Viscous friction coefficient of the pivot	1.50×10^{-3}	$\text{kg} \cdot \text{m}^2 \cdot \text{s}^{-1}$
g	Acceleration of gravity	9.80	$\text{m} \cdot \text{s}^{-2}$

$$G_c : \begin{cases} \dot{x}(t) = A_c x(t) + B_c u(t), \\ y(t) = C_c x(t). \end{cases}$$

Here, $x = [\, z \; \theta \; \dot{z} \; \dot{\theta} \,]^\mathrm{T}$, $u = i$, $y = [\, z \; \theta \,]^\mathrm{T}$ and

$$A_c = \begin{bmatrix} 0 & 0 & 1 & 0 \\ 0 & 0 & 0 & 1 \\ 0 & -m^2 l^2 g\,W & -F(J_p + ml^2)W & cml\,W \\ 0 & mgl(M + m + \frac{J_c}{r^2})W & Fml\,W & -c(M + m + \frac{J_c}{r^2})W \end{bmatrix},$$

$$B_c = \begin{bmatrix} 0 \\ 0 \\ a(J_p + ml^2)W \\ -mla\,W \end{bmatrix}, \quad C_c = \begin{bmatrix} 1 & 0 & 0 & 0 \\ 0 & 1 & 0 & 0 \end{bmatrix},$$

$$W := \left((M + m + \tfrac{J_c}{r^2})(J_p + ml^2) - m^2 l^2 \right)^{-1}.$$

$$(5.49)$$

The physical meanings of the parameters in (5.49) as well as their values are shown in Table 5.1.

As shown in Table 5.1, we assume that the parameter a, denoting the gain from i to the force applied to the cart, is uncertain and varies over $[1.90, 2.80]$. Let us denote by $G_{c,0}$ the plant corresponding to $a = a_0 = 2.35$, the center of the variation. Furthermore, we denote by $W := 0.45/2.35$ the maximum relative error of the parameter a with respect to a_0. Then, since the parameter a enters only and linearly in the input matrix B, we see that the block-diagram of G_c can be described by Fig. 5.2. In Fig. 5.2, Δ represents the uncertainty of the parameter a and given by $\Delta := \{\Delta \in \mathbb{R} : \Delta \in [-1, 1]\}$.

Our goal here is to design a (discrete-time) controller that satisfies the following control specifications:

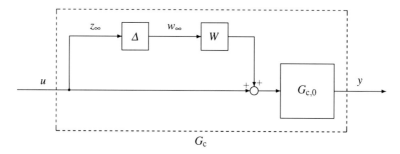

Fig. 5.2 Block-diagram of G_c

[S1] Stabilizes the closed-loop system under the variation of the parameter $a \in$ [1.90, 2.80].

[S2] Achieves good initial response, i.e., drives the state to the origin immediately with favorable transient response.

We formulate this controller synthesis problem as an multiobjective H_2/H_∞ control problem. The generalized plants (in continuous-time domain) can be derived as follows:

$$P_{c,\infty} : \begin{cases} \dot{x}(t) = A_c x(t) + B_c W w_\infty(t) + B_c u(t), \\ z_\infty(t) = u(t), \\ y(t) = C_c x(t), \end{cases} \tag{5.50}$$

$$P_{c,2} : \begin{cases} \dot{x}(t) = A_c x(t) + B_{c,w_2} w_2(t) + B_c u(t), \\ z_2(t) = C_{c,z_2} x(t) + D_{c,z_2,u} u(t), \\ y(t) = C_c x(t) \end{cases} \tag{5.51}$$

Here, the signals (w_∞, z_∞) in the generalized plant $P_{c,\infty}$ are chosen such that if $\|P_{c,\infty,K}\|_\infty < 1$ is achieved for some controller K, then the robust stability under the variation $a \in [1.90, 2.80]$ is attained. This is a direct consequence from Fig. 5.2 and the small-gain theorem.

On the other hand, in the generalized plant $P_{c,2}$, we let

$$B_{c,w_2} = \begin{bmatrix} 0 \\ \frac{\pi}{36} \\ 0 \\ 0 \end{bmatrix}, \quad C_{c,z_2} = \begin{bmatrix} 1 & 0 & 0 & 0 \\ 0 & 1 & 0 & 0 \\ 0 & 0 & 1 & 0 \\ 0 & 0 & 0 & 1 \\ 0 & 0 & 0 & 0 \end{bmatrix}, \quad D_{c,z_2,u} = \begin{bmatrix} 0 \\ 0 \\ 0 \\ 0 \\ 1 \end{bmatrix}.$$

We construct $P_{c,2}$ for the improvement of the initial response. The choice of B_{c,w_2} reflects that the initial state is given by $z(0) = 0$, $\theta(0) = \pi/36$ [rad]= 5 [°], $\dot{z}(0) = 0$ and $\dot{\theta}(0) = 0$. On the other hand, the choice of C_{c,z_2} and $D_{c,z_2,u}$ implies that $z_2 = [\, x^T \ u\,]^T$. By minimizing $\|P_{c,2,K}\|_2$ by appropriately designing an output-feedback controller K, we can expect that the state (starting from the initial state $x(0) = B_{c,w_2}$) goes to the origin immediately achieving good transient response.

In practice, we design discrete-time controllers and hence we need to discretize the generalized plants $P_{c,\infty}$ and $P_{c,2}$. For simplicity, we discretize $P_{c,\infty}$ with zero-order hold and ideal sampler under sampling period $T = 0.01$ [sec] and obtain

$$P_\infty : \begin{cases} x(k+1) = Ax(k) \quad + B_{w_\infty}w_\infty(k) + B_u u(k), \\ z_\infty(k) \quad = u(k), \\ y(k) \quad = C_c x(k). \end{cases} \tag{5.52}$$

where

$$A := \exp(A_c T),$$
$$B_{w_\infty} := \int_0^T \exp(A_c(T - \tau))B_{c,w_\infty}d\tau, \quad B_u := \int_0^T \exp(A_c(T - \tau))B_c d\tau.$$

On the other hand, since we focus on the initial response with respect to $x(0) = B_{c,w_2}$ when we build $P_{c,2}$, it is natural to choose the discretize plant P_2 given by

$$P_2 : \begin{cases} x(k+1) = Ax(k) \quad + B_{c,w_2}w_2(k) \quad + B_u u(k), \\ z_2(k) \quad = C_{c,z_2}x(k) + D_{c,z_2,u}u(k), \\ y(k) \quad = C_c x(k). \end{cases} \tag{5.53}$$

For the generalized plants P_∞ and P_2, we solve the discrete-time multiobjective H_2/H_∞ control problem formulated by

$$\inf_K \quad \|P_{2,K}\|_2 \quad \text{subject to} \quad \|P_{\infty,K}\|_\infty < 1. \tag{5.54}$$

To solve this problem, we apply the following two approaches by means of the standard LMIs and SV-LMIs, respectively.

[A] Solve the standard LMI-based SDP (5.47) by letting $\gamma_\infty = 1$ and $\widehat{D}_K = 0$.
[B] Solve the SV-LMI-based SDP (5.48) by letting $\gamma_\infty = 1$ and $\widehat{D}_K = 0$.

Here, in both approaches, we let $\widehat{D}_K = D_K = 0$, so that we can obtain strictly proper controllers. This is often an essential requirement when considering controller implementation in practice.

Remark 5.1 For stable numerical computation, we apply a scaling to the LMI (5.37a)–(5.37c) when solving the SDP (5.47). Namely, we replace the (5,5) term of (5.37a) by $-\gamma_2 I_{n_{w_2}}$, the (4,4) term of (5.37b) by $-\gamma_2 I_{n_{w_2}}$, and the right-hand side of (5.37c) by γ_2. This replacement is just a scaling and the resulting LMI condition is essentially equivalent to the original one. We apply a similar scaling to (5.39a)–(5.39c) when solving the SDP (5.48).

Before solving the multiobjective H_2/H_∞ control problem (5.54), we confirm that this problem is whether well-posed or not. In this problem setting, the H_∞ optimal controller $K^\star_\infty$ with respect to the generalized plant P_∞ satisfies $\|P_{\infty,K^\star_\infty}\|_2 = 0.4341$ and $\|P_{2,K^\star_\infty}\|_2 = 177.1631$. Therefore, the problem (5.54) is feasible. On the other hand, the H_2 optimal controller $K^\star_2$ with respect to the generalized plant P_2 satisfies $\|P_{\infty,K^\star_2}\|_2 = 1.2641$ and $\|P_{2,K^\star_2}\|_2 = 4.0022$. Namely, the H_2 optimal controller does not satisfy the H_∞ constraint and hence cannot be optimal for the problem (5.54). It follows that the problem (5.54) is nontrivial.

By solving the standard LMI-based SDP (5.47), we obtained $\gamma^\star_X = 25.9947$ and a controller K_X given by

$$K_X = \left[\begin{array}{cccc|cc} -0.8002 & -0.1920 & 0.0072 & -0.0077 & 27.4810 & -1.4077 \\ 0.1918 & 0.4315 & 0.0463 & -0.0595 & 12.8163 & -0.1641 \\ -0.0063 & 0.0159 & 0.9960 & 0.0098 & -0.3459 & 0.7807 \\ 0.0076 & -0.0596 & 0.0041 & -0.0861 & 0.3542 & 0.0729 \\ \hline -27.5175 & 12.8195 & -0.3728 & 0.3517 & 0 & 0 \end{array}\right]. \quad (5.55)$$

This controller achieves $\|P_{2,K_X}\|_2 = 16.6828$ and $\|P_{\infty,K_X}\|_\infty = 0.2934$. On the other hand, by solving the SV-LMI-based SDP (5.48), we obtained $\gamma^\star_G = 16.4882$ and a controller K_G given by

$$K_G = \left[\begin{array}{cccc|cc} -0.2187 & -0.1193 & 0.4804 & 0.0989 & 17.4810 & -2.7011 \\ 0.1024 & 0.9802 & 0.1019 & 0.0290 & 1.4546 & 0.5067 \\ -0.4827 & 0.0707 & 0.6034 & -0.1555 & -5.0435 & 0.6824 \\ -0.0666 & 0.0129 & -0.1032 & -0.2393 & 0.5533 & 2.0994 \\ \hline -17.7012 & 1.3936 & -5.0688 & 0.3486 & 0 & 0 \end{array}\right]. \quad (5.56)$$

This controller achieves $\|P_{2,K_G}\|_2 = 9.6964$ and $\|P_{\infty,K_G}\|_\infty = 0.3209$. We see that the SV-LMI approach is definitely effective in conservatism reduction since $\|P_{2,K_G}\|_2 < \|P_{2,K_X}\|_2$. However, it is also true that the SV-LMI approach is still conservative since the resulting controller K_G satisfies H_∞ constraint with a plenty of room as in $\|P_{\infty,K_G}\|_\infty = 0.3209$.

In Figs. 5.3, 5.5 and 5.7, we show the plots of $z[m]$, $\theta[°]$ and $u[A]$ when we use the controller K_X. Figures 5.9 and 5.11 are the enlargements of 5.5 and 5.7 respectively. Similarly, Figs. 5.4, 5.6 and 5.8 show the plots of $z[m]$, $\theta[°]$ and $u[A]$ when we use the controller K_G. Figures 5.10 and 5.12 are the enlargements of Figs. 5.6 and 5.8 respectively. First of all, we see from these figures that K_X and K_G successfully stabilizes the plant over the variation of the parameter $a \in [1.90, 2.80]$. More importantly, by comparing Figs. 5.3 and 5.4, we see that K_G is definitely better than K_X with respect to the response of $z[m]$. Figures 5.9 and 5.10 show that K_X and K_G achieves almost the same performance with respect to the response of $\theta[°]$. On the other hand, Figs. 5.11 and 5.12 show that the peak of the input $u[A]$ with K_G is higher than that with K_X. Still, the input $u[A]$ with K_G converges to zero faster than that with K_X. It is also true that the input $u[A]$ with K_G is more smooth than that with K_X. To summarize, we can conclude that K_G achieves better performance

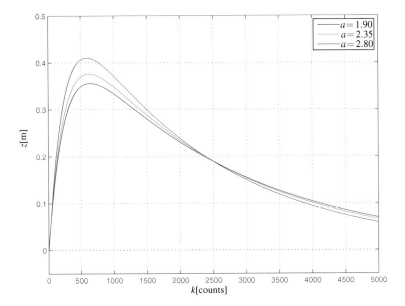

Fig. 5.3 The response of z [m] under K_X

than K_X overall. We expect that this achievement is due to the fact that K_G achieves much smaller H_2 norm than K_X.

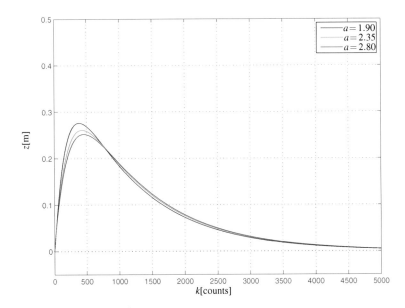

Fig. 5.4 The response of z [m] under K_G

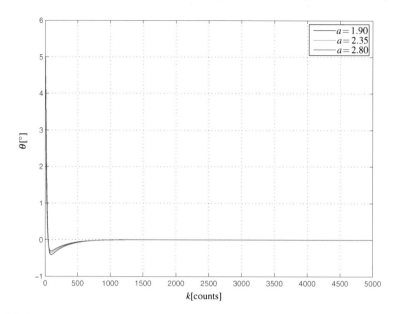

Fig. 5.5 The response of $\theta[°]$ under K_X

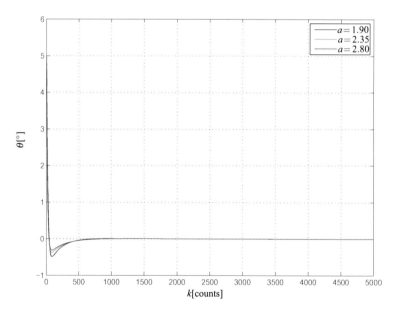

Fig. 5.6 The response of $\theta[°]$ under K_G

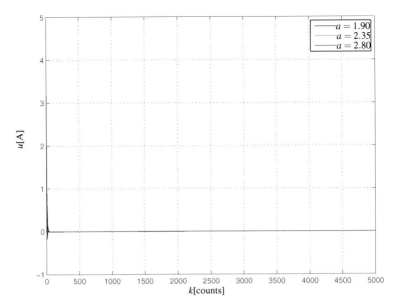

Fig. 5.7 The response of control input u[A] under K_X

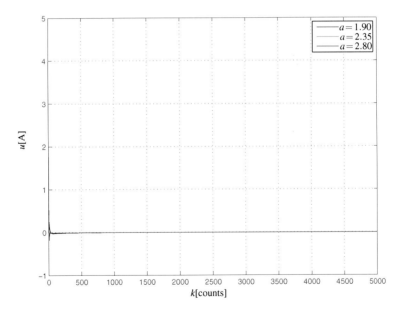

Fig. 5.8 The response of control input u[A] under K_G

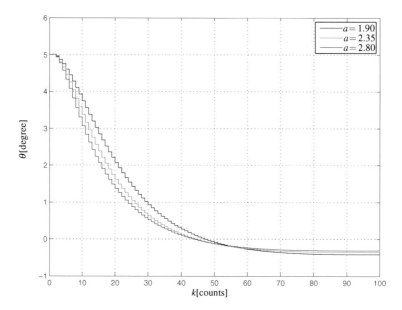

Fig. 5.9   Enlargement of Fig. 5.5

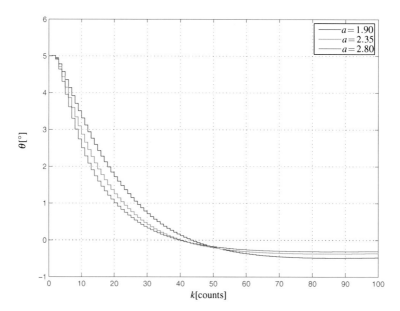

Fig. 5.10   Enlargement of Fig. 5.6

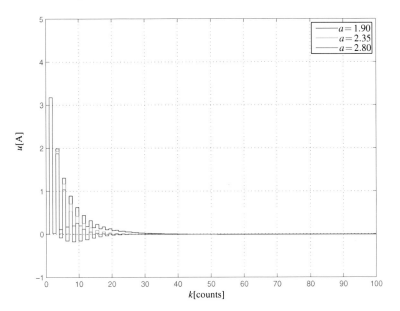

Fig. 5.11 Enlargement of Fig. 5.7

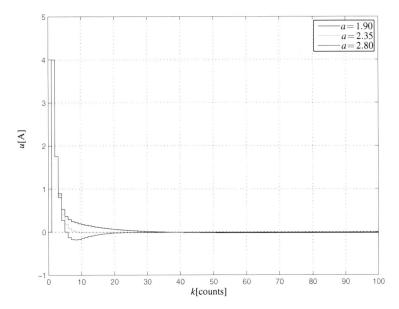

Fig. 5.12 Enlargement of Fig. 5.8

5.7 Conclusion

In this chapter, we illustrated the usefulness of SV-LMIs in multiobjective controller synthesis problems of discrete-time systems. The advantage of SV-LMIs over standard LMIs has been shown straightforwardly by the fact that the SV-LMIs allow us to employ distinct (noncommon) Lyapunov variables for each design specification. Even in the output-feedback cases, the advantage of SV-LMIs is ensured along with the linearizing change-of-variables technique tailored for SV-LMIs.

We have restricted our attention to discrete-time multiobjective controller synthesis problems, since in the continuous-time system cases the achievement in the line of SV-LMIs is restricted. To this date, it has been recognized that

- "single-shot" SV-LMI condition that can be used for H_∞ synthesis, state-feedback or output-feedback, is not available in a sound fashion. Namely, if we pursue the necessary and sufficient synthesis condition along the line of SV-LMIs, we have to employ a line-search parameter [2, 6];
- for some specific multiobjective control problems, we can ensure the effectiveness of SV-LMIs over the standard ones almost the same way as in the discrete-time system cases. One of such examples is the multiobjective H_2/D-stability synthesis problem. In this problem setting, we seek for a controller that minimizes the H_2 norm of a given generalized plant under closed-loop pole location constraints. In this case, the "recovery property" of SV-LMIs hold in a different (or say, slightly involved) fashion than (iii) of Theorems 5.1 and 5.2. Still, by using the recovery property, we can ensure that SV-LMI-based synthesis conditions yield better (no worse) controllers than standard LMI-based ones [2, 3].

References

1. de Oliveira MC, Geromel JC, Bernussou J (2002) Extended H_2 and H_∞ norm characterizations and controller parametrizations for discrete-time systems. Int J Control 75:666–679
2. Ebihara Y (2002) LMI-based Multiobjective Controller Design with Non-Common Lyapunov Variables. PHD Dissertation, Kyoto University
3. Ebihara Y, Hagiwara T (2004) New dilated LMI characterizations for continuous-time multiobjective controller synthesis. Automatica 40(11):2003–2009
4. Scherer CW, Gahinet P, Chilali M (1997) Multiobjective output-feedback control via LMI optimization. IEEE Trans Autom Control 42(7):896–911
5. Masubuchi I, Ohara A, Suda N (1998) LMI-based controller synthesis: a unified formulation and solution. Int J Robust Nonlinear Control 8(8):669–686
6. Shimomura T, Fujii T (2005) Multiobjective control via successive over-bounding of quadratic terms. Int J Robust Nonlinear Control 15(6):363–381

Chapter 6
Static Output-Feedback Synthesis

6.1 Introduction

One of the most challenging open problems in linear control theory is the synthesis of static output-feedback (SOF) controllers that meet desired performances and/or robustness specifications [1]. The static or reduced fixed-order dynamic output-feedback problem is therefore always an active research area in the control literature.

In the recent years, many attempts have been made to give efficient numerical procedures to solve related problems [2–9]. In [10], a numerical comparison was performed and classification into three categories (nonlinear programming, parametric optimization, and convex programming approaches) was proposed. Even if these three classes may overlap, it gives a clear picture of the situation at that moment. Since then, new developments have been witnessed and the literature has been enriched by numerous contributions on the so-called nonlinear programming approach [11–16] while pure convex programming methods or parametric optimization methods (which could be merged in a unique class) were scarcely still considered [17, 18]. The main reason mainly relies on the use of new powerful numerical tools based on nonsmooth optimization for the solution of static output-feedback stabilization problems or static output-feedback design with closed-loop performance guarantees [13, 19, 20]. The algorithms based on these techniques may be considered as the most numerically efficient ones at the moment as reported in different dedicated publications [21–23].

The SV-LMI approach for SOF controller synthesis can be categorized as a convex programming approach. Due to the inherent difficulties of SOF controller synthesis, pure SV-LMI synthesis condition is of course out of reach. Still, the SV-LMI leads to efficient and effective iterative methods as explicated in this chapter.

© Springer-Verlag London 2015
Y. Ebihara et al., *S-Variable Approach to LMI-Based Robust Control*,
Communications and Control Engineering, DOI 10.1007/978-1-4471-6606-1_6

6.2 A New Parametrization for Stabilizing SOF Synthesis

6.2.1 Statement of the Problem

Let the state-space model of the system be given by its minimal state-space realization:

$$\begin{cases} \dot{x}(t) = Ax(t) + Bu(t) \\ y(t) = Cx(t) \end{cases} \tag{6.1}$$

where $x \in \mathbb{R}^n$ is the state vector, $u \in \mathbb{R}^m$ is the control vector and $y \in \mathbb{R}^r$ is the output vector. All matrices are assumed to be of appropriate dimensions and it is assumed throughout the chapter that $\text{rank}(B) = m$ and $\text{rank}(C) = r$.

The model (6.1) is stabilizable by static output feedback (SOF) if there exists a gain matrix $K_{\text{SOF}} \in \mathbb{R}^{m \times r}$ such that the closed-loop matrix $A + BK_{\text{SOF}}C$ is asymptotically stable ($A + BK_{\text{SOF}}C$ is a Hurwitz matrix). Moreover, the system is stabilizable by state-feedback (SF) if there exists a gain matrix $K_{\text{SF}} \in \mathbb{R}^{n \times r}$ such that the matrix $A + BK_{\text{SF}}$ is Hurwitz. This last property is a special case of the former and corresponds to full state information output feedback ($C = I$) and has found a tractable solution through convex optimization and LMI formalism as we have seen in the previous chapters. Let us define the set of stabilizing state-feedback matrices:

$$\mathcal{K}_{\text{SF}} = \left\{ K_{\text{SF}} \in \mathbb{R}^{m \times n} \;:\; \lambda(A + BK_{\text{SF}}) \in \mathbb{C}_- \right\} \tag{6.2}$$

and the set of stabilizing static output-feedback matrices:

$$\mathcal{K}_{\text{SOF}} = \left\{ K_{\text{SOF}} \in \mathbb{R}^{m \times r} \;:\; \lambda(A + BK_{\text{SOF}}C) \in \mathbb{C}_- \right\} \tag{6.3}$$

The first problem considered in this chapter is to build non trivial sets of stabilizing SOF for (6.1).

Problem 6.1 (SOF Stabilization) Given the model (6.1), build a non trivial subset $\mathcal{K}_{\text{SOF}}^{\text{subs}} \subset \mathcal{K}_{\text{SOF}}$.

6.2.2 Parametrization of $\mathcal{K}_{\text{SOF}}$

Let us introduce the parametrization of stabilizing SOF matrices that has been first proposed in [17]. To this end, define the following notation:

$$M(P) = \begin{bmatrix} 0 & B^T P \\ PB & PA + A^T P \end{bmatrix} = \begin{bmatrix} B^T & 0 \\ A^T & I \end{bmatrix} \begin{bmatrix} 0 & P \\ P & 0 \end{bmatrix} \begin{bmatrix} B & A \\ 0 & I \end{bmatrix}. \tag{6.4}$$

A necessary and sufficient condition for the existence of a stabilizing SOF for (6.1) is given in the following theorem.

Theorem 6.1 [17] *There exists $K_{SOF} \in \mathcal{K}_{SOF}$ for the model (6.1) if and only if there exist a stabilizing SF matrix $K_{SF} \in \mathcal{K}_{SF}$, a matrix $P \in \mathbb{S}_{++}^n$ and matrices $F \in \mathbb{R}^{m \times m}$, $Z \in \mathbb{R}^{m \times r}$ such that*

$$L(P, K_{SF}, Z, F) = M(P) + \text{He}\left(\begin{bmatrix} I \\ -K_{SF}^T \end{bmatrix} \begin{bmatrix} F & ZC \end{bmatrix}\right) \prec 0. \qquad (6.5)$$

Moreover,

$$K_{SOF} = -F^{-1}Z \in \mathcal{K}_{SOF}. \qquad (6.6)$$

Proof Note that SOF stabilizability of (6.1) ($\exists\, K_{SOF} \in \mathcal{K}_{SOF}$) is equivalent to the existence of a matrix $K_{SOF} \in \mathbb{R}^{m \times r}$ and a matrix $P \in \mathbb{S}_{++}^n$ solutions to the following Lyapunov inequality:

$$\begin{aligned}
(A &+ BK_{SOF}C)^T P + P(A + BK_{SOF}C) \\
&= \begin{bmatrix} C^T K_{SOF}^T & I \end{bmatrix} M(P) \begin{bmatrix} K_{SOF}C \\ I \end{bmatrix} \\
&= \begin{bmatrix} I \\ -C^T K_{SOF}^T \end{bmatrix}^{\perp} M(P) \begin{bmatrix} I \\ -C^T K_{SOF}^T \end{bmatrix}^{\perp T} \prec 0.
\end{aligned} \qquad (6.7)$$

Applying Elimination lemma (Lemma 1.2), this is also equivalent to the existence of matrices $K_{SOF} \in \mathbb{R}^{m \times r}$, $F_1 \in \mathbb{R}^{n \times m}$, $F_2 \in \mathbb{R}^{m \times m}$ and a matrix $P \in \mathbb{S}_{++}^n$ solutions to the following matrix inequality:

$$M(P) + \text{He}\left(\begin{bmatrix} F_1 \\ F_2 \end{bmatrix} \begin{bmatrix} I & -K_{SOF}C \end{bmatrix}\right) \prec 0. \qquad (6.8)$$

The matrix F_1 is always invertible since the block $(1, 1)$ of (6.8) reads $F_1 + F_1^T \prec 0$. Factorizing F_1 in (6.8) leads to

$$M(P) + \text{He}\left(\begin{bmatrix} I \\ F_2 F_1^{-1} \end{bmatrix} \begin{bmatrix} F_1 & -F_1 K_{SOF}C \end{bmatrix}\right) \prec 0. \qquad (6.9)$$

With the notations $Z = -F_1 K_{SOF}$, $F = F_1$ and $K_{SF} = -F_1^{-T} F_2^T$, we retrieve inequality (6.5) and the parametrization above. Finally, $K_{SF} = -F_1^{-T} F_2^T \in \mathcal{K}_{SF}$ by noting that

$$\begin{bmatrix} K_{SF}^T & I \end{bmatrix} (6.5) \begin{bmatrix} K_{SF} \\ I \end{bmatrix} = \begin{bmatrix} K_{SF}^T & I \end{bmatrix} M(p) \begin{bmatrix} K_{SF} \\ I \end{bmatrix} \tag{6.10}$$
$$= (A + BK_{SF})^T P + P(A + BK_{SF}) \prec 0. \qquad \square$$

Even if the existence condition is expressed in terms of solution to a Bilinear Matrix Inequality (BMI), this parametrization has some interesting characteristics that will be useful for defining efficient numerical procedure for SOF synthesis. First, thanks to the introduction of S-variables F_1 and F_2 that have merged into F and K_{SF} in (6.5), there is a decoupling between $K_{SOF} \in \mathcal{K}_{SOF}$ and the Lyapunov certificate P. Moreover, the matrix K_{SF} at the origin of the nonconvexity of condition (6.5) must not be arbitrarily chosen but it has to be a stabilizing SF for (6.1), i.e., $K_{SF} \in \mathcal{K}_{SF}$. As we have seen in previous chapters, the computation of a stabilizing SF for (6.1) is a convex problem that may be solved via the following LMI optimization problem [24]:

$$\begin{aligned} \min_{X,R} \quad & \text{trace}(X) \\ \text{subject to trace}(X) &> \alpha \\ X &\succ 0 \\ AX + XA^T + BY + Y^T B^T &\prec 0 \end{aligned} \tag{6.11}$$

for some $\alpha > 0$ and where $K_{SF} = YX^{-1} \in \mathcal{K}_{SF}$.

For a given $K_{SF} \in \mathcal{K}_{SF}$, let us define the LMI convex set $\mathcal{L}_{SOF}^{K_{SF}}$ by

$$\mathcal{L}_{SOF}^{K_{SF}} = \{ K_{SOF} = -F^{-1}Z : $$
$$(P, Z, F) \in \mathbb{S}_{++}^n \times \mathbb{R}^{m \times r} \times \mathbb{R}^{m \times m}, \ L(P, K_{SF}, Z, F) \prec 0 \}. \tag{6.12}$$

Then, this set is a convex parametrization that approximates the set of all stabilizing SOF $\mathcal{K}_{SOF}$. Note that this convex approximation may be empty ($\mathcal{L}_{SOF}^{K_{SF}} = \emptyset$) for a given $K_{SF} \in \mathcal{K}_{SF}$ even if $\mathcal{K}_{SOF} \neq \emptyset$. However, a complete parametrization of $\mathcal{K}_{SOF}$ is obtained when K_{SF} covers the whole continuum of $\mathcal{K}_{SF}$.

$$\mathcal{K}_{SOF} = \left\{ \bigcup_i^\infty \mathcal{L}_{SOF}^{K_{SF}^i} \mid K_{SF}^i \in \mathcal{K}_{SF} \right\} \tag{6.13}$$

In fact, the problem of finding a stabilizing SOF amounts to find a triplet composed of $(K_{SOF}, K_{SF}, P) \in \mathcal{K}_{SOF} \times \mathcal{K}_{SF} \times \mathbb{S}_{++}^n$ verifying (6.5), meaning that a common Lyapunov certificate P has to be found for K_{SF} and K_{SOF}. This last interpretation is reminiscent of a necessary and sufficient condition of SOF stabilizability proposed in [25].

Corollary 6.1 *The model* (6.1) *is stabilizable via SOF if and only if there exists a triplet* $(P, K_{SOF}, K_{SF}) \in \mathbb{S}_{++}^n \times \mathbb{R}^{m \times r} \times \mathbb{R}^{m \times n}$ *such that*

$$\begin{bmatrix} C^T K_{SOF}^T & I \end{bmatrix} M(P) \begin{bmatrix} K_{SOF} C \\ I \end{bmatrix} \prec 0$$

$$\begin{bmatrix} K_{SF}^T & I \end{bmatrix} M(P) \begin{bmatrix} K_{SF} \\ I \end{bmatrix} \prec 0 \tag{6.14}$$

Remark 6.1 Note that the parametrization (6.5) of $\mathcal{K}_{SOF}$ has an equivalent one for $\mathcal{K}_{SF}$ since:

$$
\begin{aligned}
L(P, K_{SF}, Z, F) &= L'(P, K_{SOF}, H, F) \\
&= M(P) + He\left(\begin{bmatrix} F \\ H \end{bmatrix} \begin{bmatrix} I & -K_{SOF}C \end{bmatrix} \right) \prec 0
\end{aligned}
\tag{6.15}
$$

where $K_{SF} = -F^{-T} H^T \in \mathcal{K}_{SF}$. Given $K_{SOF} \in \mathcal{K}_{SOF}$, one can easily get a convex parametrization approximating the set of all stabilizing SF. This seems of no consequence since the complete set $\mathcal{K}_{SF}$ may be parameterized by the LMI convex set defined in (6.11). Nevertheless, this alternate parametrization will appear useful in the sequel when looking for sets of SOF.

For a given $K_{SOF} \in \mathcal{K}_{SOF}$, the LMI convex set $\mathscr{L}_{SF}^{K_{SOF}}$ defined by:

$$
\begin{aligned}
\mathscr{L}_{SF}^{K_{SOF}} = \big\{ &K_{SF} = -F^{-T} H^T : \\
&(P, H, F) \in \mathbb{S}_{++}^n \times \mathbb{R}^{n \times m} \times \mathbb{R}^{m \times m} : L'(P, K_{SOF}, H, F) \prec 0 \big\}
\end{aligned}
\tag{6.16}
$$

is a convex parametrization that approximates the set of all stabilizing SF $\mathcal{K}_{SF}$.

To make (6.12) completely effective, a precise description of the subset $\mathcal{K}_{SF}^g \subset \mathcal{K}_{SF}$ defined by

$$\mathcal{K}_{SF}^g = \left\{ K_{SF} \in \mathcal{K}_{SF} : \mathscr{L}_{SOF}^{K_{SF}} \neq \emptyset \right\} \tag{6.17}$$

would be necessary. The idea is therefore to find at least one element belonging to this set by generating a sufficient number of random samples in $\mathcal{K}_{SF}$ via Hit-and-Run techniques.

6.2.3 Related Parametrizations

6.2.3.1 A Slight Variation with Elimination Lemma

In [26], the formulation (6.14) is applied to the design of structured state feedback (decentralized or multiobjective control laws), while in [27] some numerical experiments are given regarding SOF stabilization.

Theorem 6.2 [26] *There exists $K_{\text{SOF}} \in \mathcal{K}_{\text{SOF}}$ for the model (6.1) if and only if there exist matrices $G \in \mathbb{R}^{m \times m}$, $W \in \mathbb{R}^{m \times r}$, $F_1 \in \mathbb{R}^{(2m+n) \times r}$, $F_2 \in \mathbb{R}^{(2m+n) \times m}$ and a matrix $P \in \mathbb{S}_{++}^n$ such that*

$$\left[\begin{bmatrix} 0 \\ G \\ WC \end{bmatrix} \begin{bmatrix} G & WC \end{bmatrix} \\ M(P) \right] + \text{He}\left(\begin{bmatrix} -F_2 & F_2 & -F_1 \end{bmatrix} \right) \prec 0 \tag{6.18}$$

under the rank constraint

$$\text{rank}\left(\begin{bmatrix} F_1 & F_2 \end{bmatrix} \right) \leq m. \tag{6.19}$$

Moreover,

$$K_{\text{SOF}} = -G^{-1} W \in \mathcal{K}_{\text{SOF}}. \tag{6.20}$$

Proof Following Corollary 6.1, the SOF stabilizability of (6.1) is equivalent to the existence of a triplet $(P, K_{\text{SOF}}, K_{\text{SF}}) \in \mathbb{S}_{++}^n \times \mathbb{R}^{m \times r} \times \mathbb{R}^{m \times n}$ such that

$$\begin{bmatrix} I \\ -C^T K_{\text{SOF}}^T \end{bmatrix}^\perp M(P) \begin{bmatrix} I \\ -C^T K_{\text{SOF}}^T \end{bmatrix}^{\perp T} \prec 0$$

$$\begin{bmatrix} I \\ -K_{\text{SF}}^T \end{bmatrix}^\perp M(P) \begin{bmatrix} I \\ -K_{\text{SF}}^T \end{bmatrix}^{\perp T} \prec 0 \tag{6.21}$$

Applying Elimination Lemma (Lemma 1.2), this is also equivalent to the existence of matrices $K_{\text{SOF}} \in \mathbb{R}^{m \times r}$, $K_{\text{SF}} \in \mathbb{R}^{m \times n}$, $G \in \mathbb{R}^{m \times m}$ and a matrix $P \in \mathbb{S}_{++}^n$ solutions to the following matrix inequality:

$$M(P) + \text{He}\left(\begin{bmatrix} I \\ -K_{\text{SF}}^T \end{bmatrix} G \begin{bmatrix} I & -K_{\text{SOF}} C \end{bmatrix} \right) \prec 0. \tag{6.22}$$

Factorizing (6.22) as

$$\begin{bmatrix} I & -K_{\text{SF}} \\ I & 0 \\ 0 & I \end{bmatrix}^T \left[\begin{bmatrix} 0 \\ G^T \\ -C^T K_{\text{SOF}}^T G^T \end{bmatrix} \begin{bmatrix} G & -G K_{\text{SOF}} C \end{bmatrix} \\ M(P) \right] \begin{bmatrix} I & -K_{\text{SF}} \\ I & 0 \\ 0 & I \end{bmatrix}$$

$$= \begin{bmatrix} -I \\ I \\ -K_{\text{SF}}^T \end{bmatrix}^\perp \left[\begin{bmatrix} 0 \\ G^T \\ -C^T K_{\text{SOF}}^T G^T \end{bmatrix} \begin{bmatrix} G & -G K_{\text{SOF}} C \end{bmatrix} \\ M(P) \right] \begin{bmatrix} -I \\ I \\ -K_{\text{SF}}^T \end{bmatrix}^{\perp T} \prec 0 \tag{6.23}$$

and applying again Elimination Lemma, we get equivalently the existence of matrices $W \in \mathbb{R}^{m \times r}$, $F_1 \in \mathbb{R}^{(2m+n) \times r}$, $F_2 \in \mathbb{R}^{(2m+n) \times n}$, $G \in \mathbb{R}^{m \times m}$ and a matrix $P \in \mathbb{S}^n_{++}$ solutions to

$$\left[\begin{bmatrix} 0 \\ G^T \\ C^T W^T \end{bmatrix} \begin{bmatrix} G & WC \end{bmatrix} \\ M(P) \right] + \mathrm{He}\left(\begin{bmatrix} -F_2 & F_2 & -F_1 \end{bmatrix} \right) \prec 0 \qquad (6.24)$$

where $W = -GK_{\mathrm{SOF}}$ and $F_1 = F_2 K_{\mathrm{SF}}$. This last inequality gives the rank constraint on the matrix $\begin{bmatrix} F_1 & F_2 \end{bmatrix}$. $\qquad\qquad\square$

Even though the rank constraint (6.19) is nonconvex and hence hard to handle in numerical computation, we can construct an effective iterative method [26] by means of the alternating projection method [9].

6.2.3.2 Polynomial Matrix Interpretation of S-variables

When dealing with SOF synthesis problems by means of SV-LMIs, "polynomial matrix interpretation" of S-variables along with [28] is useful.

Theorem 6.3 [28] *Let* $D(s) = sI_n - F$ *be a stable polynomial matrix in continuous-time system sense. If there exist matrices* $P \in \mathbb{S}^n_{++}$, $K \in \mathbb{R}^{r \times m}$ *such that*

$$\begin{bmatrix} 0 & P \\ P & 0 \end{bmatrix} - D^T N - N^T D \prec 0 \qquad (6.25)$$

where $D = \begin{bmatrix} I_n & -F \end{bmatrix}$ *and* $N = \begin{bmatrix} I_n & -(A + BKC) \end{bmatrix}$, *then* (6.1) *is stabilizable via SOF and* $K \in \mathscr{K}_{\mathrm{SOF}}$.

Proof The proof follows from lemma 8 in [28] by choosing the polynomial matrices $N(s) = sI_n - (A + BKC)$ and $D(s) = sI_n - F$ with $D(s)$ stable. $\qquad\square$

LMI condition (6.25) is only a sufficient condition for static output-feedback stabilization of the model (6.1) since $D(s)$, called the central polynomial matrix, is assumed to be given and stable. In [28, 29], the choice of an appropriate $D(s)$ (i.e., the choice of an appropriate F) is left to the designer and it is recommended that the set of zeros of $D(s)$ (i.e., eigenvalues of F) should *somehow match the expected closed-loop spectrum* as a rule of thumb. This sufficient condition developed in a polynomial framework has a nice interpretation in terms of S-variables.

First, note that the LMI (6.25) reads $\exists P \in \mathbb{S}^n_{++}, \exists K \in \mathbb{R}^{r \times m}$ such that

$$\begin{bmatrix} 0 & P \\ P & 0 \end{bmatrix} + \mathrm{He}\left(\begin{bmatrix} -I_n \\ F^T \end{bmatrix} \begin{bmatrix} I_n & -(A + BKC) \end{bmatrix} \right) \prec 0 \qquad (6.26)$$

for a given Hurwitz matrix $F \in \mathbb{R}^{n \times n}$. Looking back to inequality (6.7) rewritten as

$$
\begin{aligned}
& (A + BK_{\text{SOF}}C)^T P + P(A + BK_{\text{SOF}}C) \\
&= \begin{bmatrix} (A + BK_{\text{SOF}}C)^T & I \end{bmatrix} \begin{bmatrix} 0 & P \\ P & 0 \end{bmatrix} \begin{bmatrix} A + BK_{\text{SOF}}C \\ I \end{bmatrix} \\
&= \begin{bmatrix} I \\ -(A + BK_{\text{SOF}}C)^T \end{bmatrix}^{\perp} \begin{bmatrix} 0 & P \\ P & 0 \end{bmatrix} \begin{bmatrix} I \\ -(A + BK_{\text{SOF}}C) \end{bmatrix}^{\perp T} \prec 0.
\end{aligned}
\tag{6.27}
$$

By Elimination Lemma, this condition can be rewritten equivalently as $\exists \, P \in \mathbb{S}^n_{++}$, $\exists \, K_{\text{SOF}} \in \mathbb{R}^{r \times m}$, $\exists \, (F_1, F_2) \in \mathbb{R}^{n \times n} \times \mathbb{R}^{n \times n}$ such that

$$
\begin{bmatrix} 0 & P \\ P & 0 \end{bmatrix} + \text{He}\left(\begin{bmatrix} F_1 \\ F_2 \end{bmatrix} \begin{bmatrix} I_n & -(A + BK_{\text{SOF}}C) \end{bmatrix} \right) \prec 0.
\tag{6.28}
$$

Then, it is clear that (6.26) is nothing but (6.28) with $K = K_{\text{SOF}}$, $F_1 = -I_n$ and $F_2 = F^T$ with $F \in \mathbb{R}^{n \times n}$ selected as a Hurwitz matrix.

Since (6.28) gives another necessary and sufficient condition for SOF stabilization of model (6.1), the above discussion leads us to the next corollary on the interpretation of S-variables F_1 and F_2 in terms of matrix polynomials.

Corollary 6.2 *There exists $K_{\text{SOF}} \in \mathcal{K}_{\text{SOF}}$ for the model (6.1) if and only if there exist a matrix $K_{\text{SOF}} \in \mathbb{R}^{r \times m}$, a matrix $P \in \mathbb{S}^n_{++}$ and a stable matrix pencil $D(s) = SF_1 + F_2$ such that (6.28) is satisfied.*

The sufficient LMI condition given in Theorem 6.3 is clearly a relaxation of the necessary and sufficient condition given in Corollary 6.2. The stable matrix pencil $D(s)$ is chosen *a priori* with a particular form $F_1 = I_n$ and $F_2 = F$. Unfortunately, choosing an appropriate stable matrix pencil $D(s)$ is not an easy task in general. First, the spectrum of $D(s)$ must belong to the open left half plane of the complex plane without any additional information about its precise location. More surprising, *the state-space realization* of this pencil plays also a great role in the efficiency of the sufficient condition given by Theorem 6.3 as it is illustrated in the following simple example.

Example 6.1 [30] The model of the longitudinal motion of an F4E fighter aircraft is considered. The input is the elevator position, the output is the pitch rate, while the aircraft model is obtained via a linearization around the flight conditions: Mach 0.5, 5,000 ft:

$$
G(s) = \frac{-185.4s - 163.8}{s^3 + 15.84s^2 + 22.00s - 52.75}
\tag{6.29}
$$

It is easy to see that this open-loop unstable plant can be stabilized via SOF by any $K \in [0.3220, +\infty)$.

A minimal controllability canonical realization is built from $G(s)$. Then, as in [29], let us choose the stable central polynomial matrix $D(s) = sI - F$ with

$$F = \begin{bmatrix} -0.5588 & 0 & 0 \\ 0 & -7.6410 & -11.85 \\ 0 & 11.85 & -7.6410 \end{bmatrix}, \tag{6.30}$$

where $\det(D(s)) = s^3 + 15.84s^2 + 207.4s + 111.1$ and hence the spectrum of $D(s)$ is $\{-0.5588, -7.6410 \pm j11.85\}$.

Under the choice (6.30), the LMI condition (6.25) turns out to be infeasible. However, if we choose a minimal controllability canonical realization for $D(s)$, the sufficient condition of Theorem 6.3 gives the solution $K = 0.9991$ as a stabilizing SOF. Note that it is not always necessary to resort to a canonical controllability realization for $D(s)$ and $G(s)$ since the following choice

$$F = \begin{bmatrix} -17.8504 & -2.0472 & -1.7739 \\ -33.7792 & -59.6725 & -18.8929 \\ -8.7978 & -20.0044 & -30.1611 \end{bmatrix}$$

$$A_G = \begin{bmatrix} -15.84 & -2.75 & 3.297 \\ 8 & 0 & 0 \\ 0 & 2 & 0 \end{bmatrix}, \quad B_G = \begin{bmatrix} 8 \\ 0 \\ 0 \end{bmatrix},$$

$$C_G = \begin{bmatrix} 0 & -2.897 & -1.28 \end{bmatrix} \tag{6.31}$$

where the spectrum of $D(s)$ is given by -71.0145, $-18.3348 \pm j0.6502$, gives $K = 1.0075$ as solution to the SOF stabilization problem.

6.3 H_∞ and H_2 SOF Synthesis

6.3.1 Statement of the SOF Performance Synthesis Problem

Let the state-space model of the system be given by its minimal state-space realization:

$$\begin{cases} \dot{x}(t) = Ax(t) + B_1 w(t) + Bu(t), \\ z(t) = C_1 x(t) + D_{11} w(t) + D_{12} u(t), \quad P(s) := \begin{bmatrix} A & B_1 & B \\ C_1 & D_{11} & D_{12} \\ C & D_{21} & 0 \end{bmatrix} \\ y(t) = Cx(t) + D_{21} w(t) \end{cases} \tag{6.32}$$

where $x \in \mathbb{R}^n$ is the state vector, $u \in \mathbb{R}^m$ is the control vector, $y \in \mathbb{R}^r$ is the output vector, $w \in \mathbb{R}^{m_2}$ is the disturbance vector and $z \in \mathbb{R}^{r_2}$ is the controlled output vector. All matrices are assumed to be of appropriate dimensions and it is assumed that rank$(B) = m$ and rank$(C) = r$.

In general, when evaluating algorithms for SOF design, one important issue is not only to find stabilizing SOF but rather to find adequate candidates for optimizing some prespecified performance. For $K_{SOF} \in \mathscr{K}_{SOF}$ and $K_{SF} \in \mathscr{K}_{SF}$, consider the following transfer function matrices

$$T(s, K_{SOF}) = \left[\begin{array}{c|c} A + BK_{SOF}C & B_1 + BD_{21} \\ \hline C_1 + D_{12}K_{SOF}C & D_{11} + D_{12}K_{SOF}D_{21} \end{array} \right],$$

$$T_s(s, K_{SF}) = \left[\begin{array}{c|c} A + BK_{SF} & B_1 \\ \hline C_1 + D_{12}K_{SF} & D_{11} \end{array} \right]. \tag{6.33}$$

Let $\|T(s, K_*)\|_2$ and $\|T(s, K_*)\|_\infty$ be, respectively, the H_2 norm and the H_∞ norm of the transfer matrix $T(s, K_*)$. The generic performance synthesis problem to be addressed is stated as follows.

Problem 6.2 (*Optimal SOF control*) Find $K_{SOF}^* \in \mathscr{K}_{SOF}$ such that

$$K_{SOF}^* = \arg \left\{ \min_{K_{SOF} \in \mathscr{K}_{SOF}} \|T(s, K_{SOF})\|_* \right\} \tag{6.34}$$

where $* = 2$ or ∞.

This problem, which has been dealt with for the first time in [31] in the context of optimal H_2 SOF control, is known to be a difficult one for which no convex formulation exists yet for $* = 2$ or ∞. The problem is supposed to have many local minima and it is believed that any reformulation will also exhibit local minima. To overcome this particular problem, many approaches use heuristics to make the optimization problem more tractable. In [17], an efficient numerical cross-decomposition procedure based on a new parametrization of $\mathscr{K}_{SOF}$ has given promising results in the H_2 case. It is proposed to build on these previous results by adding a randomized step that will enforced the obtained results. Note that in [17], the H_2 optimal SOF control problem was addressed but H_∞ optimal SOF control may be considered in the same framework. In the sequel, when considering the H_2 SOF optimal control problem, it is assumed that $D_{11} = 0$ and $D_{21} = 0$.

6.3.2 H_2 Optimal Synthesis via SOF

6.3.2.1 Parametrizations for H_2 Optimal Control

First, let us extend the Parametrization of Theorem 6.1 to give a solution to the Problem (6.34). To this end, define

$$N(P_2) = M(P_2) + \left[D_{12}\ C_1 \right]^T \left[D_{12}\ C_1 \right] \tag{6.35}$$

Then, we can state the next theorem.

Theorem 6.4 *The H_2 optimal SOF for* (6.32) *is given by $K^*_{SOF_2} = -F^{*-1}Z^*$ where the triplet* $(P_2^*, Z^*, F^*) \in \mathbb{S}^{+*} \times \mathbb{R}^{m \times r} \times \mathbb{R}^{m \times m}$ *is the global optimal solution of the non convex optimization problem:*

$$\min_{P_2, Z, F, K_{SF_2}} \text{trace}(B_1^T P_2 B_1)$$
$$\text{subject to} \quad P_2 \succ 0,$$
$$L_2(P_2, K_{SF_2}, Z, F) = N(P_2) + \text{He}\left(\begin{bmatrix} I \\ -K^T_{SF_2} \end{bmatrix} \begin{bmatrix} F & ZC \end{bmatrix}\right) \prec 0.$$
$$(6.36)$$

When dealing with the simpler case of H_2 SF optimal control (problem (6.36) for which $C = I$), the exact H_2 SF optimal control may be computed via the solution of the following SDP:

$$\min_{X_2, Y, W} \text{trace}(W)$$
$$\text{subject to} \quad X_2 \succ 0,$$
$$\begin{bmatrix} -W & C_1 X_2 + D_{12} Y \\ X_2 C_1^T + Y^T D_{12}^T & -X_2 \end{bmatrix} \prec 0, \quad (6.37)$$
$$\text{He}\{A X_2 + B Y\} + B_1^T B_1 \prec 0.$$

From the solutions of this SDP, the optimal H_2 SF is given by $K_{SF} = Y X_2^{-1}$.

6.3.2.2 Numerical Procedures and Algorithms

In [17], Parametrization of Theorem 6.4 is used to derive a coordinate-descent cross-decomposition algorithm allowing to compute an H_2 suboptimal SOF. This algorithm is recalled for sake of clarity.

Algorithm 1

1. (Initialization step—k=1): choose a stabilizing SF gain $K^0_{SF_2} \in \mathscr{K}_{SF}$.
2. (Step k—first part): for fixed $K^k_{SF_2}$, solve the following LMI relaxation of minimization problem of (6.36):

$$\gamma^2_{k,1} = \min_{P_2, Z, F} \text{trace}(B_1^T P_2 B_1)$$
$$\text{subject to} \quad P_2 \succ 0, \quad L_2(P_2, K^k_{SF_2}, Z, F) \prec 0. \quad (6.38)$$

At the optimum, fix $Z_k = Z$ and $F_k = F$.

3. (Step k—second part): for fixed Z_k and F_k, solve the following LMI relaxation of minimization problem of (6.36):

$$\gamma_{k,2}^2 = \min_{P_2, K_{SF_2}} \quad \mathrm{trace}(B_1^T P_2 B_1)$$
$$\text{subject to } P_2 \succ 0, \ L_2(P_2, K_{SF_2}, Z_k, F_k) \prec 0. \tag{6.39}$$

At the optimum, fix $K_{SF_2}^k = K_{SF_2}$.

4. (Termination step), if $\gamma_{k,1} - \gamma_{k,2} < \varepsilon$, then stop, $K_{SOF} = -F_k^{-1} Z_k$, otherwise $k \leftarrow k + 1$ and go to step 2.

Note that this algorithm always generates a non increasing sequence of H_2 suboptimal costs:

$$\cdots \geq \gamma_{k-1,2} \geq \gamma_{k,1} \geq \gamma_{k,2} \geq \gamma_{k+1,1} \geq \cdots$$

Remark 6.2 The matrix K_{SF_2} in (6.36) must belong to the set:

$$\left\{ K_{SF} \in \mathcal{K}_{SF} : \exists \, P_2 \in \mathbb{S}^{+*} \text{such that} \atop \mathrm{He}\{P_2(A + BK_{SF})\} + \mathrm{Sq}\{(C_1 + D_{12}K_{SF})^T\} \prec 0 \right\}. \tag{6.40}$$

This is not restrictive since for $K_{SF} \in \mathcal{K}_{SF}$, it is always possible to find a matrix $P \succ 0$ such that:

$$\mathrm{He}\{P(A + BK_{SF})\} + \mathrm{Sq}\{(C_1 + D_{12}K_{SF})^T\} \prec 0 \tag{6.41}$$

by choosing a matrix $P \succ W_o$ where W_o is the observability gramian defined as the solution of the Lyapunov equation:

$$\mathrm{He}\{W_o(A + BK_{SF})\} + \mathrm{Sq}\{(C_1 + D_{12}K_{SF})^T\} = 0. \tag{6.42}$$

This algorithm generally works well in practice and allows to get suboptimal H_2 SOF in a moderate CPU time when the initializing $K_{SF_2}^0 \in \mathcal{K}_{SF}$ is appropriately chosen, i.e., $\mathcal{L}_{SOF}^{K_{SF_2}^0} \neq \emptyset$. Unfortunately, as may be seen in Tables 6.1 and 6.2 (detailed explanations on the results of these Tables will be given in Sect. 6.6), this is not always the case and one has to resort to more efficient alternatives. Noting that an initial stabilizing SOF may be computed by Theorem 6.1, an alternate Parametrization can be proposed as follows:

$$L_2'(P_2, K_{SOF}, H, F) = N(P_2) + \mathrm{He}\left(\begin{bmatrix} F \\ H \end{bmatrix} \begin{bmatrix} I & -K_{SOF}C \end{bmatrix} \right) \prec 0 \tag{6.43}$$

where $K_{SF} = -F^{-T}H^T \in \mathcal{K}_{SF}$. It leads to an alternate coordinate-descent cross-decomposition algorithm.

Table 6.1 Numerical results for the COMPl$_e$ib library

Ex.	n_x	n_u	n_y	OLS	n_{SOF}^1/n_{SF}
AC1	5	3	3	OLMS	240/1000
AC2	5	3	3	OLMS	334/1000
AC5	4	2	2	OLNS	283/1000
AC9	10	4	5	OLNS	34/1000
AC10	55	2	2	OLNS	0/1000
AC11	5	2	4	OLNS	996/1000
AC12	4	3	4	OLNS	997/1000
AC13	28	3	4	OLNS	39/1000
AC14	40	3	4	OLNS	13/1000
AC18	10	2	2	OLNS	17/1000
HE1	4	2	1	OLNS	93/1000
HE3	8	4	6	OLNS	12/1000
HE4	8	4	6	OLNS	1000/1000
HE5	4	2	2	OLNS	13/1000
HE6	20	4	6	OLNS	321/1000
HE7	20	4	6	OLNS	327/1000
DIS2	3	2	2	OLNS	842/1000
DIS4	6	4	6	OLNS	1000/1000
DIS5	4	2	2	OLNS	725/1000
JE2	21	3	3	OLMS	94/1000
JE3	24	3	6	OLMS	44/1000
REA1	4	2	3	OLNS	999/1000
REA2	4	2	2	OLNS	554/1000
REA3	12	1	3	OLNS	965/1000

Algorithm 2

1. (Initialization step—k = 1): choose a stabilizing SOF gain $K_{SOF} \in \mathcal{K}_{SOF}$.
2. (Step k—first part): for fixed K_{SOF}^k, solve the following LMI relaxation of minimization problem of (6.36):

$$\gamma_{k,1}^2 = \min_{P_2, Z, F} \text{trace}(B_1^T P_2 B_1)$$
$$\text{under } P_2 \succ 0, \qquad\qquad (6.44)$$
$$L_2'(P_2, K_{SOF}^k, H, F) \prec 0$$

At the optimum, fix $H_k = H$ and $F_k = F$.

3. (Step k—second part): for fixed H_k and F_k, solve the following LMI relaxation of minimization problem of (6.36):

Table 6.2 Numerical results for the COMPl$_e$ib library

Ex.	n_x	n_u	n_y	OLS	n_{SOF}^1/n_{SF}
WEC1	10	3	4	OLNS	775/1000
BDT2	82	4	4	OLMS	43/1000
IH	21	11	10	OLMS	63/1000
CSE2	60	2	30	OLNS	10/10
PAS	5	1	3	OLMS	236/1000
TF1	7	2	4	OLMS	81/1000
TF2	7	2	3	OLMS	189/1000
TF3	7	2	3	OLMS	6/1000
NN1	3	1	2	OLNS	629/1000
NN2	2	1	1	OLMS	1000/1000
NN5	7	1	2	OLNS	84/1000
NN6	9	1	4	OLNS	983/1000
NN7	9	1	4	OLNS	620/1000
NN9	5	3	2	OLNS	7/1000
NN12	6	2	2	OLNS	28/1000
NN13	6	2	2	OLNS	77/1000
NN14	6	2	2	OLNS	44/1000
NN15	3	2	2	OLMS	821/1000
NN16	8	4	4	OLMS	61/1000
NN17	3	2	1	OLNS	125/1000
HF2D10	5	2	3	OLNS	991/1000
HF2D11	5	2	3	OLNS	993/1000
HF2D14	5	2	4	OLNS	1000/1000
HF2D15	5	2	4	OLNS	1000/1000
HF2D16	5	2	4	OLNS	998/1000
HF2D17	5	2	4	OLNS	1000/1000
HF2D18	5	2	2	OLNS	755/1000
TMD	6	2	4	OLNS	654/1000
FS	5	1	3	OLNS	977/1000

$$\gamma_{k,2}^2 = \min_{P_2, K_{SOF}} \text{trace}(B_1^T P_2 B_1)$$
$$\text{under} \quad P_2 \succ 0 \qquad\qquad (6.45)$$
$$L_2'(P_2, K_{SOF}, H_k, F_k) \prec 0$$

At the optimum, fix $K_{SOF}^k = K_{SOF}$.

4. (Termination step), if $\gamma_{k,1} - \gamma_{k,2} < \varepsilon$, then stop, $K_{SOF} = K_{SOF}^k$, otherwise $k \leftarrow k + 1$ and go to step 2.

Once again, Algorithm 2 generally works well in practice except when $\mathscr{L}_{\mathrm{SF}}^{K_{\mathrm{SOF}}^0} = \emptyset$. The idea is to use potentialities of randomized algorithms to generate nontrivial sets of stabilizing SF for Algorithm 1 and sets of stabilizing SOF for Algorithm 2 in which it will be easier to find at least one instance K_{SF}^0 (resp. K_{SOF}^0) such that $\mathscr{L}_{\mathrm{SOF}}^{K_{\mathrm{SF}}^0} \neq \emptyset$ (resp. $\mathscr{L}_{\mathrm{SF}}^{K_{\mathrm{SOF}}^0} \neq \emptyset$).

6.3.3 H_∞ Optimal Synthesis via SOF

In this section, we state the results for H_∞ Optimal Synthesis via SOF by following the same lines as the previous Sect. 6.3.2. Consequently, only an abridged version giving the main differences are presented.

6.3.3.1 Parametrization and Algorithms for H_∞ Optimal Control

First, we extend the Parametrization of Theorem 6.1 to give a solution to the problem (6.34) in the optimal H_∞ SOF control case. Let us define:

$$
\begin{aligned}
N_\infty(P_\infty) &= \begin{bmatrix} I & 0 & 0 \\ A & B_1 & B \end{bmatrix}^T \begin{bmatrix} 0 & P_\infty \\ P_\infty & 0 \end{bmatrix} \begin{bmatrix} I & 0 & 0 \\ A & B_1 & B \end{bmatrix} \\
&+ \begin{bmatrix} C_1 & D_{11} & D_{12} \\ 0 & I & 0 \end{bmatrix}^T \begin{bmatrix} I & 0 \\ 0 & -\gamma^2 I \end{bmatrix} \begin{bmatrix} C_1 & D_{11} & D_{12} \\ 0 & I & 0 \end{bmatrix}.
\end{aligned} \tag{6.46}
$$

Then, we can state the next results.

Theorem 6.5 *The H_∞ optimal SOF for (6.32) is given by $K_{\mathrm{SOF}_\infty}^* = -F^{-1*}Z^*$ where the triplet $(P_\infty^*, Z^*, F^*) \in \mathbb{S}^{+*} \times \mathbb{R}^{m \times r} \times \mathbb{R}^{m \times m}$ is the global optimal solution of the non convex optimization problem (6.47):*

$$
\begin{aligned}
&\min_{P_\infty, Z, F, K_{\mathrm{SF}_\infty}} \gamma^2 \\
&\text{subject to} \quad P_\infty \succ 0, \\
&\qquad N_\infty(P_\infty) + \mathrm{He}\left(\begin{bmatrix} K_{\mathrm{SF}_\infty}^T \\ K_{w_\infty}^T \\ -I \end{bmatrix} \begin{bmatrix} ZC & ZD_{21} & F \end{bmatrix} \right) \prec 0.
\end{aligned} \tag{6.47}
$$

A coordinate-descent algorithm very similar to Algorithm 1 may be designed. The only difference is that the initialization of S-variables K_{SF_∞} and K_{w_∞} can be made reasonably by solving the H_∞ optimal full-information problem given by

$$\min_{X_\infty, R, K_w} \gamma^2$$

subject to $X_\infty \succ 0$,

$$\begin{bmatrix} \mathrm{He}\{AX_\infty + BY\} & B_1 + BK_w & (C_1 + D_{12}R)' \\ (B_1 + BK_w)' & -\gamma^2 I & (D_{11} + D_{12}K_w)' \\ (C_1 + D_{12}R) & (D_{11} + D_{12}K_w) & -I \end{bmatrix} \prec 0 \qquad (6.48)$$

where $C = \begin{bmatrix} I & 0 \end{bmatrix}^T$ and $D_{21} = \begin{bmatrix} 0 & I \end{bmatrix}'$ in (6.1) [32]. From the optimal solution of this SDP, we can let $K_{\mathrm{SF}\infty} = YX\infty^{-1}$ and $K_{w\infty} = K_w$. Since the basic principles of the algorithm have been dealt with in the previous subsection, it is not recalled here.

As in the H_2 case, an alternate parametrization may be used when $K_{\mathrm{SOF}} \in \mathcal{K}_{\mathrm{SOF}}$ is given.

$$L_\infty(P_\infty, K_{\mathrm{SOF}}, K_1, K_2, F) =$$

$$N(P_\infty) + \mathrm{He}\left(\begin{bmatrix} K_1 \\ K_2 \\ F \end{bmatrix} \begin{bmatrix} K_{\mathrm{SOF}}C & K_{\mathrm{SOF}}D_{21} & -I \end{bmatrix}\right) \prec 0. \qquad (6.49)$$

It leads to an alternate coordinate-descent cross-decomposition algorithm similar to Algorithm 2.

6.4 SOF Synthesis Under Uncertainties

6.4.1 Problem Statement and Preliminaries

Let us consider the linear time-invariant continuous-time (6.1) where we assume that A and B are uncertain and satisfy

$$M = \begin{bmatrix} A & B \end{bmatrix} \in \Omega, \quad \Omega = \mathrm{conv}\left\{\begin{bmatrix} A^{[1]} & B^{[1]} \end{bmatrix}, \ldots, \begin{bmatrix} A^{[L]} & B^{[L]} \end{bmatrix}\right\} \qquad (6.50)$$

Hence, Ω is a convex polytope of matrices for which each element may be expressed as a convex combination of the N vertices of Ω:

$$M(\theta) = \begin{bmatrix} A(\theta) & B(\theta) \end{bmatrix} = \sum_{j=1}^{N} \theta_j \begin{bmatrix} A^{[j]} & B^{[j]} \end{bmatrix}, \quad \theta \in \mathbb{E}^L. \qquad (6.51)$$

The uncertain vector of parameters θ is supposed to be time invariant, so that the realization of the model (6.50) is not known but is not time varying.

The first problem addressed in this section is to find a robustly stabilizing static output-feedback control law $u(t) = Ky(t)$ for the model (6.50), i.e., to find a single gain matrix $K \in \mathbb{R}^{r \times m}$ such that every member of the polytope:

$$\Omega_{\mathrm{bf}} = \mathrm{conv}\left\{A^{[1]} + B^{[1]}KC, \cdots, A^{[L]} + B^{[L]}KC\right\} \qquad (6.52)$$

maintains eigenvalue location in the left half plane.

It is well-known that the test of the stability of a matrix polytope is equivalent to solving an $\mathcal{N}\mathcal{P}$-hard problem [33]. Even if the particular complexity of the problem of stabilization via static output feedback is not known, one may conjecture that the related problem of robust stabilization via static output feedback of a polytope of matrices is therefore equivalent to an $\mathcal{N}\mathcal{P}$-hard problem. Moreover, due to the inherent difficulties of SOF synthesis problems (even if there is no uncertainties), the robust SOF synthesis problem is much harder than the state-feedback case.

Similarly to the state-feedback case, two different robust stability concepts have been defined to study robust stabilization via SOF:

Definition 6.1

- Ω is said to be quadratically stabilizable via static output feedback if and only if there exist a positive definite matrix $P \in \mathbb{R}^{n \times n}$ and a matrix $K \in \mathbb{R}^{r \times m}$ such that:

$$\begin{bmatrix} KC \\ I \end{bmatrix}^T M_\theta(P) \begin{bmatrix} KC \\ I \end{bmatrix} \prec 0 \quad \forall \theta \in \theta \tag{6.53}$$

where

$$M_\theta(P) := \begin{bmatrix} B(\theta)^T & 0 \\ A(\theta)^T & I \end{bmatrix} \begin{bmatrix} 0 & P \\ P & 0 \end{bmatrix} \begin{bmatrix} B(\theta) & A(\theta) \\ 0 & I \end{bmatrix}.$$

- Ω is said to be robustly stabilizable via static output feedback if and only if, for each $M(\theta) = \begin{bmatrix} A(\theta) & B(\theta) \end{bmatrix}$, there exist a positive definite matrix $P(\theta) \in \mathbb{R}^{n \times n}$ and a matrix $K \in \mathbb{R}^{r \times m}$ such that:

$$\begin{bmatrix} KC \\ I \end{bmatrix}^T M_\theta(P(\theta)) \begin{bmatrix} KC \\ I \end{bmatrix} \prec 0 \quad \forall \theta \in \theta. \tag{6.54}$$

A straightforward necessary and sufficient condition of quadratic stabilizability via SOF is the following:

Theorem 6.6 *The system* (6.1) *with polytopic-type uncertainty* (6.50) *is quadratically stabilizable via SOF if and only if there exist a positive definite matrix* $P \in \mathbb{S}_{++}^n$ *and a matrix* $K \in \mathbb{R}^{r \times m}$ *such that:*

$$\begin{bmatrix} KC \\ I \end{bmatrix}^T M_j(P) \begin{bmatrix} KC \\ I \end{bmatrix} \prec 0 \quad (\forall j \in \mathcal{I}_L) \tag{6.55}$$

where

$$M_j(P) := \begin{bmatrix} B^{[j]T} & 0 \\ A^{[j]T} & I \end{bmatrix} \begin{bmatrix} 0 & P \\ P & 0 \end{bmatrix} \begin{bmatrix} B^{[j]} & A^{[j]} \\ 0 & I \end{bmatrix}.$$

Unfortunately, this condition is useless as it stands. In [5], a necessary and sufficient condition of quadratic stabilizability via SOF is provided in terms of the intersection of a convex set and a set defined by a nonlinear real-valued function. Even if some convex programming tools may be used for numerical tractability of such a condition, the convergence of the algorithm is not ensured a priori. In [34], an LMI-based procedure is proposed for which a particular initialization is required. Moreover, extension of this approach to deal with robust guaranteed performance seems hardly possible. The next section will provide some alternatives to previous works.

Remark 6.3 In (6.50), only the matrices A and B are considered to be uncertain, while the output matrix C is supposed to be exactly known. All the results presented in this section are valid with this assumption and may be easily transposed to the case where A and C are in a polytope and B is known. Unfortunately, the more general case for which A, B and C are uncertain is out of the scope. At our knowledge, there does not exist any tractable test, (operating on the vertices of the polytope) for such a case in the literature.

6.4.2 Robust and Quadratic Stabilization via SOF

Some results based on the quadratic stability framework [17] and complete proofs are recalled. The main purpose of this subsection is then to propose a new approach encompassing the quadratic one in all cases.

6.4.2.1 Quadratic Stabilization via SOF

Theorem 6.7 [17] *The system* (6.1) *with polytopic-type uncertainty* (6.50) *is quadratically stabilizable via static output feedback if and only if there exist a positive definite matrix* $P \in \mathbb{R}^{n \times n}$, *matrices* $K_{\mathrm{SF}} \in \mathbb{R}^{m \times n}$, $Z \in \mathbb{R}^{m \times r}$, $F \in \mathbb{R}^{m \times m}$ *such that*

$$M_j(P) + \mathrm{He}\left\{ \begin{bmatrix} I \\ -K_{\mathrm{SF}}^T \end{bmatrix} \begin{bmatrix} F & ZC \end{bmatrix} \right\} \prec 0 \quad (\forall j \in \mathscr{I}_L). \tag{6.56}$$

Moreover, a quadratically stabilizing SOF is given by $K = -F^{-1}Z$.

Proof Sufficiency: It is straightforward to see that $F + F^T \prec 0$ and therefore the condition (6.56) can be rewritten as

$$M_j(P) + \mathrm{He}\left\{ \begin{bmatrix} I \\ -K_{\mathrm{SF}}^T \end{bmatrix} F \begin{bmatrix} I & -KC \end{bmatrix} \right\} \prec 0 \quad (\forall i \in \mathscr{I}_L).$$

with the notation $K = -F^{-1}Z$. Hence, multiplying each inequality by $\begin{bmatrix} C^T K^T & I \end{bmatrix}$ on the left and its transpose on the right, there exist a positive definite matrix $P \in \mathbb{R}^{n \times n}$ and a matrix $K \in \mathbb{R}^{r \times m}$ satisfying:

$$\begin{bmatrix} KC \\ I \end{bmatrix}^T M_j(P) \begin{bmatrix} KC \\ I \end{bmatrix} \prec 0 \quad (\forall i \in \mathscr{I}_L).$$

This implies quadratic stability by Theorem 6.6.

Necessity: If the system (6.1) with polytopic-type uncertainty (6.50) is quadratically stabilizable via SOF, then, from Finsler's lemma (Lemma 1.1), there exist a positive definite matrix $P \in \mathbb{R}^{n \times n}$, a matrix $K \in \mathbb{R}^{r \times m}$ and a scalar $\tau > 0$ such that

$$M_j(P) - 2\tau \begin{bmatrix} I \\ -C^T K^T \end{bmatrix} \begin{bmatrix} I \\ -C^T K^T \end{bmatrix}^T \prec 0 \quad (\forall j \in \mathscr{I}_L).$$

This implies that (6.56)) is retrieved with $K_{SF} = KC$, $F = -\tau I$ and $Z = \tau K$. $\quad\square$

Lemma 6.1 [17] *If the condition (6.56) is verified for a positive definite matrix P and matrices K_{SF}, Z, and F, then K_{SF} is necessarily a quadratically stabilizing state-feedback matrix for the system (6.1) with (6.50).*

Proof Multiplying the left of each inequality in condition (6.56) by $\begin{bmatrix} I & K'_{SF} \end{bmatrix}$ and the right by its transpose, we get

$$\mathrm{He}\{P(A^{[j]} + B^{[j]} K_{SF})\} \prec 0 \quad (\forall j \in \mathscr{I}_L).$$

These last inequalities state that K_{SF} is nothing but a quadratically stabilizing state-feedback gain. $\quad\square$

6.4.2.2 Improved Conditions via PDLF

The main advantage of the new parametrization of SOF given in the previous subsection is that it allows to decouple the computation of the SOF gain and the computation of Lyapunov matrix allowing to derive a new sufficient condition of robust SOF stabilizability based on parameter-dependent Lyapunov functions.

Theorem 6.8 *The system (6.1) with polytopic-type uncertainty (6.50) is stabilizable via SOF if there exist L positive definite matrices $P^{[j]} \in \mathbb{R}^{n \times n}$, matrices $E_1 \in \mathbb{R}^{n \times n}$, $E_2 \in \mathbb{R}^{n \times n}$, $K_{SF} \in \mathbb{R}^{m \times n}$, $Z \in \mathbb{R}^{m \times r}$, $F \in \mathbb{R}^{m \times m}$ such that*

$$\begin{bmatrix} 0 & 0 & P^{[j]} \\ 0 & 0 & 0 \\ P^{[j]} & 0 & 0 \end{bmatrix} + \mathrm{He} \left\{ \begin{bmatrix} E_1 \\ 0 \\ E_2 \end{bmatrix} \begin{bmatrix} I & -B^{[j]} & -A^{[j]} \end{bmatrix} \right\}$$
$$+ \mathrm{He} \left\{ \begin{bmatrix} 0 \\ I \\ -K_{SF}^T \end{bmatrix} \begin{bmatrix} 0 & F & ZC \end{bmatrix} \right\} \prec 0. \tag{6.57}$$

Moreover, $K = -F^{-1}Z$ is a robustly stabilizing SOF.

Proof Suppose there exist L positive definite matrices $P^{[j]}$ and matrices E_1, E_2, K_{SF}, Z, F such that (6.57) holds. Then multiplying the right of each inequality by
$$\begin{bmatrix} I & 0 \\ 0 & KC \\ 0 & I \end{bmatrix}$$ and the left by its transpose with $K = -F^{-1}Z$, we have

$$\begin{bmatrix} 0 & P^{[j]} \\ P^{[j]} & 0 \end{bmatrix} + \mathrm{He}\left\{ \begin{bmatrix} E_1 \\ E_2 \end{bmatrix} \begin{bmatrix} I & -(A^{[j]} + B^{[j]}KC) \end{bmatrix} \right\} \prec 0 \quad (\forall j \in \mathscr{I}_L).$$

Summing over the L vertices of the polytope $\mathbb{E}^L$ with weights θ_j, we have

$$\begin{bmatrix} 0 & P(\theta) \\ P(\theta) & 0 \end{bmatrix} + \mathrm{He}\left\{ \begin{bmatrix} E_1 \\ E_2 \end{bmatrix} \begin{bmatrix} I & -(A(\theta) + B(\theta)KC) \end{bmatrix} \right\} \prec 0 \quad (\forall \theta \in \mathbb{E}^L).$$

where $P(\theta) := \sum_{j=1}^{L} \theta_j P^{[j]} \succ 0 \ (\forall \theta \in \mathbb{E}^L)$. Multiplying the right of the inequality by $\begin{bmatrix} (A(\theta) + B(\theta)KC)^T & I \end{bmatrix}^T$ and the left by its transpose, we get

$$\mathrm{He}\left\{ P(\theta)(A(\theta) + B(\theta)KC) \right\} \prec 0 \quad (\forall \theta \in \mathbb{E}^L).$$

This is nothing but the condition for robust stabilizability of the polytope Ω via the static output feedback K.   □

Note that two different sets of S-variables are introduced here. K_{SF}, Z and F are used to define the new parametrization of the SOF gains, while E_1 and E_2 allows to involve parameter-dependent Lyapunov functions. An interpretation of the S-variable K_{SF} similar to the one in quadratic setup is now proposed.

Lemma 6.2 *If the condition (6.57) is verified for L positive definite matrices $P^{[j]}$ and matrices K_{SF}, Z and F, then K_{SF} is necessarily a robustly stabilizing state-feedback matrix for the system (6.1) with polytopic-type uncertainty (6.50).*

The key point of the new sufficient condition is that it ensures to always get better results than the quadratic approach. This fact is formalized in the following theorem.

Theorem 6.9 *If the system (6.1) with polytopic-type uncertainty (6.50) is quadratically stabilizable via SOF, then there exist L positive definite matrices $X^{[j]} \in \mathbb{R}^{n \times n}$, matrices $E_1 \in \mathbb{R}^{n \times n}$, $E_2 \in \mathbb{R}^{n \times n}$, $K_{SF} \in \mathbb{R}^{m \times n}$, $Z \in \mathbb{R}^{m \times r}$, $F \in \mathbb{R}^{m \times m}$ such that 6.57 holds.*

Proof Suppose the system (6.1) with polytopic-type uncertainty (6.50) is quadratically stabilizable via SOF. Then, there exists a positive definite matrix P and matrices $K_{SF} \in \mathbb{R}^{m \times n}$, $Z \in \mathbb{R}^{m \times r}$, $F \in \mathbb{R}^{m \times m}$ such that (6.56) holds. Then, there exits $\varepsilon > 0$ such that

$$M_j(P) + \mathrm{He}\left\{ \begin{bmatrix} I \\ -K_{SF}^T \end{bmatrix} \begin{bmatrix} F & ZC \end{bmatrix} \right\} + \frac{\varepsilon}{2} \begin{bmatrix} B^{[j]T} \\ A^{[j]T} \end{bmatrix} \begin{bmatrix} B^{[j]} & A^{[j]} \end{bmatrix} \prec 0 \quad (\forall j \in \mathscr{I}_L).$$

By Schur complement, this can be rewritten as

$$
\begin{bmatrix} 0 & 0 & P \\ 0 & 0 & 0 \\ P & 0 & 0 \end{bmatrix} + \mathrm{He} \left\{ \begin{bmatrix} -\varepsilon I \\ 0 \\ -P \end{bmatrix} \begin{bmatrix} I & -B^{[j]} & -A^{[j]} \end{bmatrix} \right\}
$$
$$
+ \mathrm{He} \left\{ \begin{bmatrix} 0 \\ I \\ -K_{SF}^T \end{bmatrix} \begin{bmatrix} 0 & F & ZC \end{bmatrix} \right\} \prec 0.
$$

This clearly shows that (6.57) holds with $P^{[j]} = P$ $(\forall j \in \mathscr{I}_L)$, $E_1 = -\varepsilon I$ and $E_2 = -P$. □

One of the feature of the new condition is that it may be easily extended to deal with the problem of worst-case guaranteed performance. In the next section, a procedure for worst-case guaranteed H_2 performance synthesis is derived.

6.4.3 Robust H_2 Suboptimal Control via SOF

Let us consider the system (6.32) where $D_{11} = 0$, $D_{21} = 0$. We assume that the coefficient matrices of (6.32), except for the matrix C, are affected by polytopic uncertainty as in

$$
\begin{bmatrix} A & B_1 & B \\ C_1 & D_{12} & 0 \end{bmatrix} = \sum_{j=1}^L \theta_j \begin{bmatrix} A^{[j]} & B_1^{[j]} & B^{[j]} \\ C_1^{[j]} & D_{12}^{[j]} & 0 \end{bmatrix}, \quad \theta \in \mathbb{E}^L. \tag{6.58}
$$

For a given SOF gain K, the closed-loop transfer function matrix is given by

$$
T_{zw}(s, K, \theta) = \left[\begin{array}{c|c} A(\theta) + B(\theta)KC & B_1(\theta) \\ \hline C_1(\theta) + D_{12}(\theta)KC & 0 \end{array} \right]. \tag{6.59}
$$

By letting $\mathscr{K}_{\mathrm{SOF,ROB}}$ the set of robustly stabilizing SOF gains, our goal here is to compute the optimal robust H_2 SOF gain $K^\star$ and associated the optimal robust performance $\gamma^\star$ characterized by

$$
\begin{aligned}
K^\star &:= \arg \min_{K \in \mathscr{K}_{\mathrm{SOF,ROB}}} \max_{\theta \in \theta} \|T_{zw}(s, K, \theta)\|_2, \\
\gamma^\star &:= \min_{K \in \mathscr{K}_{\mathrm{SOF,ROB}}} \max_{\theta \in \theta} \|T_{zw}(s, K, \theta)\|_2.
\end{aligned} \tag{6.60}
$$

This problem is a nonconvex and semi-infinite optimization problem and hence very hard to solve. Except by using complex approaches involving branching operations, the best one may expect is to get tractable conditions with associated numerical procedures giving a suboptimal solution to this problem. With this in mind, we study suboptimal robust H_2 SOF controller synthesis problem given as follows.

Problem 6.3 (*Suboptimal Robust H_2 SOF Controller Synthesis*) Consider the system (6.32) with polytopic-type uncertainty (6.58). For given $\gamma > 0$, find a robustly stabilizing SOF K such that

$$
\begin{aligned}
&\|T_{z_2 w_2}(s, K, \theta)\|_{2,w.c.}^2 \leq \gamma, \\
&\|T_{z_2 w_2}(s, K, \theta)\|_{2,w.c.} := \max_{\theta \in \theta} \|T_{zw}(s, K, \theta)\|_2.
\end{aligned}
\tag{6.61}
$$

Different methods to tackle this problem are now presented.

6.4.4 Quadratic Robust Suboptimal H_2 SOF Synthesis

First, the quadratic approach proposed in [17] using SV-LMIs is recalled.

Theorem 6.10 [17] *For given $\gamma_q > 0$, suppose there exists a positive definite matrix $P \in \mathbb{R}^{n \times n}$, a positive definite matrix W, matrices $K_{SF_2} \in \mathbb{R}^{m \times n}$, $Z \in \mathbb{R}^{m \times r}$, $F \in \mathbb{R}^{m \times m}$ such that*

$$
\begin{aligned}
&\text{trace}\, W < \gamma_q, \\
&-W + B_1^{[j]T} P B_1^{[j]} \prec 0, \\
&N_j(P) + \text{He}\left(\begin{bmatrix} I \\ -K_{SF_2}^T \end{bmatrix} \begin{bmatrix} F & ZC \end{bmatrix} \right) \prec 0 \quad (\forall j \in \mathscr{I}_L)
\end{aligned}
\tag{6.62}
$$

where

$$
N_j(P) = M_j(P) + \begin{bmatrix} D_{12}^{[j]} & C_1^{[j]} \end{bmatrix}^T \begin{bmatrix} D_{12}^{[j]} & C_1^{[j]} \end{bmatrix}.
$$

Then, $K = -F^{-1}Z$ is a robustly stabilizing SOF such that

$$
\|T_{zw}(s, K)\|_{2,w.c.}^2 \leq \gamma_q.
\tag{6.63}
$$

This theorem readily follows from Theorem 6.4 with simple convexity arguments related to the treatment of the polytopic uncertainty.

Related to the result in Theorem 6.10, an optimization problem may be defined:

$$
\gamma_q^* = \min_{\gamma_q, K_{SF}, X, Z, F} \gamma_q \text{ subject to (6.62).}
\tag{6.64}
$$

The insight into the genuine nature of the variable K_{SF} given in Lemma 6.1 is used in [17] to initialize a cross-decomposition algorithm in order to solve (6.64). This algorithm is not reminded here since a similar one will be presented in the next subsection.

6.4.4.1 Robust Suboptimal H_2 SOF Synthesis via PDLF

The result of Theorem 6.8 is now extended to deal with robust suboptimal H_2 SOF synthesis.

Theorem 6.11 *For given $\gamma_l > 0$, suppose there exist L positive definite matrices $P^{[j]} \in \mathbb{R}^{n \times n}$, a positive definite matrix W, matrices $E_1 \in \mathbb{R}^{n \times n}$, $E_2 \in \mathbb{R}^{n \times n}$, $G_1 \in \mathbb{R}^{m \times n}$, $G_2 \in \mathbb{R}^{n \times n}$, $K_{SF} \in \mathbb{R}^{m \times n}$, $Z \in \mathbb{R}^{m \times r}$, $F \in \mathbb{R}^{m \times m}$ such that*

$$\text{trace}(W) < \gamma_l,$$

$$\begin{bmatrix} -W & 0 \\ 0 & P^{[j]} \end{bmatrix} + \text{He}\left\{ \begin{bmatrix} G_1 \\ G_2 \end{bmatrix} \begin{bmatrix} B_1^{[j]} & -I \end{bmatrix} \right\} \prec 0,$$

$$\begin{bmatrix} 0 & 0 & P^{[j]} \\ 0 & D^{[j]T} D^{[j]} & D^{[j]T} C^{[j]} \\ P^{[j]} & C^{[j]T} D^{[j]} & C^{[j]T} C^{[j]} \end{bmatrix} + \text{He}\left\{ \begin{bmatrix} E_1 \\ 0 \\ E_2 \end{bmatrix} \begin{bmatrix} I & -B^{[j]} & -A^{[j]} \end{bmatrix} \right\} \qquad (6.65)$$

$$+\text{He}\left\{ \begin{bmatrix} 0 \\ I \\ -K_{SF}^T \end{bmatrix} \begin{bmatrix} 0 & F & ZC \end{bmatrix} \right\} \prec 0 \quad (\forall j \in \mathscr{I}_L).$$

Then, $K = -F^{-1}Z$ is a robustly stabilizing SOF such that

$$\|T_{zw}(s, K)\|_{2,w.c.}^2 \le \gamma_l. \qquad (6.66)$$

In the same way as for the quadratic case, a nonlinear optimization problem may be defined:

$$\gamma_l^* = \min_{\substack{P^{[j]}, W, K_{SF}, Z, F \\ E_1, E_2, G_1, G_2}} \gamma_l \quad \text{subject to (6.65)} \qquad (6.67)$$

Following Theorem 6.4, it is easy to prove the following lemma.

Lemma 6.3

$$\gamma^\star \le \gamma_l^* \le \gamma_q^* \qquad (6.68)$$

Remark 6.4 The method proposed in [34] may be extended in a straightforward way to the minimization of a guaranteed upper-bound for the worst-case H_2 norm. We recall that the method consists in a three-step procedure each performing an LMI optimization. The first one seeks a quadratically stabilizing state feedback as initialization which is used in the second step to get a Lyapunov function. Finally, this Lyapunov function shared by a state and an output feedback leads to the computation of the static output feedback. It is therefore clear that there are few degrees of freedom to optimize the guaranteed bound when the procedure does not simply fails.

6.4.4.2 A Coordinate Descent-Type Algorithm

The optimization problem (6.65) involve bilinear matrix inequalities in their constraints and hence the reader may think that no advance has been done (from the view point of efficient numerical computation) so far. In fact, the main advantage of the new formulation using SV-LMIs is that additional degrees of freedom allow now to define an iterative procedure based on LMI optimization converging to a local optimal solution. The set of S-variables may then be split in two. At each iteration, an LMI optimization problem is solved with respect to a set of variables while the other is frozen.

Algorithm 3

1. **Initialization:** $k = 1$
   Solve the following LMI problem for some $\varepsilon > 0$ with respect to the L positive definite matrices $X^{[j]}$ and matrices G and Y:

$$\begin{bmatrix} 0 & X^{[j]} \\ X^{[j]} & 0 \end{bmatrix} + \mathrm{He}\left\{\begin{bmatrix} A^{[j]}G + B^{[j]}Y \\ -I \end{bmatrix}\begin{bmatrix} I & -\varepsilon I \end{bmatrix}\right\} \prec 0 \ (\forall j \in \mathscr{I}_L). \quad (6.69)$$

   If it has a solution then $K_{SF,1} = YG^{-1}$ is a robust state feedback for $(A(\theta), B(\theta))$
   $(\theta \in \mathbb{E}^L)$.
2. **Step** k:

   2.1 Solve the SDP:

   $$\min \gamma_l \quad \text{subjectto (6.65) with fixed } K_{SF} = K_{SF,k}.$$

   Let the resulting optimal value be $\gamma_l^{k,1} = \gamma_l$, $F_k = F$ and $Z_k = Z$.
   2.2 Solve the SDP:

   $$\min \gamma_l \quad \text{subject to (6.65) with fixed } F = F_k \text{ and } Z = Z_k.$$

   Let the resulting optimal value be $\gamma_l^{k,2} = \gamma_l$ and $K_{SF} = K_{SF,k}$.
3. **Termination step:** If $\gamma_l^{k-1,2} - \gamma_l^{k,2} < \eta$ then stop where η is a specified small value beforehand. Let

$$K_{sub.} = -F_k^{-1}Z_k. \quad (6.70)$$

   otherwise $k \leftarrow k + 1$ and go step 2.

The initialization step is particularly interesting since we need to solve a problem of robust stabilization via state feedback of the polytope Ω. In [35], a sufficient LMI condition based on polytopic parameter-dependent Lyapunov functions is given for this problem. However, in some cases, a quadratically stabilizing state feedback may be employed to initialize the problem. Of course, in any case, no guarantee is given that the chosen initial K_{SF} will lead to a feasible second step even if the polytope Ω_2

is robustly stabilizable via SOF. Nevertheless, this procedure provides a sequence of feasible solutions and a non increasing sequence of criteria.

Theorem 6.12 *The sequence $\{\gamma_l^{k,i}\}_{k,i}$ is a monotonic non increasing sequence converging to a local minimum in rough sense.*

Another interesting point in the previous algorithm is that at each semi iteration 2.1 or 2.2, the complete set of N parameter-dependent Lyapunov functions is included in the set of decision variables and one can hope to shape this set to get better performance bounds.

6.5 Numerical Procedures

If we use those results Theorem 6.1 and its extensions within convex optimization framework, we need to fix the initial stabilizing state-feedback gain K_{SF} beforehand. Since the success/failure of SV-LMI-based conditions for SOF synthesis fully depends on the initial SF gain K_{SF}, it is natural to evaluate the effectiveness of SV-LMIs by applying them under large number of initial SF gains and examine the rate of success/failure. For generating initial SF gains, we use Hit-and-Run (HR) method as explicated in the following.

6.5.1 Hit-and-Run Synthesis Techniques for Stabilization

For given $A \in \mathbb{R}^{n \times n}$ and $B \in \mathbb{R}^{n \times m}$, we apply Hit-and-Run (HR) for generating a set of stabilizing state-feedback gains. In every step of the HR algorithm we generate matrix $D = Y/\|Y\|$ with $Y = \mathtt{randn(m,n)}$ (in MATLAB code), which is uniformly distributed on the unit sphere in the space of matrices equipped with Frobenius norm. Matrix D is a random direction in the space of $m \times n$-matrices. For given $K^0 \in \mathcal{K}_{SF}$, we call *boundary oracle* an algorithm which provides $J = \{t \in R : K^0 + tD \in \mathcal{K}_{SF}\}$. In the simplest case when $\mathcal{K}_{SF}$ is convex, the set J is the interval $(-\underline{t}, \overline{t})$ where $\overline{t} = \sup\{t : K^0 + tD \in \mathcal{K}_{SF}\}, \underline{t} = \sup\{t : K^0 - tD \in \mathcal{K}_{SF}\}$. In more general situations boundary oracle provides all intersections over the straight line $-\infty < t < +\infty$ with $K^0 + tD \in \mathcal{K}_{SF}$. The set J consists of finite number of intervals, the algorithm for calculating their end points is presented in [36]. However sometimes "brute force" approach is more simple. Introduce $f(t) = \max \Re \operatorname{eig}(F + tG)$ where $F = A + BK^0$ and $G = BD$, the end points of the intervals are solutions of the equation $f(t) = 0$ and can be found by use of standard 1D equation solvers (such as command fsolve" in Matlab). To summarize, HR method works as follows.

1. Find a starting point $K^0 \in \mathcal{K}_{SF}$; $i = 0$.
2. At the point $K^i \in \mathcal{K}_{SF}$ generate a random direction $D^i \in \mathbb{R}^{m \times r}$ uniformly distributed on the unit sphere.

3. Apply boundary oracle procedure, i.e., define the set

$$L_i = \{t \in \mathbb{R} : K^i + t D^i \in \mathcal{K}_{SF}\}.$$

4. Generate a point t_i uniformly distributed in L_i (we recall that L_i is, in general, a finite set of intervals), and compute a new point

$$K^{i+1} = K^i + t_i D^i.$$

5. Go to step 2 and increase i.

6.5.2 Two LMI/Randomized Algorithms for SOF Stabilization

Two algorithms using complementary advantages of degrees of freedom offered by parametrization (6.5) and Hit-and-Run numerical efficiency are built in order to generate non trivial sets of stabilizing SOF. The first algorithm uses Hit-and-Run only for generating a large subset $\mathcal{K}_{SF}^{n_{SF}}$ of $\mathcal{K}_{SF}$ that will serve for checking if $\mathcal{L}_{SOF}^{K_{SF}}$ is empty.

Algorithm 4

1. Compute a stabilizing SF $K_{SF}^0 = RX^{-1} \in \mathcal{K}_{SF}$ via the solution of the LMI problem (6.11):
2. From K_{SF}^0, generate a set $\mathcal{K}_{SF}^{n_{SF}} \subset \mathcal{K}_{SF}$ which is the collection of n_{SF} samples of stabilizing SF matrices via H.R.;
3. Compute a set $\mathcal{K}_{SOF}^{n_{SOF}^1} \subset \mathcal{K}_{SOF}$ which is the collection of n_{SOF}^1 samples of stabilizing SOF matrices: $\forall\, K_{SF}^i \in \mathcal{K}_{SF}^{n_{SF}}$, if the LMI set $\mathcal{L}_{SOF}^{K_{SF}^i} \neq \emptyset$, add the solution $K_i = -F_i^{-1} Z_i$ to the collection set $\mathcal{K}_{SOF}^{n_{SOF}^1}$.

Remark 6.5 Using LMI (6.11) is not the only way to initialize Algorithm 4. K_{SF}^0 may also be computed as the LMI solution of the H_2 or H_∞ state-feedback problems (see [37] for more details).

Remark 6.6 Step 3 is performed by testing the realizability of the LMI set $\mathcal{L}_{SOF}^{K_{SF}}$ for each instance $K_{SF} \in \mathcal{K}_{SF}^{n_{SF}}$. To avoid numerical problems, this step must be done by solving the following semidefinite programming problem:

$$\begin{aligned}
\min_{P,Z,F} \ & \text{trace}(P) \quad \text{subject to} \\
& \text{trace}(P) > \alpha, \\
& P \succ 0, \\
& M(P) + \text{He} \left\{ \begin{bmatrix} I \\ -K_{SF}^T \end{bmatrix} \begin{bmatrix} F & ZC \end{bmatrix} \right\} \prec 0
\end{aligned} \tag{6.71}$$

for some $\alpha > 0$.

The interest of running this algorithm is as follows. Namely, it will be a good way to evaluate the conservatism of the convex approximation of $\mathcal{K}_{\text{SOF}}$ induced by parametrization (6.5). Indeed, the percentage $n^1_{\text{SOF}}/n_{\text{SF}}$ may be defined as a quantitative measure of both the conservatism of the convex approximation as well as the difficulty to stabilize the plant via SOF.

Remark 6.7 If K is a stabilizing SOF known *a priori*, then the set $\mathcal{L}^{KC}_{\text{SOF}} \neq \emptyset$ since the choice $Z = -FK$ will lead to the existence of $P \succ 0$ and $F \in \mathbb{R}^{m \times m}$ such that

$$
\begin{aligned}
M(P) &+ \text{He}\left\{\begin{bmatrix} I \\ -C^T K^T \end{bmatrix}\begin{bmatrix} F & -FKC \end{bmatrix}\right\} \\
&= M(P) + \text{He}\left\{\begin{bmatrix} I \\ -C^T K^T \end{bmatrix} F \begin{bmatrix} I & -KC \end{bmatrix}\right\} \prec 0.
\end{aligned}
\tag{6.72}
$$

Applying elimination lemma, this last condition is equivalent to the existence of a matrix $P \in \mathbb{S}^{+*}$ such that

$$
\text{He}\{P(A + BKC)\} \prec 0
\tag{6.73}
$$

which is obviously true since $K \in \mathcal{K}_{\text{SOF}}$

A much more numerically efficient mixed LMI/randomized algorithm generating sets of stabilizing SOF may be deduced from the previous one. It mainly avoids to run the computational burden of checking if LMI set $\mathcal{L}^{K^i_{\text{SF}}}_{\text{SOF}}$ is empty for every instance $\forall\, K^i_{\text{SF}} \in \mathcal{K}^{n_{\text{SF}}}_{\text{SF}}$ but starts a hit-and-run generation of $\mathcal{K}^{n^2_{\text{SOF}}}_{\text{SOF}}$ as soon as an initial $K_{\text{SOF}} \in \mathcal{K}_{\text{SOF}}$ is found by LMI step. It proves to be much more efficient in practice to generate sets of stabilizing SOF in almost every studied cases of the COMPl$_e$ib library.

Algorithm 5

1. Compute a stabilizing state-feedback $K^0_{\text{SF}} = RX^{-1} \in \mathcal{K}_{\text{SF}}$ via the solution of the LMI problem (6.11);
2. From K^0_{SF}, generate a set $\mathcal{K}^{n_{\text{SF}}}_{\text{SF}} \subset \mathcal{K}_{\text{SF}}$ which is the collection of n_{SF} samples of stabilizing SF matrices via H.R.;
3. Find an initial stabilizing SOF $K^0_{\text{SOF}} = -F_0^{-1} Z_0$: Check every $K^i_{\text{SF}} \in \mathcal{K}^{n_{\text{SF}}}_{\text{SF}}$ until $\mathcal{L}^{K^i_{\text{SF}}}_{\text{SOF}} \neq \emptyset$;
4. From K^0_{SOF}, compute a set $\mathcal{K}^{n^2_{\text{SOF}}}_{\text{SOF}} \subset \mathcal{K}_{\text{SOF}}$ which is the collection of n^2_{SOF} samples of stabilizing SOF matrices via H.R.

6.6 Numerical Examples

6.6.1 Results from the COMPl$_e$ib Library

The COMPl$_e$ib library is composed of different LTI models (6.1) ranging from purely academic problems to more realistic industrial examples. The underlying systems that are already open-loop asymptotically stable have not been considered here for obvious reasons. With this restriction, 53 different models have been tested, which are mainly classified in seven classes as follows.

- Aerospace models: Aircraft models (AC), helicopter models (HE), jet engine models (JE),
- Reactor models (REA),
- Decentralized interconnected systems (DIS),
- Academic tests problems (NN),
- Various applications: Wind energy conversion model (WEC), binary distillation towers (BDT) and terrain following models (TF), string of high-speed vehicles (IH), strings (CSE), piezoelectric bimorph actuator (PAS),
- Second order models: A tuned mass damper (TMD), a flexible satellite (FS),
- 2D heat flow models (HF2D).

For precise details concerning each single example and benchmark, the interested reader may read [38, 39].

6.6.1.1 Stabilization Results

The main results concerning the construction of stabilizing SOF by Algorithm 4 are presented in the Tables 6.1 and 6.2. The following notations are used in Tables 6.1 and 6.2.

- OLS stands for Open-Loop Stability,
- OLMS stands for Open-Loop Marginally Stable (a unique eigenvalue at 0 or multiple eigenvalue with 0 real part but scalar associated Jordan blocks),
- OLNS stands for Open-Loop Non Stable `max(real(eig(A)))>0`,
- $n_{\mathrm{SOF}}^1/n_{\mathrm{SF}}$ stands for the portion of stabilizing SOF found by Algorithm 4.

The results in Tables 6.1 and 6.2 can be summarized as follows.

- First note that it has been possible to stabilize every model except `AC10`. In this case, the HIFOO package that may be considered as one of the most effective tool for SOF stabilization and for optimal H_∞ SOF control is not able either to stabilize it.
- A quick look at the reference [40] clearly shows that Algorithm 4 gives far better results than this reference. The algorithm in [40] was unable to stabilize plants `AC10`, `NN1`, `NN5`, `NN7` and `NN12`.

- Hard to stabilize examples may be identified as the ones for which less than 5 % of the initializing state-feedback succeed in finding a stabilizing SOF. AC9, AC13, AC14, AC18, HE3, HE5, JE3, BDT2, TF3, NN9, NN12, NN14 are such plants for which n_{SOF}^1 obtained with Algorithm 4 is rather low. This is mainly due to the bad conditioning of numerical operations (matrix inversion) and LMI optimization rather than a failure of the method in itself. This is confirmed by the reference [39] where similar failures of SDP solvers were already noticed for some of the previous examples (AC14, AC18, JE3) for the problems of H_2 or H_∞ state feedback optimal control which are known to have a convex formulation. Moreover, the initialization $K_{SF} = KC$ where $K \in \mathcal{K}_{SOF}$ is known a priori, does not perform well for all these examples, demonstrating that the plant matrices are poorly conditioned.
- It is not a surprise to note that $n_{SOF}^1 \leq n_{SOF}^2$ since if the initialization step succeeds $n_{SOF}^2 = 1,000$ while $\max(n_{SOF}^1) = 1,000$. More surprising is the easiness to get n_{SOF}^2 stabilizing SOF for almost all examples and considering also that the complete run is really fast in general. Size seems to be a limiting factor for LMI step but not for Hit-and-Run step.

Figure 6.1 shows the population stabilizing gains with the boundary, for example, AC7. It is interesting to note that the exact shape of the set of stabilizing SOF is easily obtained with the mixed LMI/randomized algorithm, Algorithm 4 (see [41] for comparison). Similar results have been obtained for all two-parameter examples of the database COMPL$_e$ib, labeled AC4, NN1, NN5 (see below), NN17 and HE1.

Figures 6.2 and 6.3 show a comparison of the populations respectively obtained via Algorithms 4 and 5 for benchmark NN5.

Fig. 6.1 Set of stabilizing SOF generated with LMI/Randomized Algorithm 5 for AC7 example

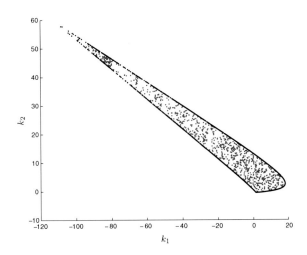

Fig. 6.2 Set of
stabilizing SOF
generated with
LMI/Randomized
algorithm, Algorithm 4, for
NN5 example

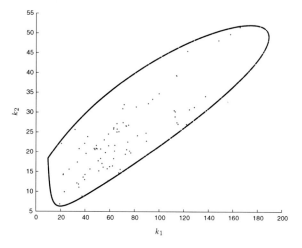

Fig. 6.3 Set of
stabilizing SOF
generated with
LMI/Randomized
algorithm, Algorithm 5, for
NN5 example

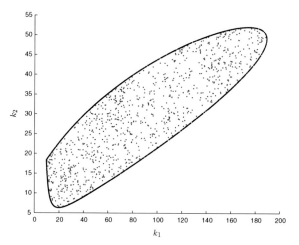

6.6.2 Examples with Uncertainties

6.6.2.1 Example 1

Algorithm 3 is now applied to the robust H_2 SOF synthesis problem of an uncertain
continuous-time system defined by the following system matrices:

$$A = \begin{bmatrix} 0 & \alpha - 1 \\ \beta & 0 \end{bmatrix}, \; B = \begin{bmatrix} \alpha \\ 1 - \beta \end{bmatrix}, \; B_1 = \begin{bmatrix} 0 \\ 1 \end{bmatrix},$$
$$C = \begin{bmatrix} 1 & 0 \end{bmatrix}, \qquad C_1 = I_2, \qquad D_{12} = \begin{bmatrix} 0 \\ 1 \end{bmatrix}.$$

The uncertain parameters are defined as $|\alpha - 0.5| \leq \gamma$, $|\beta - 0.5| \leq \gamma$. This defines a polytope of matrices (A, B) with four vertices. This example is borrowed from [42] where a quadratic state feedback stabilizing gain is computed and consequently modified for static output-feedback purposes. Note that for $\gamma = 0.5$, two of the vertices of the polytope are uncontrollable pairs. For a given value of $\gamma \in (0, 0.5)$, a necessary and sufficient condition of robust stabilizability via static output feedback is obtained analytically by

$$K < 0 \text{ and } K > \frac{\gamma - 0.5}{\gamma + 0.5}. \tag{6.74}$$

In [42], it is shown that the set of quadratic state feedback stabilizing gains is not empty if and only if $\gamma \in (0, 0.36)$. For $\gamma = 0.4$, the pair (A, B) is therefore not quadratically stabilizable via static output feedback. Using the Algorithm3 with $\varepsilon = 10^{-14}$ for the initialization step, we get a robust suboptimal H_2 SOF gain and the actual worst-case H_2 performance as in

$$K_{sub.} = -0.0108, \quad ||T_{zw}(s, K_{sub.})||_{2w.c.} = 71.17. \tag{6.75}$$

To get an idea of the conservatism of our approach, a brute force method has been used by gridding the space of uncertain parameters and the interval of admissible robust static output-feedback gains given by (6.74). Then, it turned out that

$$K^\star = -0.0572, \quad ||T_{zw}(s, K^\star)||_{2w.c.} = 41.42. \tag{6.76}$$

This example is particularly interesting since it shows that the new sufficient condition using SV-LMIs clearly outperforms the previous ones based on quadratic stability and may be a valuable alternative when this last approach fails. Obviously, the result may be rather conservative due to the particular chosen initialization.

6.6.2.2 Example 2

The example here is based on the model of the linearized equations of the VTOL helicopter borrowed from [43] and modified as in [5]. The data are given by

$$A = \begin{bmatrix} -0.0366 & 0.0271 & 0.0188 & -0.4555 \\ 0.0482 & -1.0100 & 0.0024 & -4.0208 \\ 0.1002 & a_1 & -0.7070 & a_2 \\ 0 & 0 & 1 & 0 \end{bmatrix},$$

$$B = \begin{bmatrix} -0.4422 & -0.1761 \\ b_1 & 7.5922 \\ 5.52 & -4.49 \\ 0 & 0 \end{bmatrix}, \quad C = \begin{bmatrix} 0 & 1 & 0 & 0 \end{bmatrix}.$$

Here, a_1, a_2 and b_1 denote uncertain parameters.

Table 6.3 Computation results

	GPS96	PA01	Algorithm 3
Guaranteed H_2 cost	4.5	4.66	3.9651

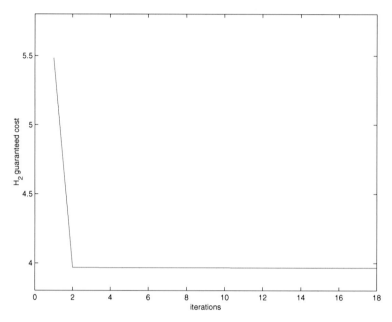

Fig. 6.4 Number of Iterations and γ_i^k in Algorithm 3

The matrices defining the H_2 performance channel are $B_1 = I_4$, $C_1 = \begin{bmatrix} I_4 \\ 0_{2\times 4} \end{bmatrix}$, and $D_{2u} = \begin{bmatrix} 0_{4\times 4} \\ I_2 \end{bmatrix}$. The quadratic approach developed respectively in [5] and [17], denoted by GPS96 and PA01, are then compared with the proposed algorithm, Algorithm 3. See Table 6.3 for the computation results. Algorithm 3 gives its result after 10 iterations, (20 LMI iterations). In the initialization we let $\varepsilon = 1$ and the resulting robust suboptimal H_2 SOF is

$$K_{sub.} = \begin{bmatrix} 1.541 \\ -10.252 \end{bmatrix}. \tag{6.77}$$

Once again, the new method gives a much better result than the previous ones based on quadratic framework. Figure 6.4 shows the convergence of Algorithm 3. It is interesting to note that after 2 iterations, the algorithm has reached a suboptimal bound very close to the final one.

References

1. Syrmos VL, Abdallah CT, Dorato P, Grigoriadis K (1997) Static output feedback: a survey. Automatica 33(2):125–137
2. Anderson BDO, Bose NK, Jury EI (1975) Output feedback stabilization and related problems—solutions via decision methods. IEEE Trans Automat Control 20(1):53–66
3. Rotea MA, Iwasaki T (1994) An alternative to the D-K iteration?. In: American control conference, Baltimore, pp 53–57
4. Iwasaki T, Skelton RE (1995) The XY-centring algorithm for the dual LMI problem: a new approach to fixed-order control design. Int J Control 62(6):1257–1272
5. Geromel JC, Peres PLD, Souza SR (1996) Convex analysis of output feedback control problems: robust stability and performance. IEEE Trans Automat Control 41(7):997–1003
6. El Ghaoui L, Oustry F, AitRami M (1997) A cone complementarity linearization algorithm for static ouput-feedback and related problems. IEEE Trans Automat Control 42(8):1171–1176
7. Geromel JC, de Souza CC, Skelton RE (1998) Static output feedback controllers: stability and convexity. IEEE Trans Automat Control 43(1):120–125
8. Iwasaki T (1999) The dual iteration for fixed-order control. IEEE Trans Automat Control 44(4):783–788
9. Grigoriadis KM, Beran EB (2000) In: El Ghaoui L, Niculescu SI (eds) Advances in linear matrix inequality methods in control, chapter 13: alternating projection algorithms for linear matrix inequalities problems with rank constraints. SIAM, Philadelphia
10. de Oliveira MC, Geromel JC (1997) Numerical comparison of output feedback design methods. In: American control conference, Albuquerque, New Mexico, June 1997
11. Leibfritz F (2001) An LMI-based algorithm for designing suboptimal static H_2/H_∞ output feedback controllers. SIAM J Control Optim 39(6):1711–1735
12. Apkarian P, Arzelier D, Henrion D, Peaucelle D (2003) Quelques perspectives pour la programmation mathèmatique en thèorie de la commande robuste. In: ROADEF, Avignon, France, pp 268–270
13. Burke JV, Henrion D, Lewis AS, Overton ML (2006) Stabilization via nonsmooth, nonconvex optimization. IEEE Trans Automat Control 51:1760–1769
14. Burke JV, Henrion D, Lewis AS, Overton ML (2006) Hifoo—a matlab package for fixed-order controller design and H-infinity optimization. In: IFAC symposium on robust control design, ROCOND06, Toulouse, France, June 2006
15. Apkarian P, Noll D (2006) Nonsmooth H_∞ synthesis. IEEE Trans Autom Control 51(1):71–86
16. Apkarian P, Noll D (2006) Controller design via nonsmooth multi-directional search. SIAM J Control Optim 44(6):1923–1949
17. Peaucelle D, Arzelier D (2001) An efficient numerical solution for H_2 static output feedback synthesis. In: European control conference, Porto, Portugal, September 2001, pp 3800–3805
18. Prempain E, Postlethwaite I (2005) Static H-infinity loop shaping control of a fly-by-wire helicopter. Automatica 41(9):1517–1528
19. Gumussoy S, Millstone M, Overton ML (2008) H_∞ strong stabilization via HIFOO, a package for fixed-order controller design. In: IEEE conference on decision and control, Cancun, Mexico, December 2008
20. Gumussoy S, Overton ML (2008) Fixed-order H_∞ controller design via HIFOO, a specialized nonsmooth optimization package. In: IEEE American control conference, pp 2750–2754, Seattle, USA, June 2008
21. Apkarian P, Noll D, Rondepierre A (2008) Mixed H_2/H_∞ control via nonsmooth optimization. SIAM J Control Optim 47(3):1516–1546
22. Henrion D (2006) Solving static output feedback problems by direct search optimization. In: Proceedings of the joint IEEE CCA/CACSD/ISIC conference, Munich, Germany, October 2006
23. Gumussoy S, Henrion D, Millstone M, Overton ML (2009) Multiobjective robust control with HIFOO 2.0. In: IFAC robust control design, ROCOND09, Haifa, Israel, June 2009

24. Bernussou J, Geromel JC, Peres PLD (1989) A linear programing oriented procedure for quadratic stabilization of uncertain systems. Syst Control Lett 13:65–72
25. Astolfi A, Colaneri P (2005) Hankel/Toeplitz matrices and the static output feedback stabilization problem. Math Control Signals Syst 17(4):231–268
26. Ebihara Y, Tokuyama K, Hagiwara T (2004) Structured controller synthesis using lmi and alternating projection method. Int J Control 77(12):1137–1147
27. Ebihara Y, Hagiwara T (2003) Structured controller synthesis using LMI and alternating projection method. In: IEEE conference on decision and control, pp 5632–5637, Maui, Hawaii, USA, December 2003
28. Henrion D, Arzelier D, Peaucelle D (2003) Positive polynomial matrices and improved LMI robustness conditions. Automatica 39(8):1479–1485
29. Henrion D, Šebek M, Kučra V (2003) Positive polynomials and robust stabilization with fixed-order controllers. IEEE Trans Automat Control 48(7):1178
30. Ackermann J (1993) Robust control: systems with uncertain physical parameters. Springler, London
31. Levine WS, Athans M (1970) On the determination of the optimal constant output feedback gains fro linear multivariable systems. IEEE Trans Autom Control 15(1):44–48
32. Zhou K, Doyle JC, Glover K (1996) Robust and opimal control. Prentice Hall, Englewood Cliffs
33. Coxson GE, Demarco CL (1991) Testing robust stability of general matrix polytopes is an NP-hard computation. In: Proceedings of the allerton conference on communication, control and computingMonticello, pp 105–106, Illinois, USA, October 1991
34. Benton RE, Smith D (2000) A non-iterative LMI-basedalgorithm for robust static output feedback stabilization. Int J Control Autom Syst 72(14):1322–1330
35. Arzelier D, Peaucelle D (2002) An iterative method for mixed H_2/H_∞ synthesis via static output-feedback. In: IEEE conference on decision and control, Las Vegas, December 2002
36. Gryazina EN, Polyak BT (2006) Stability domain in the parameter space: D-decomposition revisited. Automatica 42(1):13–26
37. Arzelier D, Gryazina EN, Peaucelle D, Polyak BT (2010) Mixed LMI/randomized methods for static output feedback control design. In: American control conference, Baltimore, June 2010
38. Leibfritz F, Lipinski W (2003) Description of the benchmark examples in compleib 1.0. Technical report, Department of Mathematics, University of Trier. www.complib.de
39. Leibfritz F, Lipinski W (2004) COMPleib 1.0: COnstraint Matrix-optimization Problem - user manual and quick reference. Technical report, Department of Mathematics, University of Trier. www.complib.de
40. Prempain E (2005) A comparison of two algorithms for the static H_∞ loop shaping problem. In IFAC World Congress, Prague, July 2005
41. Henrion D, Sebek M (2008) Plane geometry and convexity of polynomial stability regions. In: Proceedings of the international symposium on symbolic and algebraic computations (ISSAC), Hagenberg, Austria, July 2008
42. Colaneri P, Geromel JC, Locatelli A (1997) Control theory and design : an RH_2 and RH_∞ viewpoint. Academic Press, London
43. Keel L, Bhattacharyya SP, Howze JW (1988) Robust control with structured perturbations. IEEE Trans Automat Control 33:68–78

Chapter 7
Robust Performance Analysis of Discrete-Time Periodic Systems

7.1 Introduction

In this section, we analyze stability, H_2, and H_∞ performance of linear discrete-time periodic systems in terms of SV-LMIs. We consider N-periodic system G described by

$$G : \begin{cases} x(k+1) = A_k x(k) + B_k w(k), \\ z(k) \quad = C_k x(k) + D_k w(k), \end{cases} \tag{7.1}$$

where $x \in \mathbb{R}^n$ is the state, $w \in \mathbb{R}^{n_w}$ the input, and $z \in \mathbb{R}^{n_z}$ the output, respectively. For all $k \geq 0$, matrices $A_k \in \mathbb{R}^{n \times n}$, $B_k \in \mathbb{R}^{n \times n_w}$, $C_k \in \mathbb{R}^{n_z \times n}$, and $D_k \in \mathbb{R}^{n_z \times n_w}$ are all N-periodic, i.e., $A_{k+N} = A_k$, etc.

7.2 Stability Analysis of Linear Periodic Systems

7.2.1 Definition and Basic Results

The system (7.1) is said to be (asymptotically) stable if $x(k) \to 0$ $(k \to \infty)$ for any $x(0)$. It is known that the stability is characterized by the monodromy matrix Φ defined by $\Phi := \prod_{k=0}^{N-1} A_k = A_{N-1} A_{N-2}, \ldots, A_1 A_0$ [1]. Indeed, since $x(lN) = \Phi^l x(0)$, we see that the system is stable if and only if Φ is Schur stable. This result is formally stated in the next lemma.

Lemma 7.1 *The system* (7.1) *is stable if and only if the monodromy matrix* Φ *defined by* $\Phi := \prod_{k=0}^{N-1} A_k$ *is Schur stable.*

The eigenvalues of Φ is referred to *characteristic multipliers* of the system (7.1). Therefore, the stability condition of the system (7.1) can be restated equivalently that all of the characteristic multipliers belong to the open unit disk.

© Springer-Verlag London 2015
Y. Ebihara et al., *S-Variable Approach to LMI-Based Robust Control*,
Communications and Control Engineering, DOI 10.1007/978-1-4471-6606-1_7

Since the stability of the system (7.1) is characterized by the Schur stability of the matrix Φ, we can readily apply the results for stability analysis of discrete-time LTI systems. The next result is a direct consequence of the Lyapunov inequality characterizing the Schur stability of matrices.

Lemma 7.2 *The system (7.1) is stable if and only if there exists $P \succ 0$ such that*

$$-P + \Phi^T P \Phi \prec 0. \tag{7.2}$$

Note that the LMI (7.2) is not useful in several applications, since the matrix Φ is represented by multiplications among A_k ($k = 0, \ldots, N - 1$). The next theorem shows that the stability can be characterized by a set of LMIs where the matrix A_k ($k = 0, \ldots, N - 1$) are free from such multiplications.

Theorem 7.1 *The system (7.1) is stable if and only if there exist $P_k \succ 0$ ($k = 0, \ldots, N - 1$) such that*

$$-P_k + A_k^T P_{k+1} A_k \prec 0 \ (k = 0, \ldots, N - 1) \tag{7.3}$$

where $P_N = P_0$.

Proof Even though this theorem can be proved by the periodic Lyapunov Lemma [2], we give here a primitive proof for the equivalence of the conditions in Lemma 7.2 and Theorem 7.1 by means of simple matrix operation.
(7.2)$\Rightarrow$(7.3): Suppose (7.2) holds. Then, there exists $P_0 \succ 0$ such that

$$-P_0 + A_0^T \cdots A_{N-2}^T A_{N-1}^T P_0 A_{N-1} A_{N-2} \cdots A_0 \prec 0.$$

Then, there exists sufficiently small $\varepsilon_{N-1} > 0$ such that

$$-P_0 + A_0^T \cdots A_{N-2}^T P_{N-1} A_{N-2} \cdots A_0 \prec 0, \quad P_{N-1} := A_{N-1}^T P_0 A_{N-1} + \varepsilon_{N-1} I_n \succ 0.$$

Applying this procedure recursively, we see that there exist ε_k ($k = 1, \ldots, N - 1$) such that

$$-P_0 + A_0^T \cdots A_{N-2}^T P_{N-1} A_{N-2} \cdots A_0 \prec 0, \quad P_{N-1} := A_{N-1}^T P_0 A_{N-1} + \varepsilon_{N-1} I_n \succ 0,$$
$$-P_0 + A_0^T \cdots A_{N-3}^T P_{N-2} A_{N-3} \cdots A_0 \prec 0, \quad P_{N-2} := A_{N-2}^T P_{N-1} A_{N-2} + \varepsilon_{N-2} I_n \succ 0,$$
$$-P_0 + A_0^T \cdots A_{N-4}^T P_{N-3} A_{N-4} \cdots A_0 \prec 0, \quad P_{N-3} := A_{N-3}^T P_{N-2} A_{N-3} + \varepsilon_{N-3} I_n \succ 0,$$
$$\vdots$$
$$-P_0 + A_0^T P_1 A_0 \prec 0, \qquad\qquad\qquad P_1 := A_1^T P_2 A_1 + \varepsilon_1 I_n \succ 0.$$

Therefore, we see that there exist $P_k \succ 0$ ($k = 0, \ldots, N - 1$) such that

$$-P_0 + A_0^T P_1 A_0 \prec 0,$$
$$-P_k + A_k^T P_{k+1} A_k = -\varepsilon_k I_n \prec 0 \ (k = 1, \ldots, N - 1).$$

This implies (7.3).

(7.3)⇒(7.2): Suppose (7.3) holds. Then, we have from $-P_{N-1}+A_{N-1}^T P_0 A_{N-1} \prec 0$ that

$$-A_{N-2}^T P_{N-1} A_{N-2} + \left(\prod_{k=N-2}^{N-1} A_k\right)^T P_0 \left(\prod_{k=N-2}^{N-1} A_k\right) \preceq 0$$

where $\prod_{k=p}^q A_k := A_q \cdots A_{p+1} A_p$. From this nonstrict inequality and the strict inequality $-P_{N-2} + A_{N-2}^T P_{N-1} A_{N-2} \prec 0$, we have

$$-P_{N-2} + \left(\prod_{k=N-2}^{N-1} A_k\right)^T P_0 \left(\prod_{k=N-2}^{N-1} A_k\right) \prec 0.$$

Repeating the same procedure, we arrive at

$$-P_0 + \left(\prod_{k=0}^{N-1} A_k\right)^T P_0 \left(\prod_{k=0}^{N-1} A_k\right) \prec 0.$$

This completes the proof.   □

7.2.2 Robust Stability Analysis Using SV-LMIs

Suppose the matrix A_k ($k = 0, \ldots, N - 1$) in (7.1) is affected by polytopic uncertainty as in

$$\begin{bmatrix} A_0 \\ \vdots \\ A_{N-1} \end{bmatrix} = \sum_{l=1}^L \theta_l \begin{bmatrix} A_0^{[l]} \\ \vdots \\ A_{N-1}^{[l]} \end{bmatrix}, \quad \theta \in \mathbb{E}^L. \tag{7.4}$$

Here, $A_k^{[l]}$ ($k = 0, \ldots, N - 1$, $l = 1, \ldots, L$) are known matrices. The above expression entails $A_k = A_k(\theta) = \sum_{l=1}^L \theta_l A_k^{[l]}$ ($k = 0, \ldots, N - 1$) where the uncertain parameter θ is uniform over one period. One may think that the following polytopic representation, where the parameter θ depends on the time instance k, would be more natural than (7.4):

$$A_k = A_k(\theta_k) = \sum_{l=1}^L \theta_{k,l} A_k^{[l]}, \ \theta_k \in \mathbb{E}^L \quad (k = 0, \ldots, N - 1). \tag{7.5}$$

However, it is easy to confirm that (7.5) can be represented as (7.4) by appropriately reconstructing vertex matrices of the polytope. In this sense, the representation (7.4) is not restrictive. For example, in the case $N = 2$ and $L = 2$, we see that the latter form

$$\begin{bmatrix} A_0 \\ A_1 \end{bmatrix} = \begin{bmatrix} \theta_{0,1} A_0^{[1]} + \theta_{0,2} A_0^{[2]} \\ \theta_{1,1} A_1^{[1]} + \theta_{1,2} A_1^{[2]} \end{bmatrix}$$

can be rewritten as

$$\begin{bmatrix} A_0 \\ A_1 \end{bmatrix} = \theta_1 \begin{bmatrix} A_0^{[1]} \\ A_1^{[1]} \end{bmatrix} + \theta_2 \begin{bmatrix} A_0^{[1]} \\ A_1^{[2]} \end{bmatrix} + \theta_3 \begin{bmatrix} A_0^{[2]} \\ A_1^{[1]} \end{bmatrix} + \theta_4 \begin{bmatrix} A_0^{[2]} \\ A_1^{[2]} \end{bmatrix}$$

where $\theta_1 = \theta_{0,1}\theta_{1,1}$, $\theta_2 = \theta_{0,1}\theta_{1,2}$, $\theta_3 = \theta_{0,2}\theta_{1,1}$, $\theta_4 = \theta_{0,2}\theta_{1,2}$ and we can confirm $\theta \in \mathbb{E}^4$.

For the uncertain periodic system described by (7.1) and (7.4) (without input and output), our first goal is to derive LMIs to assess the robust stability. More precisely, if we define $\Phi(\theta) := \prod_{k=0}^{N-1} A_k(\theta)$, we want to assess whether $\rho(\Phi(\theta)) < 1$ holds for all $\theta \in \mathbb{E}^L$.

As noted previously, the LMI condition (7.2) is not useful for the robust stability analysis. This is because the robust LMI condition

$$-P + \Phi(\theta)^T P \Phi(\theta) \prec 0 \quad (\forall \theta \in \mathbb{E}^L)$$

involves multiplications among θ_i, and hence not easily tractable. However, if we rely on the LMI (7.3), we can obtain the next theorem providing sufficient condition for the robust stability.

Theorem 7.2 *The uncertain periodic system described by (7.1) and (7.4) is robustly stable if there exist $P_k \succ 0$ ($k = 0, \ldots, N - 1$) such that*

$$-P_k + A_k^{[l]T} P_{k+1} A_k^{[l]} \prec 0 \quad (k = 0, \ldots, N-1, \, l = 1, \ldots, L) \tag{7.6}$$

where $P_N = P_0$.

Proof Suppose (7.6) holds. Note that for each k this set of LMIs can be rewritten equivalently as

$$\begin{bmatrix} -P_k & A_k^{[l]T} P_{k+1} \\ P_{k+1} A_k^{[l]} & -P_{k+1} \end{bmatrix} \prec 0 \quad (l = 1, \ldots, L).$$

Therefore, for any $\theta \in \mathbb{E}^L$, we have

$$\sum_{l=1}^{L} \theta_l \begin{bmatrix} -P_k & A_k^{[l]T} P_{k+1} \\ P_{k+1} A_k^{[l]} & -P_{k+1} \end{bmatrix} = \begin{bmatrix} -P_k & A_k(\theta)^T P_{k+1} \\ P_{k+1} A_k(\theta) & -P_{k+1} \end{bmatrix} \prec 0$$

or equivalently

$$-P_k - A_k(\theta)^T P_{k+1} A_k(\theta) \prec 0 \ (\forall \theta \in \mathbb{E}^L).$$

Since this condition holds for $k = 0, \ldots, N - 1$, we see from Theorem 7.1 that the system is robustly stable. □

In this theorem, we see that if (7.6) holds, then $-P_0 + \Phi(\theta)^T P_0 \Phi(\theta) \prec 0$ holds for all $\theta \in \mathbb{E}^L$. Namely, the condition (7.6) ensures the robust stability of $\Phi(\theta)$ by parameter-independent Lyapunov function of the form $V(x) = x^T P_0 x$. This clearly shows that the condition (7.6) is conservative. The next theorems show that we can reduce the conservatism by using S-variable LMIs.

Theorem 7.3 [3] *The uncertain periodic system described by (7.1) and (7.4) is robustly stable if there exist* $P_k^{[l]} \succ 0$, G_k $(k = 0, \ldots, N - 1)$ *such that*

$$\begin{bmatrix} -P_k^{[l]} & 0 \\ 0 & P_{k+1}^{[l]} \end{bmatrix} + \text{He}\left\{ \begin{bmatrix} A_k^{[l]T} \\ -I \end{bmatrix} \begin{bmatrix} 0 & G_{k+1} \end{bmatrix} \right\} \prec 0 \tag{7.7}$$

$$(k = 0, \ldots, N - 1, l = 1, \ldots, L)$$

where $P_N^{[l]} = P_0^{[l]}$ $(l = 1, \ldots, L)$ *and* $G_N = G_0$. *Moreover, if the LMI (7.6) holds with* $P_k = P_k^\star$ $(k = 0, \ldots, N - 1)$, *then the LMI (7.7) holds with* $P_k^{[l]} = P_k^\star$ $(k = 0, \ldots, N - 1, l = 1, \ldots, L)$ *and* $G_k = P_k^\star$ $(k = 0, \ldots, N - 1)$.

Theorem 7.4 *The uncertain periodic system described by (7.1) and (7.4) is robustly stable if there exist* $P_k^{[l]} \succ 0$, $F_{k,j}$ $(k = 0, \ldots, N - 1, l = 1, \ldots, L, j = 1, 2)$ *such that*

$$\begin{bmatrix} -P_k^{[l]} & 0 \\ 0 & P_{k+1}^{[l]} \end{bmatrix} + \text{He}\left\{ \begin{bmatrix} A_k^{[l]T} \\ -I \end{bmatrix} \begin{bmatrix} F_{k+1,1} & F_{k+1,2} \end{bmatrix} \right\} \prec 0 \tag{7.8}$$

$$(k = 0, \ldots, N - 1, l = 1, \ldots, L)$$

where $P_N^{[l]} = P_0^{[l]}$ $(l = 1, \ldots, L)$ *and* $F_{N,j} = F_{0,j}$ $(j = 1, 2)$. *Moreover, if the LMI (7.7) holds with* $P_k^{[l]} = P_k^{[l]\star}$ $(k = 0, \ldots, N - 1, l = 1, \ldots, L)$ *and* $G_k = G_i^\star$ $(k = 0, \ldots, N - 1)$, *then the LMI (7.8) holds with* $P_k^{[l]} = P_k^{[l]\star}$ $(k = 0, \ldots, N - 1, l = 1, \ldots, L)$ *and* $\begin{bmatrix} F_{k,1} & F_{k,2} \end{bmatrix} = \begin{bmatrix} 0 & G_k^\star \end{bmatrix}$ $(k = 0, \ldots, N - 1)$.

Proof of Theorem 7.3 Similarly to the proof of Theorem 7.2, we can easily confirm that if (7.7) holds then

$$\begin{bmatrix} -P_k(\theta) & 0 \\ 0 & P_{k+1}(\theta) \end{bmatrix} + \mathrm{He}\left\{ \begin{bmatrix} A_k(\theta)^T \\ -I \end{bmatrix} \begin{bmatrix} 0 & G_{k+1} \end{bmatrix} \right\} \prec 0 \tag{7.9}$$
$$(k = 1, \ldots, N - 1, \ \forall \theta \in \mathbb{E}^L).$$

Here, we defined $P_k(\theta) := \sum_{l=1}^{L} \theta_l P_k^{[l]}$. It is clear that $P_k(\theta) \succ 0$ $(k = 0, \ldots, N - 1, \ \forall \theta \in \mathbb{E}^L)$. Multiplying $[\ I \ A_k(\theta)^T\ ]$ from the left of (7.9) and its transpose from the right, we readily obtain

$$-P_k(\theta) - A_k(\theta)^T P_{k+1}(\theta) A_k(\theta) \prec 0 \ (k = 0, \ldots, N_1, \ \forall \theta \in \mathbb{E}^L).$$

From this condition and Theorem 7.1, we see that the system is robustly stable. The second assertion readily follows from Schur complement.   □

Proof of Theorem 7.4 The first assertion follows similarly to the proof of Theorem 7.3. The second assertion obviously holds from the fact that the LMI (7.7) is a special case of (7.8) by letting $[\ F_{k,1}\ F_{k,2}\ ] = [\ 0\ G_k\ ]$ $(k = 0, \ldots, N - 1)$.   □

From these theorems, we see that the LMI (7.7) is no more conservative than the LMI (7.6), and the LMI (7.8) is no more conservative than the LMI (7.7). Note that such conservatism reduction has been achieved at the expense of the increased number of decision variables.

We derived LMIs (7.7) and (7.8) by naturally extending the LTI case results. Seemingly, there is no constructive way to derive LMIs for further conservatism reduction. However, by noting that the stability of periodic systems is characterized by the Schur stability of monodromy matrices, we can derive an LMI that outperforms those in Theorems 7.3 and 7.4.

Theorem 7.5 [4] *The uncertain periodic system described by (7.1) and (7.4) is robustly stable if there exist* $P^{[l]} \succ 0$ $(l = 1, \ldots, L)$ *and* $F \in \mathbb{R}^{Nn \times (N+1)n}$ *such that*

$$\mathcal{P}_N(P^{[l]}) + \mathrm{He}\left\{ \mathscr{A}^{[l]T} F \right\} \prec 0 \ (= 1, \ldots, L). \tag{7.10}$$

where

$$\mathcal{P}_N(P^{[l]}) := \begin{bmatrix} -P^{[l]} & 0 & 0 \\ 0 & 0_{(N-1)n} & 0 \\ 0 & 0 & P^{[l]} \end{bmatrix},$$

$$\mathscr{A}^{[l]} := \begin{bmatrix} A_0^{[l]} & -I & 0 & \cdots & & 0 \\ 0 & A_1^{[l]} & -I & 0 & & \vdots \\ \vdots & \ddots & \ddots & \ddots & \ddots & \vdots \\ \vdots & & \ddots & \ddots & & 0 \\ 0 & \cdots & \cdots & 0 & A_{N-1}^{[l]} & -I \end{bmatrix} \in \mathbb{R}^{Nn \times (N+1)n} \ (l = 1, \ldots, L). \tag{7.11}$$

Moreover, if the LMI (7.8) holds with $P_k^{[l]} = P_k^{[l]\star}$ and $[\, F_{1,i}\ F_{2,i}\,] = [\, F_{1,i}^{\star}\ F_{2,i}^{\star}\,]\ (k = 0, \ldots, N-1,\ l = 1, \ldots, L)$, then the LMI (7.10) holds with $P^{[l]} = P_0^{[l]\star}\ (l = 1, \ldots, L)$ and

$$F = \begin{bmatrix} F_{1,1}^{\star} & F_{1,2}^{\star} & 0 & \cdots & & \cdots & 0 \\ 0 & F_{2,1}^{\star} & F_{2,2}^{\star} & 0 & & & \vdots \\ \vdots & \ddots & \ddots & \ddots & & \ddots & \vdots \\ \vdots & & & \ddots & F_{N-1,1}^{\star} & F_{N-2,2}^{\star} & 0 \\ 0 & \cdots & & \cdots & 0 & F_{0,1}^{\star} & F_{0,2}^{\star} \end{bmatrix}. \tag{7.12}$$

This theorem follows from [4] with dual system representation of (7.1). The proof is given below for completeness.

Proof of Theorem 7.4 Suppose (7.10) holds. Then, for any $\theta \in \mathbb{E}^L$ we have

$$\mathscr{P}_N(P(\theta)) + \mathrm{He}\left\{ \mathscr{A}(\theta)^T F \right\} \prec 0 \tag{7.13}$$

where $P(\theta) := \sum_{l=1}^{L} \theta_l P^{[l]}$ and

$$\mathscr{A}(\theta) := \begin{bmatrix} A_0(\theta) & -I & 0 & \cdots & & 0 \\ 0 & A_1(\theta) & -I & 0 & & \vdots \\ \vdots & & \ddots & \ddots & \ddots & \vdots \\ \vdots & & & \ddots & \ddots & 0 \\ 0 & \cdots & & 0 & A_{N-1}(\theta) & -I \end{bmatrix}.$$

It is obvious that $P(\theta) \succ 0\ (\forall \theta \in \mathbb{E}^L)$. Moreover, we see that

$$\mathscr{A}(\theta)^{\perp} = \begin{bmatrix} I & A_0(\theta)^T & (A_1(\theta)A_0(\theta))^T & \cdots & \Phi(\theta)^T \end{bmatrix}^T.$$

It follows from (7.13) that

$$\mathscr{A}(\theta)^{\perp T} \mathscr{P}_N(P(\theta)) \mathscr{A}(\theta)^{\perp} \prec 0 \quad \Leftrightarrow \quad -P(\theta) + \Phi(\theta)^T P(\theta)\Phi(\theta) \prec 0.$$

This clearly shows that the system of interest is robustly stable.

We next prove the second assertion for the case $N = 2$. The proof for general cases follows similarly. From the underlying assumption, there exist $P_k^{[l]\star} \succ 0\ (k = 0, 1,\ l = 1, \ldots, L)$ and $F_{k,1}^{\star},\ F_{k,2}^{\star}\ (k = 0, 1)$ such that

$$\mathscr{D}_{0,l} := \begin{bmatrix} -P_0^{[l]\star} & 0 \\ 0 & P_1^{[l]\star} \end{bmatrix} + \mathrm{He}\left\{ \begin{bmatrix} A_0^{[l]T} \\ -I \end{bmatrix} \begin{bmatrix} F_{1,1}^\star & F_{1,2}^\star \end{bmatrix} \right\} \prec 0 \ (l = 1, \ldots, L),$$

$$\mathscr{D}_{1,l} := \begin{bmatrix} -P_1^{[l]\star} & 0 \\ 0 & P_0^{[l]\star} \end{bmatrix} + \mathrm{He}\left\{ \begin{bmatrix} A_1^{[l]T} \\ -I \end{bmatrix} \begin{bmatrix} F_{0,1}^\star & F_{0,2}^\star \end{bmatrix} \right\} \prec 0 \ (l = 1, \ldots, L).$$

$$(7.14)$$

Then, for each l $(l = 1, \ldots, L)$, we have

$$\begin{bmatrix} I & 0 & 0 \\ 0 & I & 0 \\ 0 & I & 0 \\ 0 & 0 & I \end{bmatrix}^T \begin{bmatrix} \mathscr{D}_{0,l} & 0 \\ 0 & \mathscr{D}_{1,l} \end{bmatrix} \begin{bmatrix} I & 0 & 0 \\ 0 & I & 0 \\ 0 & I & 0 \\ 0 & 0 & I \end{bmatrix} \prec 0$$

$$(7.15)$$

$$\Leftrightarrow \begin{bmatrix} -P_0^{[l]\star} & 0 & 0 \\ 0 & 0_n & 0 \\ 0 & 0 & P_0^{[l]\star} \end{bmatrix} + \mathrm{He}\left\{ \begin{bmatrix} A_0^{[l]T} & 0 \\ -I & A_1^{[l]T} \\ 0 & -I \end{bmatrix} \begin{bmatrix} F_{1,1}^\star & F_{1,2}^\star & 0 \\ 0 & F_{0,1}^\star & F_{0,2}^\star \end{bmatrix} \right\} \prec 0.$$

This shows that (7.10) for $N = 2$ holds with $P^{[l]} = P_0^{[l]\star}$ $(l = 1, \ldots, L)$ and F given by (7.12). □

From Theorem 7.5, we can conclude that the LMI (7.10) is no more conservative than the LMI (7.8). In Sect. 7.5 we show by numerical examples that the LMI (7.10) yields strictly better (less conservative) results than the LMI (7.8).

Remark 7.1 It is interesting to note that the matrix $\mathscr{A}^{[l]}$ defined by (7.11) represents the dynamics of the system (7.1) and (7.4) corresponding to the lth vertex since

$$\mathscr{A}^{[l]} \begin{bmatrix} x_0 \\ \vdots \\ x_N \end{bmatrix} = \begin{bmatrix} A_0^{[l]} & -I & 0 & \cdots & & \cdots & 0 \\ 0 & A_1^{[l]} & -I & 0 & & & \vdots \\ \vdots & & \ddots & \ddots & \ddots & & \vdots \\ \vdots & & & \ddots & \ddots & & 0 \\ 0 & \cdots & & \cdots & 0 & A_{N-1}^{[l]} & -I \end{bmatrix} \begin{bmatrix} x_0 \\ x_1 \\ \vdots \\ \vdots \\ x_N \end{bmatrix} = 0.$$

From this representation, we see that the nonzero matrix part and the lower triangular part with all zero matrices in the above equation corresponds to the causal dynamics of the system, whereas the upper triangular part with all zero matrices corresponds to the noncausal dynamics of the system. With this mind, let us recall the SV-LMI (7.10) where we employ S-variable $F \in \mathbb{R}^{Nn \times (N+1)n}$ that is full (unstructured). Therefore, from dynamical system point of view, the SV-LMI condition (7.10) can be interpreted in the way that it assess the robust stability of causal periodic system given by (7.1) and (7.4) with noncausal system whose dynamics can be represented by

$$F \begin{bmatrix} x_0 \\ \vdots \\ x_N \end{bmatrix} = 0.$$

7.3 H_2 Performance Analysis of Linear Periodic Systems

7.3.1 Lifting-Based Treatment

Let us revisit the lifting-based treatment of periodic systems [5]. As is well known, the behavior of the periodic system (7.1) can be fully captured via an LTI system that is obtained by the discrete-time system lifting. The LTI system is given by

$$\widehat{G} : \begin{cases} \xi_{\kappa+1} = \widehat{A}\xi_\kappa + \widehat{B}\widehat{w}_\kappa, \\ \widehat{z}_\kappa = \widehat{C}\xi_\kappa + \widehat{D}\widehat{w}_\kappa \end{cases} \tag{7.16}$$

where

$$\xi_\kappa = x_{\kappa N}, \quad \widehat{w}_\kappa = \begin{bmatrix} w_{(\kappa+1)N-1} \\ \vdots \\ w_{\kappa N} \end{bmatrix}, \quad \widehat{z}_\kappa = \begin{bmatrix} z_{(\kappa+1)N-1} \\ \vdots \\ z_{\kappa N} \end{bmatrix} \tag{7.17}$$

and

$$\left[\begin{array}{c|c} \widehat{A} & \widehat{B} \\ \hline \widehat{C} & \widehat{D} \end{array}\right] =$$

$$\left[\begin{array}{c|cccc} \Phi & B_{N-1} & A_{N-1}B_{N-2} & \cdots & \left(\prod_{k=1}^{N-1} A_k\right) B_0 \\ \hline C_{N-1}\prod_{k=0}^{N-2} A_k & D_{N-1} & C_{N-1}B_{N-2} & \cdots & C_{N-1}\left(\prod_{k=1}^{N-2} A_k\right) B_0 \\ C_{N-2}\prod_{k=0}^{N-3} A_k & 0 & D_{N-2} & & C_{N-2}\left(\prod_{k=1}^{N-3} A_k\right) B_0 \\ \vdots & \vdots & \ddots & \ddots & \vdots \\ C_0 & 0 & \cdots & 0 & D_0 \end{array}\right], \tag{7.18}$$

$$\prod_{k=p}^{q} A_k := A_q \cdots A_{p+1}A_p \ (q \geq p), \quad \prod_{k=p}^{q} A_k := I \ (q < p).$$

This LTI system is obtained by paying attention to the behavior of the sampled state $\xi_\kappa := x_{\kappa N}$. The one-period-ahead state $\xi_{\kappa+1}$ is determined from ξ_κ and the inputs over the period that are lifted as in $\widehat{w}_\kappa$. The lifted output $\widehat{z}_\kappa$ can be obtained from ξ_κ and $\widehat{w}_\kappa$.

As we have already seen, the original periodic system G is stable if and only if the fictitious LTI system $\widehat{G}$ is stable. Moreover, the input-to-output properties of system G is inherited to $\widehat{G}$.

7.3.2 Definition of H_2 Norm and Basic Results

As shown in [5, 6], we can define the generalized H_2 norm of the N-periodic system (7.1) as $\sqrt{||\widehat{G}||_2^2/N}$, where $||\widehat{G}||_2$ stands for the H_2 norm of the LTI system (7.16). The generalized H_2 norm corresponds to the mean of all the responses corresponding to impulsive inputs applied to each of the n_w input channels at each time k in the N-period.

Definition 7.1 For the N-periodic system G represented by (7.1), the generalized H_2 norm $||G||_2$ is defined by $||G||_2 := \sqrt{||\widehat{G}||_2^2/N}$. Here, $||\widehat{G}||_2$ stands for the H_2 norm of the LTI system (7.16).

From this definition, we can readily apply the LTI-case result to obtain an LMI for the H_2 performance analysis of periodic system G.

Theorem 7.6 *For the N-periodic system G described by (7.1) and given $\gamma > 0$, the condition $||G||_2 < \gamma$ holds if and only if there exist $P \succ 0$ and $Z \succ 0$ such that*

$$-P + \widehat{A}^T P \widehat{A} + \widehat{C}^T \widehat{C} \prec 0, \quad Z \succ \widehat{B}^T P \widehat{B} + \widehat{D}^T \widehat{D}, \quad \mathrm{trace}(Z) < N\gamma^2. \quad (7.19)$$

The LMI (7.19) involves $\widehat{A}, \widehat{B}, \widehat{C}$, and $\widehat{D}$ that are given by multiplications among the coefficient matrices A_k, B_k, C_k and D_k, see (7.18). As we have seen in the robust stability analysis, such multiplications become obstacles for robust H_2 performance analysis. The next theorem shows an LMI where such multiplications are successfully circumvented.

Theorem 7.7 *For the N-periodic system G described by (7.1) and given $\gamma > 0$, the condition $||G||_2 < \gamma$ holds if and only if there exist $P_k \succ 0$ and $Z_k \succ 0$ ($k = 0, \ldots, N-1$) such that*

$$-P_k + A_k^T P_{k+1} A_k + C_k^T C_k \prec 0, \quad Z_k \succ B_k^T P_{k+1} B_k + D_k^T D_k \ (k = 0, \ldots, N-1),$$

$$\mathrm{trace}\left(\sum_{k=0}^{N-1} Z_k\right) < N\gamma^2$$

$$(7.20)$$

where $P_N = P_0$.

Proof of Theorem 7.7 We prove the equivalence of (7.19) and (7.20) for $N = 2$ just for simplicity.

(7.19) $\Rightarrow$ (7.20) Suppose (7.19) holds. Then from (7.18) there exist $P_0^\star \succ 0$ and $Z^\star \succ 0$ with the partition

$$Z^\star = \begin{bmatrix} Z_1^\star & Z_{10}^\star \\ Z_{10}^{\star T} & Z_0^\star \end{bmatrix}$$

such that

$$-P_0^\star + A_0^T A_1^T P_0^\star A_1 A_0 + A_0^T C_1^T C_1 A_0 + C_0^T C_0 \prec 0, \qquad (7.21a)$$

$$\begin{bmatrix} Z_1^\star & Z_{10}^\star \\ Z_{10}^{\star T} & Z_0^\star \end{bmatrix} \succ [\, B_1 \ A_1 B_0 \,]^T P_0^\star [\, B_1 \ A_1 B_0 \,] + \begin{bmatrix} D_1 & C_1 B_0 \\ 0 & D_0 \end{bmatrix}^T \begin{bmatrix} D_1 & C_1 B_0 \\ 0 & D_0 \end{bmatrix},$$
$$\qquad (7.21b)$$

$$\mathrm{trace}(Z_1^\star + Z_0^\star) < 2\gamma^2. \qquad (7.21c)$$

This implies

$$-P_0^\star + A_0^T(A_1^T P_0^\star A_1 + C_1^T C_1)A_0 + C_0^T C_0 \prec 0,$$
$$Z_0^\star \succ B_0^T(A_1^T P_0^\star A_1 + C_1^T C_1)B_0 + D_0^T D_0,$$
$$Z_1^\star \succ B_1^T P_0^\star B_1 + D_1^T D_1,$$
$$\mathrm{trace}(Z_1^\star + Z_2^\star) < 2\gamma^2.$$

Then, there exists $\varepsilon_1 > 0$ such that $P_1^\star := A_1^T P_0^\star A_1 + C_1^T C_1 + \varepsilon_1 I$ satisfies

$$-P_0^\star + A_0^T P_1^\star A_0 + C_0^T C_0 \prec 0,$$
$$Z_0^\star \succ B_0^T P_1^\star B_0 + D_0^T D_0,$$
$$Z_1^\star \succ B_1^T P_0^\star B_1 + D_1^T D_1,$$
$$\mathrm{trace}(Z_1^\star + Z_2^\star) < 2\gamma^2.$$

This clearly shows that (7.20) holds with $P_k = P_k^\star$ and $Z_k = Z_k^\star$ ($k = 0, 1$).
(7.20) $\Rightarrow$ (7.19) Suppose (7.20) holds. Then there exist $P_k^\star \succ 0$ and $Z_k^\star \succ 0$ ($k = 0, 1$)
such that

$$-P_0^\star + A_0^T P_1^\star A_0 + C_0^T C_0 \prec 0,$$
$$-P_1^\star + A_1^T P_0^\star A_1 + C_1^T C_1 \prec 0,$$
$$Z_0^\star \succ B_0^T P_1^\star B_0 + D_0^T D_0,$$
$$Z_1^\star \succ B_1^T P_0^\star B_1 + D_1^T D_1,$$
$$\mathrm{trace}(Z_1^\star + Z_2^\star) < 2\gamma^2.$$

Then, eliminating $P_1^\star$ from the first two inequalities, we have (7.21a), and hence the
fist LMI in (7.19) holds with $P = P_0^\star$. On the other hand, if we define

$$\begin{bmatrix} Z_1 & Z_{10} \\ Z_{10}^T & Z_0 \end{bmatrix} := \widehat{B}^T P_0^\star \widehat{B} + \widehat{D}^T \widehat{D}$$

$$= [\, B_1 \ A_1 B_0 \,]^T P_0^\star [\, B_1 \ A_1 B_0 \,] + \begin{bmatrix} D_1 & C_1 B_0 \\ 0 & D_0 \end{bmatrix}^T \begin{bmatrix} D_1 & C_1 B_0 \\ 0 & D_0 \end{bmatrix},$$
$$\qquad (7.22)$$

we see that

$$Z^\star := \begin{bmatrix} Z_1^\star & Z_{10} \\ Z_{10}^T & Z_0^\star \end{bmatrix} \succ \begin{bmatrix} Z_1 & Z_{10} \\ Z_{10}^T & Z_0 \end{bmatrix}$$

since

$$Z_0^\star \succ B_0^T P_1^\star B_0 + D_0^T D_0 \succ B_0^T (A_1^T P_0^\star A_1 + C_1^T C_1) B_0 + D_0^T D_0 = Z_0,$$
$$Z_1^\star \succ B_1^T P_0^\star B_1 + D_1^T D_1 = Z_1.$$

By observing that $\text{trace}(Z^\star) = \text{trace}(Z_1^\star + Z_2^\star) < 2\gamma^2$, we can conclude that (7.19) is satisfied with $P = P_0^\star$ and $Z = Z^\star$.   □

7.3.3 Robust H_2 Performance Analysis Using SV-LMIs

Consider the case where the periodic system (7.1) is subject to the polytopic uncertainty of the form

$$M_k := \begin{bmatrix} A_k & B_k \\ C_k & D_k \end{bmatrix}, \quad M_k^{[l]} = \begin{bmatrix} A_k^{[l]} & B_k^{[l]} \\ C_k^{[l]} & D_k^{[l]} \end{bmatrix} \quad (l = 1, \ldots, L),$$

$$\begin{bmatrix} M_0 \\ \vdots \\ M_{N-1} \end{bmatrix} = \sum_{l=1}^{L} \theta_l \begin{bmatrix} M_0^{[l]} \\ \vdots \\ M_{N-1}^{[l]} \end{bmatrix}, \quad \theta \in \mathbb{E}^L. \tag{7.23}$$

Here, $M_k^{[l]}$ ($k = 0, \ldots, N - 1$, $l = 1, \ldots, L$) are known matrices that define the vertices of the polytope. This is an extension of the representation (7.4), and hence θ is uniform over one period. We emphasize that the parameter vector θ is time-invariant, and does not depend on the time-instance k. We adopt this specific polytopic representation with the same reason as before. In the sequel we denote by G_θ the periodic system given by (7.1) and (7.23) corresponding to $\theta \in \mathbb{E}^L$.

 Our goal in this subsection is to analyze the worst-case H_2 norm defined by

$$\gamma_2^\star := \max_{\theta \in \mathbb{E}^L} \|G_\theta\|_2.$$

In the case where G_θ is not robustly stable under the variation $\theta \in \mathbb{E}^L$, we interpret $\gamma_2^\star = \infty$. Since exact computation of $\gamma_2^\star$ is hard, we are interested in constructing tractable and less conservative methods for the upper bound computation of $\gamma_2^\star$. The SV-LMIs are effective in reducing conservatism.

 Before moving to the SV-LMIs, we show the next basic result obtained straightforwardly from (7.20).

Theorem 7.8 [7] *The uncertain periodic system described by (7.1) and (7.23) is robustly stable and $\gamma_2^\star < \gamma$ holds if there exist $P_k, Z_k \succ 0$ ($k = 0, \ldots, N - 1$) such* *that*

$$-P_k + A_k^{[l]T} P_{k+1} A_k^{[l]} + C_k^{[l]T} C_k^{[l]} \prec 0,$$
$$Z_k \succ B_k^{[l]T} P_{k+1} B_k^{[l]} + D_k^{[l]T} D_k^{[l]} \quad (k = 0, \ldots, N-1, \; l = 1, \ldots, L),$$
$$\text{trace}\left(\sum_{k=0}^{N-1} Z_k\right) < N\gamma^2 \tag{7.24}$$

where $P_N = P_0$.

This is a natural extension of Theorem 7.2 for the robust H_2 performance analysis. On the other hand, along the same line as we derived SV-LMI conditions (7.7) and (7.8) for robust stability analysis, we can obtain the next results.

Theorem 7.9 [7] *The uncertain periodic system described by* (7.1) *and* (7.23) *is robustly stable and* $\gamma_2^\star < \gamma$ *holds if there exist* $P_k^{[l]} \succ 0$, $Z_k \succ 0$, G_k $(k = 0, \ldots, N-1, \; l = 1, \ldots, L)$ *such that*

$$\begin{bmatrix} -P_k^{[l]} + C_k^{[l]T} C_k^{[l]} & 0 \\ 0 & P_{k+1}^{[l]} \end{bmatrix} + \text{He}\left\{ \begin{bmatrix} A_k^{[l]T} \\ -I \end{bmatrix} \begin{bmatrix} 0 & G_{k+1} \end{bmatrix} \right\} \prec 0,$$
$$\begin{bmatrix} -Z_k + D_k^{[l]T} D_k^{[l]} & 0 \\ 0 & P_{k+1}^{[l]} \end{bmatrix} + \text{He}\left\{ \begin{bmatrix} B_k^{[l]T} \\ -I \end{bmatrix} \begin{bmatrix} 0 & G_{k+1} \end{bmatrix} \right\} \prec 0,$$
$$(k = 0, \ldots, N-1, \; l = 1, \ldots, L), \tag{7.25}$$
$$\text{trace}\left(\sum_{k=0}^{N-1} Z_k\right) < N\gamma^2$$

where $P_N^{[l]} = P_0^{[l]}$ $(l = 1, \ldots, L)$ *and* $G_N = G_0$. *Moreover, if the LMI* (7.24) *holds with* $P_k = P_k^\star$ $(k = 0, \ldots, N-1)$, *then the LMI* (7.25) *holds with* $P_k^{[l]} = P_k^\star$ $(k = 0, \ldots, N-1, \; l = 1, \ldots, L)$ *and* $G_k = P_k^\star$ $(k = 0, \ldots, N-1)$.

Theorem 7.10 [7] *The uncertain periodic system described by* (7.1) *and* (7.23) *is robustly stable and* $\gamma_2^\star < \gamma$ *holds if there exist* $P_k^{[l]} \succ 0$, $Z_k \succ 0$, $F_{k,j}$, $H_{k,j}$ $(k = 0, \ldots, N-1, \; l = 1, \ldots, L, \; j = 1, 2)$ *such that*

$$\begin{bmatrix} -P_k^{[l]} + C_k^{[l]T} C_k^{[l]} & 0 \\ 0 & P_{k+1}^{[l]} \end{bmatrix} + \text{He}\left\{ \begin{bmatrix} A_k^{[l]T} \\ -I \end{bmatrix} \begin{bmatrix} F_{k+1,1} & F_{k+1,2} \end{bmatrix} \right\} \prec 0,$$
$$\begin{bmatrix} -Z_k + D_k^{[l]T} D_k^{[l]} & 0 \\ 0 & P_{k+1}^{[l]} \end{bmatrix} + \text{He}\left\{ \begin{bmatrix} B_k^{[l]T} \\ -I \end{bmatrix} \begin{bmatrix} H_{k+1,1} & H_{k+1,2} \end{bmatrix} \right\} \prec 0,$$
$$(k = 0, \ldots, N-1, \; l = 1, \ldots, L), \tag{7.26}$$
$$\text{trace}\left(\sum_{k=0}^{N-1} Z_k\right) < N\gamma^2$$

where $P_N^{[l]} = P_0^{[l]}$ $(l = 1, \ldots, L)$ and $F_{N,j} = F_{0,j}$, $H_{N,j} = H_{0,j}$ $(j = 1, 2)$. Moreover, if the LMI (7.25) holds with $P_k^{[l]} = P_k^{[l]*}$ $(k = 0, \ldots, N-1, l = 1, \ldots, L)$ and $G_k = G_i^{\star}$ $(k = 0, \ldots, N-1)$, then the LMI (7.26) holds with $P_k^{[l]} = P_k^{[l]*}$ $(k = 0, \ldots, N-1, l = 1, \ldots, L)$, $[F_{k,1} \ F_{k,2}] = [0 \ G_k^{\star}]$ $(k = 0, \ldots, N-1)$ and $[H_{k,1} \ H_{k,2}] = [0 \ G_k^{\star}]$ $(k = 0, \cdots, N-1)$.

To compare the effectiveness of the LMI conditions given by Theorems 7.8–7.10, define the SDPs

$$\overline{\gamma_{2}^{\star}}_{qs} := \inf_{P_k, Z_k} \gamma \quad \text{subject to} \quad (7.24),$$
$$\overline{\gamma_{2}^{\star}}_{G} := \inf_{P_k, Z_k, G_k} \gamma \quad \text{subject to} \quad (7.25),$$
$$\overline{\gamma_{2}^{\star}}_{F} := \inf_{P_k, Z_k, F_{k,1}, F_{k,2}, H_{k,1}, H_{k,2}} \gamma \quad \text{subject to} \quad (7.26).$$

Then, we can see from these theorems that

$$\gamma_2^{\star} \le \overline{\gamma_2^{\star}}_F \le \overline{\gamma_2^{\star}}_G \le \overline{\gamma_2^{\star}}_{qs} \tag{7.27}$$

holds. Namely, by using the SV-LMI conditions (7.25) and (7.26) we can obtain no more conservative results for the upper bound computation of the worst-case H_2 norm $\gamma_2^{\star}$.

For robust stability analysis of uncertain periodic systems we derived LMI (7.10) that outperforms LMIs (7.7) and (7.8). Such LMI for robust H_2 performance analysis was obtained in [8, 9] under more general setting enabling memory-type robust controller synthesis. To state the robust H_2 performance analysis condition, let us define $\mathscr{A}^{[l]}$ $(l = 1, \ldots, L)$ by (7.11) as well as

$$\mathscr{C}^{[l]} := [\ \mathscr{C}_u^{[l]} \ 0 \ ] \in \mathbb{R}^{Nn_z \times (N+1)n}, \quad \mathscr{C}_u^{[l]} := \text{diag}(C_0^{[l]}, \ldots, C_{N-1}^{[l]}) \in \mathbb{R}^{Nn_z \times Nn},$$

$$\overline{\mathscr{A}\mathscr{B}}_k^{[l]} := \begin{bmatrix} B_k^{[l]} & -I & 0 & \cdots & \cdots & 0 \\ 0 & A_{k+1}^{[l]} & \ddots & \ddots & & \vdots \\ \vdots & & \ddots & \ddots & & \vdots & \vdots \\ \vdots & & & \ddots & \ddots & \vdots & 0 \\ 0 & \cdots & \cdots & 0 & A_{N-1}^{[l]} & -I \end{bmatrix} \in \mathbb{R}^{(N-k)n \times ((N-k)n + n_w)},$$

$$\overline{\mathscr{C}\mathscr{D}}_k^{[l]} := \begin{bmatrix} D_k^{[l]} & 0 & \cdots & \cdots & 0 \\ 0 & C_{k+1}^{[l]} & \ddots & & \vdots \\ \vdots & \ddots & \ddots & \ddots & \vdots \\ 0 & \cdots & 0 & C_{N-1}^{[l]} & 0 \end{bmatrix} \in \mathbb{R}^{(N-k)n_z \times ((N-k)n_z + n_w)}.$$

Then, we can state the next result.

Theorem 7.11 [8] *The uncertain periodic system described by (7.1) and (7.23) is robustly stable and $\gamma_2^* < \gamma$ holds if there exist $P^{[l]} \succ 0$, $Z_k \succ 0$, F, F_k ($k = 0, \ldots, N-1$, $l = 1, \ldots, L$) such that*

$$\mathscr{P}_N(P^{[l]}) + \mathscr{C}^{[l]T}\mathscr{C}^{[l]} + \mathrm{He}\{\mathscr{A}^{[l]T}F\} \prec 0, \tag{7.28a}$$

$$\mathscr{Z}_k(P^{[l]}, Z_k) + \overline{\mathscr{C}\mathscr{D}}_k^{[l]T}\overline{\mathscr{C}\mathscr{D}}_k + \mathrm{He}\{\overline{\mathscr{A}\mathscr{B}}_k^{[l]T}F_k\} \prec 0 \ (k = 0, \ldots, N-1), \tag{7.28b}$$

$$\mathrm{trace}\left(\sum_{k=0}^{N-1} Z_k\right) < N\gamma^2. \tag{7.28c}$$

where $\mathscr{P}_N(\cdot)$ is defined by (7.11) and

$$\mathscr{Z}_k(P^{[l]}, Z_k) := \begin{bmatrix} -Z_k & 0 & 0 \\ 0 & 0_{(N-k-1)n} & 0 \\ 0 & 0 & P^{[l]} \end{bmatrix}.$$

Moreover, if the LMI (7.8) holds with $P_k^{[l]} = P_k^{[l]\star}$, $[\, F_{k,1} \ F_{k,2} \,] = [\, F_{k,1}^\star \ F_{k,2}^\star \,]$, and $[\, H_{k,1} \ H_{k,2} \,] = [\, H_{k,1}^\star \ H_{k,2}^\star \,]$ ($k = 0, \ldots, N-1$, $l = 1, \ldots, L$), then the LMI (7.10) holds with $P^{[l]} = P_0^{[l]\star}$ ($l = 1, \ldots, L$) and

$$F = \begin{bmatrix} F_{1,1}^\star & F_{1,2}^\star & 0 & \cdots & & & 0 \\ 0 & F_{2,1}^\star & F_{2,2}^\star & 0 & & & \vdots \\ \vdots & \ddots & \ddots & \ddots & & \ddots & \vdots \\ \vdots & & & \ddots & F_{N-1,1}^\star & F_{N-2,2}^\star & 0 \\ 0 & \cdots & & & 0 & F_{0,1}^\star & F_{0,2}^\star \end{bmatrix} \in \mathbb{R}^{Nn \times (N+1)n},$$

$$F_k = \begin{bmatrix} H_{k+1,1}^\star & H_{k+1,2}^\star & 0 & \cdots & & & 0 \\ 0 & F_{k+2,1}^\star & F_{k+2,2}^\star & 0 & & & \vdots \\ \vdots & & \ddots & \ddots & \ddots & & \vdots \\ \vdots & & & & \ddots & \ddots & \vdots \\ & & & \ddots & F_{N-1,1}^\star & F_{N-1,2}^\star & 0 \\ 0 & \cdots & & & 0 & F_{0,1}^\star & F_{0,2}^\star \end{bmatrix} \in \mathbb{R}^{(N-k)n \times ((N-k)n+n_w)}$$

$$(k = 0, \ldots, N-1). \tag{7.29}$$

Let us confirm the validity of Theorem 7.11 for the simplest case $N = 2$. In this case the LMI (7.28a)–(7.28c) can be written explicitly as

$$\mathscr{P}_2(P^{[l]}) + \mathrm{Sq}\left\{\begin{bmatrix} C_0^{[l]} & 0 & 0 \\ 0 & C_1^{[l]} & 0 \end{bmatrix}^T\right\} + \mathrm{He}\left\{\begin{bmatrix} A_0^{[l]} & -I & 0 \\ 0 & A_1^{[l]} & -I \end{bmatrix}^T F\right\} \prec 0, \tag{7.30a}$$

$$\mathscr{Z}_0(P^{[l]}, Z_0) + \mathrm{Sq}\left\{\begin{bmatrix} D_0^{[l]} & 0 & 0 \\ 0 & C_1^{[l]} & 0 \end{bmatrix}^T\right\} + \mathrm{He}\left\{\begin{bmatrix} B_0^{[l]} & -I & 0 \\ 0 & A_1^{[l]} & -I \end{bmatrix}^T F_0\right\} \prec 0,$$

(7.30b)

$$\mathscr{Z}_1(P^{[l]}, Z_1) + \mathrm{Sq}\left\{\begin{bmatrix} D_1^{[l]} & 0 \end{bmatrix}^T\right\} + \mathrm{He}\left\{\begin{bmatrix} B_1^{[l]} & -I \end{bmatrix}^T F_1\right\} \prec 0, \qquad (7.30c)$$

$$\mathrm{trace}\,(Z_0 + Z_1) < 2\gamma^2. \qquad (7.30d)$$

To verify the first statement, suppose (7.30) holds. Then, for every $\theta \in \mathbb{E}^L$ we have

$$\mathscr{P}_2(P(\theta)) + \mathrm{Sq}\left\{\begin{bmatrix} C_0(\theta) & 0 & 0 \\ 0 & C_1(\theta) & 0 \end{bmatrix}^T\right\} + \mathrm{He}\left\{\begin{bmatrix} A_0(\theta) & -I & 0 \\ 0 & A_1(\theta) & -I \end{bmatrix}^T F\right\} \prec 0, \quad (7.31a)$$

$$\mathscr{Z}_0(P(\theta), Z_0) + \mathrm{Sq}\left\{\begin{bmatrix} D_0(\theta) & 0 & 0 \\ 0 & C_1(\theta) & 0 \end{bmatrix}^T\right\} + \mathrm{He}\left\{\begin{bmatrix} B_0(\theta) & -I & 0 \\ 0 & A_1(\theta) & -I \end{bmatrix}^T F_0\right\} \prec 0,$$

(7.31b)

$$\mathscr{Z}_1(P(\theta), Z_1) + \mathrm{Sq}\left\{\begin{bmatrix} D_1(\theta) & 0 \end{bmatrix}^T\right\} + \mathrm{He}\left\{\begin{bmatrix} B_1(\theta) & -I \end{bmatrix}^T F_1\right\} \prec 0, \qquad (7.31c)$$

$$\mathrm{trace}\,(Z_0 + Z_1) < 2\gamma^2 \qquad (7.31d)$$

where $P(\theta) := \sum_{l=1}^L \theta_l P^{[l]}$. By eliminating F from (7.31a) noting

$$\begin{bmatrix} A_0(\theta) & -I & 0 \\ 0 & A_1(\theta) & -I \end{bmatrix}^{\perp} = \begin{bmatrix} I \\ A_0(\theta) \\ A_1(\theta)A_0(\theta) \end{bmatrix},$$

we have

$$-P(\theta) + A_0(\theta)^T A_1(\theta)^T P(\theta) A_1(\theta) A_0(\theta) + A_0(\theta)^T C_1(\theta)^T C_1(\theta) A_0(\theta) + C_0(\theta)^T C_0(\theta) \prec 0$$

or equivalently,

$$-P(\theta) + A_0(\theta)^T \left(A_1(\theta)^T P(\theta) A_1(\theta) + C_1(\theta)^T C_1(\theta) \right) A_0(\theta) + C_0(\theta)^T C_0(\theta) \prec 0.$$

Similarly, by eliminating F_0 from (7.31b) and F_1 from (7.31c), we can obtain

$$Z_0 \succ B_0(\theta)^T \left(A_1(\theta)^T P(\theta) A_1(\theta) + C_1(\theta)^T C_1(\theta) \right) B_0(\theta) + D_0(\theta)^T D_0(\theta),$$
$$Z_1 \succ B_1(\theta)^T P(\theta) B_1(\theta) + D_1(\theta)^T D_1(\theta).$$

Then, there exists $\varepsilon > 0$, which is uniform over $\theta \in \mathbb{E}^L$, such that $P_1(\theta) := A_1(\theta)^T P(\theta) A_1(\theta) + C_1(\theta)^T C_1(\theta) + \varepsilon_1 I$ satisfies

$$-P_0(\theta) + A_0(\theta)^T P_1(\theta) A_0(\theta) + C_0(\theta)^T C_0(\theta) \prec 0,$$
$$-P_1(\theta) + A_1(\theta)^T P_0(\theta) A_1(\theta) + C_1(\theta)^T C_1(\theta) \prec 0,$$
$$Z_0 \succ B_0(\theta)^T P_1(\theta) B_0(\theta) + D_0(\theta)^T D_0(\theta),$$
$$Z_1 \succ B_1(\theta)^T P_0(\theta) B_1(\theta) + D_1(\theta)^T D_1(\theta).$$

where we defined $P_0(\theta) := P(\theta)$. From this set of inequalities, (7.31d), and Theorem 7.7, we have $\|G_\theta\|_2 < \gamma$ ($\forall \theta \in \mathbb{E}^L$) and therefore $\gamma_2^\star < \gamma$.

On the other hand, to verify the second statement, let us write explicitly the LMI condition (7.26) as

$$\begin{bmatrix} -P_0^{[l]\star} + C_0^{[l]T} C_0^{[l]} & 0 \\ 0 & P_1^{[l]\star} \end{bmatrix} + \mathrm{He} \left\{ \begin{bmatrix} A_0^{[l]T} \\ -I \end{bmatrix} [F_{1,1}^\star \; F_{1,2}^\star] \right\} \prec 0, \tag{7.32a}$$

$$\begin{bmatrix} -P_1^{[l]\star} + C_1^{[l]T} C_1^{[l]} & 0 \\ 0 & P_0^{[l]\star} \end{bmatrix} + \mathrm{He} \left\{ \begin{bmatrix} A_1^{[l]T} \\ -I \end{bmatrix} [F_{0,1}^\star \; F_{0,2}^\star] \right\} \prec 0, \tag{7.32b}$$

$$\begin{bmatrix} -Z_0^\star + D_0^{[l]T} D_0^{[l]} & 0 \\ 0 & P_1^{[l]\star} \end{bmatrix} + \mathrm{He} \left\{ \begin{bmatrix} B_0^{[l]T} \\ -I \end{bmatrix} [H_{1,1}^\star \; H_{1,2}^\star] \right\} \prec 0, \tag{7.32c}$$

$$\begin{bmatrix} -Z_1^\star + D_1^{[l]T} D_0^{[l]} & 0 \\ 0 & P_0^{[l]\star} \end{bmatrix} + \mathrm{He} \left\{ \begin{bmatrix} B_1^{[l]T} \\ -I \end{bmatrix} [H_{0,1}^\star \; H_{0,2}^\star] \right\} \prec 0, \tag{7.32d}$$

$$\mathrm{trace} \, (Z_1^\star + Z_2^\star) < 2\gamma^2. \tag{7.32e}$$

Then, applying the same procedure as (7.15)–(7.32a) and (7.32b) and eliminating $P_1^{[l]}$, we can confirm that (7.31a) holds with $P^{[l]} = P_0^{[l]\star}$ and

$$F = \begin{bmatrix} F_{1,1}^\star & F_{1,2}^\star & 0 \\ 0 & F_{0,1}^\star & F_{0,2}^\star \end{bmatrix}.$$

Similarly, eliminating $P_1^{[l]}$ from (7.32b) and (7.32c) we see (7.31b) holds with $P^{[l]} = P_0^{[l]\star}$ and

$$F_0 = \begin{bmatrix} H_{1,1}^\star & H_{1,2}^\star & 0 \\ 0 & F_{0,1}^\star & F_{0,2}^\star \end{bmatrix}.$$

It is obvious from the equivalence of (7.32c) and (7.31c) that (7.31c) holds with

$$F_1 = [H_{0,1}^\star \; H_{0,2}^\star].$$

To write down explicitly the effectiveness of the LMI (7.28a)–(7.28c), define the SDP

$$\overline{\gamma_2^*}_{\text{lift}} := \inf_{P^{[l]}, F, F_k} \gamma \quad \text{subject to} \quad (7.28).$$

Then, we can see from (7.27) and Theorem 7.11 that the next relation holds.

$$\gamma_2^* \leq \overline{\gamma_2^*}_{\text{lift}} \leq \overline{\gamma_2^*}_F \leq \overline{\gamma_2^*}_G \leq \overline{\gamma_2^*}_{\text{qs}}. \tag{7.33}$$

7.4 H_∞ Performance Analysis of Linear Periodic Systems

7.4.1 Definition and Basic Results

For the periodic system G given by (7.1), its l_2 induced-norm is defined by

$$\|G\|_{l_2/l_2} := \sup_{\|w\|_{l_2}=1} \|z\|_{l_2}.$$

Since the input-output behavior of the lifted LTI system $\hat{G}$ is equivalent to that of G, it is obvious that $\|G\|_{l_2/l_2} = \|\hat{G}\|_{l_2/l_2}$. It is well known that the latter l_2 induced-norm defined in the time-domain coincides with the H_∞ norm $\|\hat{G}\|_\infty$ defined in the frequency-domain as in

$$\|\hat{G}\|_\infty := \max_{\phi \in [0, 2\pi]} \|\widehat{G}(e^{j\phi})\|.$$

In view of these facts, we define the H_∞ norm of the periodic system G by $\|\hat{G}\|_\infty$.

Definition 7.2 For the N-periodic system G represented by (7.1), the H_∞ norm $\|G\|_\infty$ is defined by $\|G\|_\infty := \|\hat{G}\|_\infty$. Here, $\|\widehat{G}\|_\infty$ stands for the H_∞ norm of the LTI system $\widehat{G}$ given by (7.16).

Similarly to the LTI case, the H_∞ norm $\|G\|_\infty$ is a reasonable measure to assess the performance of the periodic system (7.1). From Definition 7.2 and KYP-lemma for discrete-time LTI systems, the next theorem readily follows.

Theorem 7.12 For the N-periodic system G described by (7.1) and given $\gamma > 0$, the condition $\|G\|_\infty < \gamma$ holds if and only if there exist $P \succ 0$ such that

$$\begin{bmatrix} -P + \widehat{C}^T \widehat{C} & \widehat{A}^T P & \widehat{C}^T \widehat{D} \\ P\widehat{A} & -P & P\widehat{B} \\ \widehat{D}^T \widehat{C} & \widehat{B}^T P & \widehat{D}^T \widehat{D} - \gamma^2 I \end{bmatrix} \prec 0. \tag{7.34}$$

On the other hand, the following theorem characterizes the H_∞ norm of the periodic system G without resorting to the lifted LTI system $\widehat{G}$. This result is often refereed to periodic KYP lemma [5].

Theorem 7.13 [5] *For the N-periodic system G described by* (7.1) *and given* $\gamma > 0$, *the condition* $\|G\|_\infty < \gamma$ *holds if and only if there exist* $P_k \succ 0$ $(k = 0, \ldots, N-1)$ *such that*

$$\begin{bmatrix} -P_k + C_k^T C_k & A_k^T P_{k+1} & C_k^T D_k \\ P_{k+1} A_k & -P_{k+1} & P_{k+1} B_k \\ D_k^T C_k & B_k^T P_{k+1} & D_k^T D_k - \gamma^2 I \end{bmatrix} \prec 0 \ (k = 0, \ldots, N-1) \quad (7.35)$$

where $P_N = P_0$.

It is worth noting how the LMIs (7.34) and (7.35) are related each other. To see the relation clearly, let us consider the case $N = 2$. Then, (7.35) can be written (after permutation of rows and columns and Schur complement) as

$$\begin{bmatrix} -P_0 + C_0^T C_0 & C_0^T D_0 \\ D_0^T C_0 & D_0^T D_0 - \gamma^2 I \end{bmatrix} + \begin{bmatrix} A_0^T \\ B_0^T \end{bmatrix} P_1 \begin{bmatrix} A_0^T \\ B_0^T \end{bmatrix}^T \prec 0, \quad (7.36a)$$

$$\begin{bmatrix} -P_1 + C_1^T C_1 & C_1^T D_1 & A_1^T P_0 \\ D_1^T C_1 & D_1^T D_1 - \gamma^2 I & B_1^T P_0 \\ P_0 A_1 & P_0 B_1 & -P_0 \end{bmatrix} \prec 0. \quad (7.36b)$$

By applying Schur complement to (7.36b) we have

$$C_1^T C_1 + \begin{bmatrix} C_1^T D_1 & A_1^T P_0 \end{bmatrix} \begin{bmatrix} D_1^T D_1 - \gamma^2 I & B_1^T P_0 \\ P_0 B_1 & -P_0 \end{bmatrix}^{-1} \begin{bmatrix} D_1^T C_1 \\ P_0 A_1 \end{bmatrix} \prec P_1. \quad (7.37)$$

Substituting this inequality to (7.36a), we obtain

$$\begin{bmatrix} -P_0 + C_0^T C_0 & C_0^T D_0 \\ D_0^T C_0 & D_0^T D_0 - \gamma^2 I \end{bmatrix}$$

$$+ \begin{bmatrix} A_0^T \\ B_0^T \end{bmatrix} \left(C_1^T C_1 + \begin{bmatrix} C_1^T D_1 & A_1^T P_0 \end{bmatrix} \begin{bmatrix} D_1^T D_1 - \gamma^2 I & B_1^T P_0 \\ P_0 B_1 & -P_0 \end{bmatrix}^{-1} \begin{bmatrix} D_1^T C_1 \\ P_0 A_1 \end{bmatrix} \right) \begin{bmatrix} A_0^T \\ B_0^T \end{bmatrix}^T \prec 0 \quad (7.38)$$

or equivalently,

$$\begin{bmatrix} -P_0 + A_0^T C_1^T C_1 A_0 + C_0^T C_0 & A_0^T C_1^T C_1 B_0 + C_0^T D_0 & A_0^T C_1^T D_1 & A_0^T A_1^T P_0 \\ * & B_0^T C_1^T C_1 B_0 + D_0^T D_0 - \gamma^2 I & B_0^T C_1^T D_1 & B_0^T A_1^T P_0 \\ * & * & D_1^T D_1 - \gamma^2 I & B_1^T P_0 \\ * & * & * & -P_0 \end{bmatrix} \prec 0. \quad (7.39)$$

From permutation we readily obtain

$$\begin{bmatrix} -P_0 + A_0^T C_1^T C_1 A_0 + C_0^T C_0 & A_0^T A_1^T P_0 & A_0^T C_1^T D_1 & A_0^T C_1^T C_1 B_0 + C_0^T D_0 \\ * & -P_0 & P_0 B_1 & P_0 A_1 B_0 \\ * & * & D_1^T D_1 - \gamma^2 I & D_1^T C_1 B_0 \\ * & * & * & B_0^T C_1^T C_1 B_0 + D_0^T D_0 - \gamma^2 I \end{bmatrix} \prec 0. \quad (7.40)$$

This clearly shows that (7.34) for $N = 2$ holds with $P = P_0$.

On the other hand, suppose (7.34) for $N = 2$ holds with $P = P_0$. Then, we can trace back from (7.40) and arrive at (7.38). Then, there exists $\varepsilon_1 > 0$ such that

$$P_1 := \left(C_1^T C_1 + \left[C_1^T D_1 \ A_1^T P_0 \right] \begin{bmatrix} D_1^T D_1 - \gamma^2 I & B_1^T P_0 \\ P_0 B_1 & -P_0 \end{bmatrix}^{-1} \begin{bmatrix} D_1^T C_1 \\ P_0 A_1 \end{bmatrix} \right) + \varepsilon_1 I,$$

(7.41a)

$$\begin{bmatrix} -P_0 + C_0^T C_0 & C_0^T D_0 \\ D_0^T C_0 & D_0^T D_0 - \gamma^2 I \end{bmatrix} + \begin{bmatrix} A_0^T \\ B_0^T \end{bmatrix} P_1 \begin{bmatrix} A_0^T \\ B_0^T \end{bmatrix}^T \prec 0.$$

(7.41b)

By Schur complement, the former equality (interpreted as inequality without ε_1) can be rewritten as

$$\begin{bmatrix} -P_1 + C_1^T C_1 & C_1^T D_1 & A_1^T P_0 \\ D_1^T C_1 & D_1^T D_1 - \gamma^2 I & B_1^T P_0 \\ P_0 A_1 & P_0 B_1 & -P_0 \end{bmatrix} \prec 0.$$

(7.42)

The inequalities (7.41b) and (7.42) clearly show that (7.36) holds.

7.4.2 Robust H_∞ Performance Analysis Using SV-LMIs

Consider the case where the periodic system (7.1) is subject to the polytopic uncertainty of the form (7.23). Our goal in this subsection is to analyze the worst case H_∞ norm defined by

$$\gamma_\infty^\star := \max_{\theta \in \mathbb{E}^L} \| G_\theta \|_\infty.$$

Similarly to the robust H_2 performance analysis, we interpret $\gamma_\infty^\star = \infty$ if G_θ is not robustly stable under the variation $\theta \in \mathbb{E}^L$. The exact computation of $\gamma_\infty^\star$ is again hard, and hence we compute its upper bounds. For the upper bound computation the next theorem readily follows from Theorem 7.13. The validity of this theorem can be seen from simple convexity arguments on the polytope.

Theorem 7.14 *The uncertain periodic system described by (7.1) and (7.23) is robustly stable and $\gamma_\infty^\star < \gamma$ holds if there exist $P_k \succ 0$ $(k = 0, \ldots, N - 1)$ such that*

$$\begin{bmatrix} -P_k + C_k^{[l]T} C_k^{[l]} & A_k^{[l]T} P_{k+1} & C_k^{[l]T} D_k^{[l]} \\ P_{k+1} A_k^{[l]} & -P_{k+1} & P_{k+1} B_k^{[l]} \\ D_k^{[l]T} C_k^{[l]} & B_k^{[l]T} P_{k+1} & D_k^{[l]T} D_k^{[l]} - \gamma^2 I \end{bmatrix} \prec 0$$

$$(k = 0, \ldots, N - 1, \ l = 1, \ldots, L)$$

(7.43)

where $P_N = P_0$.

This approach is conservative due to the use of common (or fixed) Lyapunov matrix P. The conservatism can be reduced by SV-LMIs given in the next two theorems.

Theorem 7.15 *The uncertain periodic system described by (7.1) and (7.23) is robustly stable and $\gamma_\infty^\star < \gamma$ holds if there exist $P_k^{[l]} \succ 0$, G_k $(k = 0, \ldots, N - 1,\ l = 1, \ldots, L)$ such that*

$$
\begin{bmatrix} -P_k^{[l]} + C_k^{[l]T} C_k^{[l]} & 0 & C_k^{[l]T} D_k^{[l]} \\ 0 & P_{k+1}^{[l]} & 0 \\ D_k^{[l]T} C_k^{[l]} & 0 & D_k^{[l]T} D_k^{[l]} - \gamma^2 I \end{bmatrix} + \mathrm{He}\left\{ \begin{bmatrix} A_k^{[l]T} \\ -I \\ B_k^{[l]T} \end{bmatrix} \begin{bmatrix} 0 & G_{k+1} & 0 \end{bmatrix} \right\} \prec 0
$$
$$(k = 0, \ldots, N - 1,\ l = 1, \ldots, L)$$

$$(7.44)$$

where $P_N^{[l]} = P_0^{[l]}$ $(l = 1, \ldots, L)$ and $G_N = G_0$. Moreover, if the LMI (7.43) holds with $P_k = P_k^\star$ $(k = 0, \ldots, N - 1)$, then the LMI (7.44) holds with $P_k^{[l]} = P_k^\star$ $(k = 0, \ldots, N - 1,\ l = 1, \ldots, L)$ and $G_k = P_k^\star$ $(k = 0, \ldots, N - 1)$.

Theorem 7.16 *The uncertain periodic system described by (7.1) and (7.23) is robustly stable and $\gamma_\infty^\star < \gamma$ holds if there exist $P_k^{[l]} \succ 0$, $F_{k,j}$ $(k = 0, \ldots, N - 1,\ l = 1, \ldots, L,\ j = 1, 2, 3)$ such that*

$$
\begin{bmatrix} -P_k^{[l]} + C_k^{[l]T} C_k^{[l]} & 0 & C_k^{[l]T} D_k^{[l]} \\ 0 & P_{k+1}^{[l]} & 0 \\ D_k^{[l]T} C_k^{[l]} & 0 & D_k^{[l]T} D_k^{[l]} - \gamma^2 I \end{bmatrix} + \mathrm{He}\left\{ \begin{bmatrix} A_k^{[l]T} \\ -I \\ B_k^{[l]T} \end{bmatrix} \begin{bmatrix} F_{k+1,1} & F_{k+1,2} & F_{k+1,3} \end{bmatrix} \right\} \prec 0
$$
$$(k = 0, \ldots, N - 1,\ l = 1, \ldots, L)$$

$$(7.45)$$

where $P_N^{[l]} = P_0^{[l]}$ $(l = 1, \ldots, L)$ and $F_{N,j} = F_{0,j}$ $(j = 1, 2, 3)$. Moreover, if the LMI (7.44) holds with $P_k^{[l]} = P_k^{[l]\star}$ $(k = 0, \ldots, N - 1,\ l = 1 \ldots, L)$ and $G_k = G_k^\star$ $(k = 0, \ldots, N - 1)$, then the LMI (7.45) holds with $P_k^{[l]} = P_k^{[l]\star}$ $(k = 0, \ldots, N - 1,\ l = 1 \ldots, L)$ and $[\, F_{k,1}\ F_{k,2}\ F_{k,3}\,] = [\, 0\ G_k^\star\ 0\,]$ $(k = 0, \ldots, N - 1)$.

To compare the effectiveness of the LMI conditions given by Theorems 7.14–7.16, define the SDPs

$$
\begin{aligned}
\overline{\gamma_{\infty\mathrm{qs}}^\star} &:= \inf_{P_k} \gamma \quad \text{subject to} \quad (7.43), \\
\overline{\gamma_{\infty G}^\star} &:= \inf_{P_k^{[l]}, G_k} \gamma \quad \text{subject to} \quad (7.44), \\
\overline{\gamma_{\infty F}^\star} &:= \inf_{P_k^{[l]}, F_{k,1}, F_{k,2}, F_{k,3}} \gamma \quad \text{subject to} \quad (7.45).
\end{aligned}
$$

$$(7.46)$$

Then, it is a direct consequence from these theorems that

$$\gamma_\infty^\star \le \overline{\gamma_{\infty F}^\star} \le \overline{\gamma_{\infty G}^\star} \le \overline{\gamma_{\infty\mathrm{qs}}^\star} \tag{7.47}$$

holds. Namely, by using the SV-LMI conditions (7.44) and (7.45) we can obtain no more conservative results than the quadratic-stability-based approach (7.43) for the upper bound computation of the worst-case H_∞ norm $\gamma_\infty^\star$.

We next show the LMI condition corresponding to (7.10) and (7.28a)–(7.28c). To this end, define

$$\mathscr{B} := \mathrm{diag}(B_0^{[l]}, \ldots, B_{N-1}^{[l]}) \in \mathbb{R}^{Nn \times Nn_w}, \quad \mathscr{D} := \mathrm{diag}(D_0^{[l]}, \ldots, D_{N-1}^{[l]}) \in \mathbb{R}^{Nn_z \times Nn_w}.$$

Then, we can show that the next theorem holds.

Theorem 7.17 [8] *The uncertain periodic system described by (7.1) and (7.23) is robustly stable and $\gamma_\infty^\star < \gamma$ holds if there exist $P^{[l]} \succ 0$ $(k = 0, \ldots, N-1, l = 1, \ldots, L)$ and $F \in \mathbb{R}^{Nn \times ((N+1)n+Nn_w)}$ such that*

$$\mathscr{P}_{\infty,N}(P^{[l]}, \gamma) + \mathrm{Sq}\left\{[\ \mathscr{C}^{[l]}\ \mathscr{D}^{[l]}\ ]^T\right\} + \mathrm{He}\left\{[\ \mathscr{A}^{[l]}\ \mathscr{B}^{[l]}\ ]^T F\right\} \prec 0 \qquad (7.48)$$

where

$$\mathscr{P}_{N,\infty}(P^{[l]}, \gamma) := \begin{bmatrix} -P^{[l]} & 0 & 0 & 0 \\ 0 & 0_{(N-1)n} & 0 & 0 \\ 0 & 0 & P^{[l]} & 0 \\ 0 & 0 & 0 & -\gamma^2 I \end{bmatrix}.$$

Moreover, if the LMI (7.8) holds with $P_k^{[l]} = P_k^{[l]\star}$ and $[\ F_{k,1}\ F_{k,2}\ F_{k,3}\ ] = [\ F_{k,1}^\star\ F_{k,2}^\star\ F_{k,3}^\star\ ]$, then the LMI (7.10) holds with $P^{[l]} = P_0^{[l]\star}$ $(l = 1, \ldots, L)$ and

$$F = \begin{bmatrix} F_{1,1}^\star & F_{1,2}^\star & 0 & \cdots & \cdots & 0 & F_{1,3}^\star & 0 & \cdots & \cdots & 0 \\ 0 & F_{2,1}^\star & F_{2,2}^\star & 0 & & \vdots & 0 & F_{2,3}^\star & \ddots & & \vdots \\ \vdots & \ddots & \ddots & \ddots & \ddots & \vdots & \vdots & \ddots & \ddots & \ddots & \vdots \\ \vdots & & \ddots & \ddots & \ddots & 0 & \vdots & & \ddots & \ddots & 0 \\ 0 & \cdots & \cdots & 0 & F_{0,1}^\star & F_{0,2}^\star & 0 & \cdots & \cdots & 0 & F_{0,3}^\star \end{bmatrix} \in \mathbb{R}^{Nn \times ((N+1)n+Nn_w)}.$$

$$(7.49)$$

This theorem follows from [8] and dual system representation of the periodic system (7.1). Still we illustrate here its validity for the case $N = 2$. To verify the first assertion, suppose (7.48) holds. Then, for every $\theta \in \mathbb{E}^L$ we have

$$\mathscr{P}_{\infty,2}(P(\theta)), \gamma) + \mathrm{Sq}\left\{[\ \mathscr{C}(\theta)\ \mathscr{D}(\theta)\ ]^T\right\} + \mathrm{He}\left\{[\ \mathscr{A}(\theta)\ \mathscr{B}(\theta)\ ]^T F\right\} \prec 0 \quad (7.50)$$

where $P(\theta) := \sum_{l=1}^{L} \theta_l P^{[l]}$ and

$$\mathscr{P}_{\infty,2}(P(\theta),\gamma) = \begin{bmatrix} -P(\theta) & 0 & 0 & 0 \\ 0 & 0_{n,n} & 0 & 0 \\ 0 & 0 & P(\theta) & 0 \\ 0 & 0 & 0 & -\gamma^2 I \end{bmatrix},$$

$$[\,\mathscr{A}(\theta)\ \mathscr{B}(\theta)\,] = \begin{bmatrix} A_0(\theta) & -I & 0 & B_0(\theta) & 0 \\ 0 & A_1(\theta) & -I & 0 & B_1(\theta) \end{bmatrix},$$

$$[\,\mathscr{C}(\theta)\ \mathscr{D}(\theta)\,] = \begin{bmatrix} C_0(\theta) & 0 & 0 & D_0(\theta) & 0 \\ 0 & C_1(\theta) & 0 & 0 & D_1(\theta) \end{bmatrix}.$$

Since

$$[\,\mathscr{A}(\theta)\ \mathscr{B}(\theta)\,]^\perp = \begin{bmatrix} I & 0 & 0 \\ A_0(\theta) & 0 & B_0(\theta) \\ A_1(\theta)A_0(\theta) & B_1(\theta) & A_1(\theta)B_0(\theta) \\ 0 & 0 & I \\ 0 & I & 0 \end{bmatrix}$$

holds, we see

$$[\,\mathscr{A}(\theta)\ \mathscr{B}(\theta)\,]^{\perp T}\mathscr{P}_{\infty,2}(P(\theta),\gamma)[\,\mathscr{A}(\theta)\ \mathscr{B}(\theta)\,]^\perp$$
$$= \begin{bmatrix} -P(\theta) & 0 \\ 0 & -\gamma^2 I \end{bmatrix}$$
$$+ \begin{bmatrix} (A_1(\theta)A_0(\theta))^T \\ [\,B_1(\theta)\ A_1(\theta)B_0(\theta)]^T \end{bmatrix} P(\theta)\left[\,A_1(\theta)A_0(\theta)\ [\,B_1(\theta)\ A_1(\theta)B_0(\theta)]\,\right]$$
$$= \begin{bmatrix} -P(\theta) & 0 \\ 0 & -\gamma^2 I \end{bmatrix} + \begin{bmatrix} \widehat{A}(\theta)^T \\ \widehat{B}(\theta)^T \end{bmatrix} P \begin{bmatrix} \widehat{A}(\theta)^T \\ \widehat{B}(\theta)^T \end{bmatrix}^T,$$

$$[\,\mathscr{A}(\theta)\ \mathscr{B}(\theta)\,]^{\perp T}\mathrm{Sq}\left\{[\,\mathscr{C}(\theta)\ \mathscr{D}(\theta)\,]^T\right\}[\,\mathscr{A}(\theta)\ \mathscr{B}(\theta)\,]^\perp$$
$$= \mathrm{Sq}\left\{ \begin{bmatrix} C_1(\theta)A_0(\theta) & D_1(\theta) & C_1(\theta)B_0(\theta) \\ C_0(\theta) & 0 & D_0(\theta) \end{bmatrix}^T \right\}$$
$$= \mathrm{Sq}\left\{[\,\widehat{C}(\theta)\ \widehat{D}(\theta)\,]\right\}.$$

It follows from (7.50) that

$$\begin{bmatrix} -P(\theta) + \widehat{C}(\theta)^T\widehat{C}(\theta) & \widehat{C}(\theta)^T\widehat{D}(\theta) \\ \widehat{D}(\theta)^T\widehat{C}(\theta) & \widehat{D}(\theta)^T\widehat{D}(\theta) - \gamma^2 I \end{bmatrix} + \begin{bmatrix} \widehat{A}(\theta)^T \\ \widehat{B}(\theta)^T \end{bmatrix} P(\theta) \begin{bmatrix} \widehat{A}(\theta)^T \\ \widehat{B}(\theta)^T \end{bmatrix}^T \prec 0.$$
$$(7.51)$$

This clearly shows that $\|G_\theta\|_\infty < \gamma$ and hence $\gamma^\star_\infty < \gamma$.

To see that the second assertion holds for $N = 2$, suppose the set of the lth LMI in (7.45) holds with $P_k^{[l]} = P_k^{[l]\star}$ ($k = 0, 1$) and $F_{k,j} = F^\star_{k,j}$ ($k = 0, 1,\ j = 1, 2, 3$). The corresponding LMIs can be written explicitly as

$$\mathscr{D}_{0,l} := \begin{bmatrix} -P_0^{[l]} + C_0^{[l]T} C_0^{[l]} & 0 & C_0^{[l]T} D_0^{[l]} \\ 0 & P_1^{[l]} & 0 \\ D_0^{[l]T} C_0^{[l]} & 0 & D_0^{[l]T} D_0^{[l]} - \gamma^2 I \end{bmatrix} + \mathrm{He}\left\{ \begin{bmatrix} A_0^{[l]T} \\ -I \\ B_0^{[l]T} \end{bmatrix} \begin{bmatrix} F_{1,1}^\star & F_{1,2}^\star & F_{1,3}^\star \end{bmatrix} \right\} \prec 0,$$

$$\tag{7.52}$$

$$\mathscr{D}_{1,l} := \begin{bmatrix} -P_1^{[l]} + C_1^{[l]T} C_1^{[l]} & 0 & C_1^{[l]T} D_1^{[l]} \\ 0 & P_0^{[l]} & 0 \\ D_1^{[l]T} C_1^{[l]} & 0 & D_1^{[l]T} D_1^{[l]} - \gamma^2 I \end{bmatrix} + \mathrm{He}\left\{ \begin{bmatrix} A_1^{[l]T} \\ -I \\ B_1^{[l]T} \end{bmatrix} \begin{bmatrix} F_{0,1}^\star & F_{0,2}^\star & F_{0,3}^\star \end{bmatrix} \right\} \prec 0.$$

$$\tag{7.53}$$

Then, for each l $(l = 1, \ldots, L)$, we have

$$\begin{bmatrix} I & 0 & 0 & 0 & 0 \\ 0 & I & 0 & 0 & 0 \\ 0 & 0 & 0 & I & 0 \\ 0 & I & 0 & 0 & 0 \\ 0 & 0 & I & 0 & 0 \\ 0 & 0 & 0 & 0 & I \end{bmatrix}^T \begin{bmatrix} \mathscr{D}_{0,l} & 0 \\ 0 & \mathscr{D}_{1,l} \end{bmatrix} \begin{bmatrix} I & 0 & 0 & 0 & 0 \\ 0 & I & 0 & 0 & 0 \\ 0 & 0 & 0 & I & 0 \\ 0 & I & 0 & 0 & 0 \\ 0 & 0 & I & 0 & 0 \\ 0 & 0 & 0 & 0 & I \end{bmatrix} \prec 0$$

$$\Leftrightarrow \begin{bmatrix} -P_0^{[l]\star} & 0 & 0 & 0 & 0 \\ 0 & 0_n & 0 & 0 & 0 \\ 0 & 0 & P_0^{[l]\star} & 0 & 0 \\ 0 & 0 & 0 & -\gamma^2 I & 0 \\ 0 & 0 & 0 & 0 & -\gamma^2 I \end{bmatrix} + \mathrm{Sq}\left\{ \begin{bmatrix} C_0 & 0 & 0 & D_0 & 0 \\ 0 & C_1 & 0 & 0 & D_1 \end{bmatrix}^T \right\}$$

$$\tag{7.54}$$

$$+ \mathrm{He}\left\{ \begin{bmatrix} A_0^{[l]T} & 0 \\ -I & A_1^{[l]T} \\ 0 & -I \\ B_0^{[l]T} & 0 \\ 0 & B_1^{[l]T} \end{bmatrix} \begin{bmatrix} F_{1,1}^\star & F_{1,2}^\star & 0 & F_{1,3}^\star & 0 \\ 0 & F_{0,1}^\star & F_{0,2}^\star & 0 & F_{0,3}^\star \end{bmatrix} \right\} \prec 0.$$

This shows that (7.48) for $N = 2$ holds with $P^{[l]} = P_0^{[l]\star}$ $(l = 1, \ldots, L)$ and F given by (7.49).

To write down explicitly the effectiveness of the LMI (7.48), define the SDP

$$\overline{\gamma_{\infty \text{lift}}^\star} := \inf_{P^{[l]}, F} \gamma \quad \text{subject to} \quad (7.48). \tag{7.55}$$

Then, we can see from (7.47) and Theorem 7.17 that the next relation holds.

$$\gamma_\infty^\star \le \overline{\gamma_{\infty \text{lift}}^\star} \le \overline{\gamma_{\infty F}^\star} \le \overline{\gamma_{\infty G}^\star} \le \overline{\gamma_{\infty \text{qs}}^\star}. \tag{7.56}$$

7.5 Numerical Examples

In this section, we illustrate the effectiveness of the LMI conditions introduced in the previous sections by numerical examples. The examples include robust stability analysis of uncertain LPTV systems and robust H_∞ performance analysis

Table 7.1 Computed results
for the robust stability
analysis

	Lower bounds	Number of LMI variables
$\underline{\delta}_{qs}$	1.2192	$Nn(n+1)/2 = 9$
$\underline{\delta}_G$	1.5195	$LNn(n+1)/2 + Nn^2 = 39$
$\underline{\delta}_F$	1.6283	$LNn(n+1)/2 + 2Nn^2 = 51$
$\underline{\delta}_{lift}$	1.6800	$Ln(n+1)/2 + Nn(N+1)n = 57$

of uncertain LTI systems. In the following numerical computation, we solved SDPs
with MATLAB R2011b and SeDuMi [10] and YALMIP [11] on a PC with Intel(R)
Core(TM)2 Extreme CPU X9770 3.20 GHz.

7.5.1 Robust Stability Analysis

Let us consider a 3-periodic uncertain LPTV system described by (7.1) and (7.4).
We assume that $L = 3$ in (7.4) and

$$
\begin{aligned}
\left[A_0^{[1]} \ A_1^{[1]} \ A_2^{[1]} \right] &= \delta \begin{bmatrix} 0.8 & 0.4 & 0.6 & -0.1 & 0.0 & -0.4 \\ -0.2 & -0.9 & 0.5 & -0.7 & 0.4 & 0.5 \end{bmatrix}, \\
\left[A_0^{[2]} \ A_1^{[2]} \ A_2^{[2]} \right] &= \delta \begin{bmatrix} -0.7 & 0.3 & 0.2 & -0.4 & 0.7 & 0.1 \\ -0.5 & 0.5 & 0.0 & -0.9 & -0.5 & 0.1 \end{bmatrix}, \\
\left[A_0^{[3]} \ A_1^{[3]} \ A_2^{[3]} \right] &= \delta \begin{bmatrix} -0.4 & 0.3 & 0.0 & 0.1 & 0.5 & -0.2 \\ 0.2 & -0.4 & 0.0 & 0.3 & -0.2 & -0.3 \end{bmatrix},
\end{aligned}
\tag{7.57}
$$

where $\delta > 0$. Our goal here is to determine the maximal value of δ, denoted by δ_{max},
such that the LPTV system is stable for all $\theta \in \mathbb{E}^3$ and $0 \le \delta < \delta_{max}$. We note that
this example is borrowed from [4].

To compute the lower bounds of δ_{max}, we carry out a bisection search over δ
by means of the LMIs (7.6)–(7.8) and (7.10). The resulting lower bounds, denoted
by $\underline{\delta}_{qs}$, $\underline{\delta}_G$, $\underline{\delta}_F$, and $\underline{\delta}_{lift}$, respectively, are shown in Table 7.1, accompanied by the
number of LMI variables. We also show the eigenvalue plots of $\Phi(\theta)$ over $\theta \in \mathbb{E}^3$
for $\delta = \underline{\delta}_{qs} = 1.2192$, $\delta = \underline{\delta}_G = 1.5195$, $\delta = \underline{\delta}_F = 1.6283$, and $\delta = \underline{\delta}_{lift} = 1.6800$
in Figs. 7.1, 7.2, 7.3 and 7.4, respectively.

From Fig. 7.1, we can infer that the lower bound $\underline{\delta}_{qs} = 1.2192$ is very conservative, and Fig. 7.2 shows that $\delta = \underline{\delta}_G = 1.5195$ is still far from exact value.
The lower bound $\delta = \underline{\delta}_F = 1.6283$ outperforms those preceding results, but
Fig. 7.3 clearly shows room for further improvement. On the other hand, in Fig. 7.4
we see that eigenvalues of $\Phi(\theta)$ come close to the stability boundary. Indeed the
value 1.6800 can be numerically verified to be exact, with the worst-case parameter
$\theta_{wc} = [\ 0.8490 \quad 0.1510 \quad 0.0000\ ]$ achieving $\rho(\Phi(\theta_{wc})) = 1$. See [4] for the details
of the exactness verification.

Fig. 7.1 Eigenvalue plots of $\Phi(\theta)$ over $\theta \in \mathbb{E}^3$ $(\delta = \underline{\delta}_{qs} = 1.2192)$

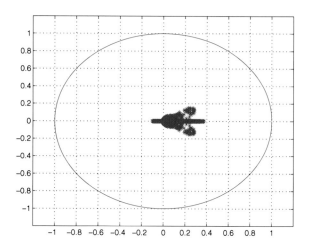

Fig. 7.2 Eigenvalue plots of $\Phi(\theta)$ over $\theta \in \mathbb{E}^3$ $(\delta = \underline{\delta}_G = 1.5195)$

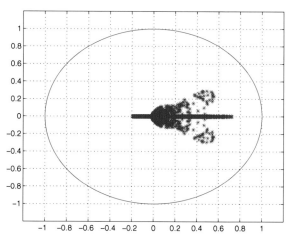

7.5.2 Robust H_∞ Performance Analysis

We next consider the robust H_∞ performance analysis problem of uncertain periodic systems. We consider the case where $N = 3$ and $L = 3$ in (7.23) and the corresponding vertex matrices are given by

$$C_k^{[l]} = \begin{bmatrix} 0.8\ 0.1 \end{bmatrix}, \quad D_k^{[l]} = 0.3 \quad (k = 0, 1, 2, \ l = 1, 2, 3),$$

Fig. 7.3 Eigenvalue plots of $\Phi(\theta)$ over $\theta \in \mathbb{E}^3$ $(\delta = \underline{\delta}_F = 1.6283)$

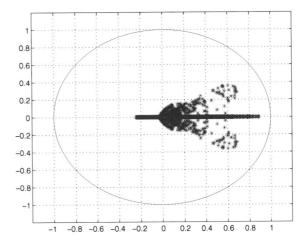

Fig. 7.4 Eigenvalue plots of $\Phi(\theta)$ over $\theta \in \mathbb{E}^3$ $(\delta = \underline{\delta}_{\text{lift}} = 1.6800)$

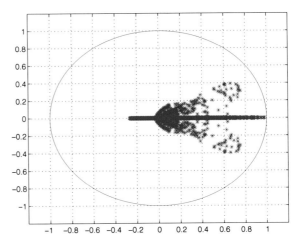

$$
\begin{bmatrix} B_0^{[1]} & B_1^{[1]} & B_2^{[1]} \end{bmatrix} = \begin{bmatrix} 0.2 & 0.3 & 0.4 \\ -0.5 & -0.5 & -0.5 \end{bmatrix},
$$
$$
\begin{bmatrix} B_0^{[2]} & B_1^{[2]} & B_2^{[2]} \end{bmatrix} = \begin{bmatrix} 0.3 & 0.4 & 0.5 \\ -0.4 & -0.4 & -0.4 \end{bmatrix}, \tag{7.58}
$$
$$
\begin{bmatrix} B_0^{[3]} & B_1^{[3]} & B_2^{[3]} \end{bmatrix} = \begin{bmatrix} 0.7 & 0.8 & 0.9 \\ -0.3 & -0.3 & -0.3 \end{bmatrix},
$$

and $A_k^{[l]}$ $(k = 0, 1, 2, \; l = 1, 2, 3)$ are given by (7.57) with $\delta = 1.2$.

To compare the effectiveness of the LMIs shown in Theorems 7.14–7.17, we solved the SDPs given in (7.46) and (7.55) and computed the upper bounds of the worst-case H_∞ norm $\gamma_\infty^\star$ denoted by $\overline{\gamma_{\infty\text{qs}}^\star}$, $\overline{\gamma_{\infty G}^\star}$, $\overline{\gamma_{\infty F}^\star}$, and $\overline{\gamma_{\infty\text{lift}}^\star}$, respectively. The results are shown in Table 7.2 accompanied by the number of LMI variables.

Table 7.2 Computed upper bounds of the worst-case H_∞ norm

	Upper bounds	Number of LMI variables
$\overline{\gamma^\star_{\infty qs}}$	15.5148	$Nn(n+1)/2 = 9$
$\overline{\gamma^\star_{\infty G}}$	1.4090	$LNn(n+1)/2 + Nn^2 = 39$
$\overline{\gamma^\star_{\infty F}}$	1.2728	$LNn(n+1)/2 + Nn(2n+n_w) = 57$
$\overline{\gamma^\star_{\infty lift}}$	1.2224	$Ln(n+1)/2 + Nn((N+1)n + Nn_w) = 75$

Table 7.3 H_∞ norm of G_θ on each vertex

θ	$\|G_\theta\|_\infty$
$[\,1\,0\,0\,]^T$	0.7788
$[\,0\,1\,0\,]^T$	1.2224
$[\,0\,0\,1\,]^T$	1.1891

From Table 7.2 we can confirm that the relation (7.56) holds among the values $\overline{\gamma^\star_{\infty qs}}$, $\overline{\gamma^\star_{\infty G}}$, $\overline{\gamma^\star_{\infty F}}$, and $\overline{\gamma^\star_{\infty lift}}$. In particular, the value $\overline{\gamma^\star_{\infty qs}}$ is large, implying that the quadratic-stability-based approach given in (7.43) is quite conservative. On the other hand, the conservatism is drastically reduced by SV-LMI approach as shown by $\overline{\gamma^\star_{\infty G}}$, $\overline{\gamma^\star_{\infty F}}$ and $\overline{\gamma^\star_{\infty lift}}$. Of course the conservatism reduction has been obtained at the expense of the increased number of variables.

To examine how the computed values are close to exact $\gamma^\star_\infty$, we show the H_∞ norm of G_θ for each vertex in Table 7.3. From this table, we see that $\|G_\theta\|_\infty = 1.2224$ for $\theta = [\,0\,1\,0\,]^T$. Therefore, $\overline{\gamma^\star_{\infty lift}}$ computed from Theorem 7.17 is numerically verified to be exact.

References

1. Bittanti S, Colaneri P (1996) Analysis of discrete-time linear periodic systems. In: Leondes CT (ed) Control and dynamic systems. Academic Press, New York, pp 313–339
2. Bittanti S, Bolzern P, Colaneri P (1985) The extended periodic Lyapunov lemma. Automatica 21:603–605
3. De Souza CE, Trofino A (2000) An LMI approach to stabilization of linear discrete-time periodic systems. Int J Control 73:696–709
4. Ebihara Y, Peaucelle D, Arzelier D (2010) Analysis of uncertain discrete-time linear periodic systems based on system lifting and LMIs. Eur J Control 16(5):532–544
5. Bittanti S, Colaneri P (2009) Periodic systems: filtering and control. Springer, London
6. Zhang C, Zhang J, Furuta K (1997) Analysis of H_2 and H_∞ performance of discrete periodically time-varying controllers. Automatica 33(4):619–634
7. Farges C, Peaucelle D, Arzelier D, Daafouz J (2007) Robust H_2 performance analysis and synthesis of linear polytopic discrete-time periodic systems via LMIs. Syst Control Lett 56:159–166
8. Ebihara Y, Peaucelle D, Arzelier D (2008) Periodically time-varying dynamical controller synthesis for polytopic-type uncertain discrete-time linear systems. In: Proceedings of the conference on decision and control, pp 5438–5443

9. Ebihara Y (2013) Periodically time-varying memory state-feedback for robust H_2 control of uncertain discrete-time linear systems. Asian J Control 15(2):409–419
10. Sturm JF (1999) Using SeDuMi 1.02, a MATLAB toolbox for optimization over symmetric cones. Optim Methods Softw 11–12:625–653
11. Löfberg J (2004) YALMIP: a toolbox for modeling and optimization in MATLAB. In: Proceedings of the IEEE computer aided control system design, pp 284–289

Chapter 8
Robust Controller Synthesis of Periodic Discrete-Time Systems

8.1 Introduction

Consider discrete-time linear N-periodic system P described by

$$P : \begin{cases} x(k+1) = A_k x(k) + B_k w(k) + E_k u(k), \\ z(k) \;\;\;= C_k x(k) + D_k w(k) + F_k u(k), \end{cases} \tag{8.1}$$

where $x \in \mathbb{R}^n$ is the state, $w \in \mathbb{R}^{n_w}$ the disturbance input, $z \in \mathbb{R}^{n_z}$ the performance output, and $u \in \mathbb{R}^{n_u}$ the control input, respectively. As in the previous section, matrices $A_k \in \mathbb{R}^{n \times n}$, $B_k \in \mathbb{R}^{n \times n_w}$, $C_k \in \mathbb{R}^{n_z \times n}$, $D_k \in \mathbb{R}^{n_z \times n_w}$, $E_k \in \mathbb{R}^{n \times n_u}$, $F_k \in \mathbb{R}^{n_z \times n_u}$, are all N-periodic for all $k \geq 0$.

For this LPTV system, we design periodic static state-feedback controller K of the form

$$K : u_k = K_k x_k \tag{8.2}$$

where $K_k \in \mathbb{R}^{n_u \times n}$ are N-periodic for all $k \geq 0$, i.e., $K_{k+N} = K_k$. The resulting closed-loop system can be rewritten as:

$$P_K : \begin{cases} x(k+1) = (A_k + E_k K_k) x(k) + B_k w(k), \\ z(k) \;\;\;= (C_k + F_k K_k) x(k) + D_k w(k). \end{cases} \tag{8.3}$$

Our goal in this section is to provide LMI conditions for the synthesis of K such that P_K becomes stable, and further, $\|P_K\|_2$ or $\|P_K\|_\infty$ is minimized. In particular, in the case where P is affected by polytopic uncertainties, we show that SV-LMI conditions are effective for conservatism reduction.

© Springer-Verlag London 2015
Y. Ebihara et al., *S-Variable Approach to LMI-Based Robust Control*,
Communications and Control Engineering, DOI 10.1007/978-1-4471-6606-1_8

8.2 Dual System

In this section, we quickly review the notion of dual system [1] for periodic systems. In Chap. 7, we consider stability and performance analysis of N-periodic system G described by

$$G : \begin{cases} x(k+1) = A_k x(k) + B_k w(k), \\ z(k) \quad = C_k x(k) + D_k w(k) \end{cases} \tag{8.4}$$

where $A_k \in \mathbb{R}^{n \times n}$, $B_k \in \mathbb{R}^{n \times n_w}$, $C_k \in \mathbb{R}^{n_z \times n}$, and $D_k \in \mathbb{R}^{n_z \times n_w}$ are all N-periodic. For this LPTV system G, its *dual system*, denoted by G^{d}, is defined by

$$G^{\mathrm{d}} : \begin{cases} x^{\mathrm{d}}(k+1) = A_k^{\mathrm{d}} x^{\mathrm{d}}(k) + B_k^{\mathrm{d}} w^{\mathrm{d}}(k), \\ z^{\mathrm{d}}(k) \quad = C_k^{\mathrm{d}} x^{\mathrm{d}}(k) + D_k^{\mathrm{d}} w^{\mathrm{d}}(k) \end{cases} \tag{8.5}$$

where $A_k^{\mathrm{d}} \in \mathbb{R}^{n \times n}$, $B_k^{\mathrm{d}} \in \mathbb{R}^{n \times n_z}$, $C_k^{\mathrm{d}} \in \mathbb{R}^{n_w \times n}$, and $D_k^{\mathrm{d}} \in \mathbb{R}^{n_w \times n_z}$ are all N-periodic and defined by

$$A_k^{\mathrm{d}} := A_{N-k-1}^T, \quad B_k^{\mathrm{d}} := C_{N-k-1}^T, \quad C_k^{\mathrm{d}} := B_{N-k-1}^T, \quad D_k^{\mathrm{d}} := D_{N-k-1}^T. \tag{8.6}$$

It is known that the stability and input–output system gains of the dual system G^{d} coincide with those of the original system G. To see this, let us apply the discrete-time system lifting to G^{d} and obtain

$$\widehat{G^{\mathrm{d}}} : \begin{cases} \xi_{\kappa+1}^{\mathrm{d}} = \widehat{A^{\mathrm{d}}} \xi_\kappa^{\mathrm{d}} + \widehat{B^{\mathrm{d}}} \widehat{w}_\kappa^{\mathrm{d}}, \\ \widehat{z}_\kappa^{\mathrm{d}} = \widehat{C^{\mathrm{d}}} \xi_\kappa^{\mathrm{d}} + \widehat{D^{\mathrm{d}}} \widehat{w}_\kappa^{\mathrm{d}} \end{cases} \tag{8.7}$$

where

$$\xi_\kappa^{\mathrm{d}} = x_{\kappa N}^{\mathrm{d}}, \quad \widehat{w}_\kappa^{\mathrm{d}} = \begin{bmatrix} w_{\kappa N}^{\mathrm{d}} \\ \vdots \\ w_{(\kappa+1)N-1}^{\mathrm{d}} \end{bmatrix}, \quad \widehat{z}_\kappa^{\mathrm{d}} = \begin{bmatrix} z_{\kappa N}^{\mathrm{d}} \\ \vdots \\ z_{(\kappa+1)N-1}^{\mathrm{d}} \end{bmatrix} \tag{8.8}$$

and

$$\left[\begin{array}{c|c} \widehat{A^{\mathrm{d}}} & \widehat{B^{\mathrm{d}}} \\ \hline \widehat{C^{\mathrm{d}}} & \widehat{D^{\mathrm{d}}} \end{array} \right] =$$

$$\left[\begin{array}{c|cccc} \prod_{k=0}^{N-1} A_k^{\mathrm{d}} & \left(\prod_{k=1}^{N-1} A_k^{\mathrm{d}}\right) B_0^{\mathrm{d}} & \cdots & A_{N-1}^{\mathrm{d}} B_{N-2}^{\mathrm{d}} & B_{N-1}^{\mathrm{d}} \\ \hline C_0^{\mathrm{d}} & D_0^{\mathrm{d}} & 0 & \cdots & 0 \\ \vdots & \vdots & \ddots & \ddots & \vdots \\ C_{N-2}^{\mathrm{d}}(\prod_{k=0}^{N-3} A_k^{\mathrm{d}}) & C_{N-2}^{\mathrm{d}}\left(\prod_{k=1}^{N-3} A_k^{\mathrm{d}}\right) B_0^{\mathrm{d}} & \cdots & D_{N-2}^{\mathrm{d}} & 0 \\ C_{N-1}^{\mathrm{d}}(\prod_{k=0}^{N-2} A_k^{\mathrm{d}}) & C_{N-1}^{\mathrm{d}}\left(\prod_{k=1}^{N-2} A_k^{\mathrm{d}}\right) B_0^{\mathrm{d}} & \cdots & C_{N-1}^{\mathrm{d}} B_{N-2}^{\mathrm{d}} & D_{N-1}^{\mathrm{d}} \end{array}\right], \tag{8.9}$$

$$\prod_{k=p}^{q} A_k^{\mathrm{d}} := A_q^{\mathrm{d}} \cdots A_{p+1}^{\mathrm{d}} A_p^{\mathrm{d}} \ (q \ge p), \quad \prod_{k=p}^{q} A_k^{\mathrm{d}} := I \ (q < p).$$

Here, in contrast to (7.17), we stack the input and output signals in reversing order in (8.8). Note that the stability and input-output properties of system G^{d} is inherited to $\widehat{G}^{\mathrm{d}}$.

By comparing (7.18) and (8.9) and noting (8.6), we can confirm that

$$\widehat{A^{\mathrm{d}}} = \widehat{A}^T, \quad \widehat{B^{\mathrm{d}}} = \widehat{C}^T, \quad \widehat{C^{\mathrm{d}}} = \widehat{B}^T, \quad \widehat{D^{\mathrm{d}}} = \widehat{D}^T.$$

Hence, the LTI system $\widehat{G}^{\mathrm{d}}$ is the dual system of the LTI system $\widehat{G}$. It follows that $\widehat{G}$ is stable if and only if $\widehat{G}^{\mathrm{d}}$ is stable. Moreover, $\|\widehat{G}\|_2 = \|\widehat{G}^{\mathrm{d}}\|_2$ and $\|\widehat{G}\|_\infty = \|\widehat{G}^{\mathrm{d}}\|_\infty$ hold. We therefore conclude that G is stable if and only if G^{d} is stable. Moreover, we have $\|G\|_2 = \|G^{\mathrm{d}}\|_2$ and $\|G\|_\infty = \|G^{\mathrm{d}}\|_\infty$.

The necessity of introducing dual system is exactly the same as in the LTI case but recalled here for clarity. If we apply LMI results given in Chap. 7 to the closed-loop system P_K given by (8.3), the coefficient matrices $A_k + E_k K_k$ and $C_k + F_k K_k$ has multiplication with Lyapunov matrices or S-variables from left-hand side. These bilinear terms cannot be linearized by change of variables because of the interruption by E_k and F_k. On the other hand, if we consider the dual system representation, the coefficient matrices $A_k + E_k K_k$ and $C_k + F_k K_k$ are multiplied by Lyapunov matrices or S-variables from right-hand side, which enables change of variables for linearization. We clarify the technique for the linearizing change of variables more concretely in the subsequent sections.

8.3 Stabilizing Periodic State-Feedback Controller Synthesis

8.3.1 Basic Results

For LPTV system G given by (8.1), consider the synthesis of periodic state-feedback controller K of the form (8.2) such that the closed-loop system (8.3) becomes stable.

From Theorem 7.1 and the results reviewed for dual system, we can readily obtain the next theorem.

Theorem 8.1 *For LPTV system G given by (8.1), the following statements are equivalent:*

(i) *There exists a periodic state-feedback controller K of the form (8.2) such that the closed-loop system P_K becomes stable.*
(ii) *There exist $X_k \succ 0$ and K_k $(k = 0, \ldots, N - 1)$ such that*

$$-X_{k+1} + (A_k + E_k K_k) X_k (A_k + E_k K_k)^T \prec 0 \quad (k = 0, \ldots, N - 1). \quad (8.10)$$

where $X_N = X_0$.
(iii) *There exist $X_k \succ 0$ and Y_k $(k = 0, \ldots, N - 1)$ such that*

$$\begin{bmatrix} -X_{k+1} & A_k X_k + E_k Y_k \\ (A_k X_k + E_k Y_k)^T & -X_k \end{bmatrix} \prec 0 \quad (k = 0, \ldots, N - 1). \quad (8.11)$$

where $X_N = X_0$.

Moreover, if the system of LMIs (8.11) is feasible, then a stabilizing controller of the form (8.2) is given by

$$K_k = Y_k X_k^{-1} \quad (k = 0, \ldots, N - 1). \quad (8.12)$$

Proof We fist prove the equivalence of (i) and (ii). From Theorem 7.2 and the dual system representation given by (8.6), we see that (i) holds if and only if there exist $P_k \succ 0$ and K_k $(k = 0, \ldots, N - 1)$ such that

$$-P_k + (A_{N-k-1} + E_{N-k-1} K_{N-k-1}) P_{k+1} (A_{N-k-1} + E_{N-k-1} K_{N-k-1})^T \prec 0$$

$$(k = 0, \ldots, N - 1).$$

If we define $k' := N - k - 1$ and $X_{k'} := P_{N-k'}$, we have

$$-X_{k'+1} + (A_{k'} + E_{k'} K_{k'}) X_{k'} (A_{k'} + E_{k'} K_{k'}) \prec 0 \quad (k' = 0, \ldots, N - 1)$$

where $X_0 = P_N = P_0$ and $X_N = P_0$ and hence $X_N = X_0$. This proves (i)$\Leftrightarrow$(ii).

The equivalence of (ii) and (iii) follows immediately. From Schur complement, we see that (ii) is equivalent to

$$\begin{bmatrix} -X_{k+1} & (A_k + E_k K_k) X_k \\ X_k (A_k + E_k K_k)^T & -X_k \end{bmatrix} \prec 0 \quad (k = 0, \ldots, N-1).$$

By applying change of variables

$$Y_k := K_k X_k \quad (k = 0, \ldots, N-1), \tag{8.13}$$

the above system of LMIs is equivalent to (8.11). The last assertion of (iii) is a direct consequence of this change of variables. We note that the reverse operation (8.12) to obtain K_k ($k = 0, \ldots, N-1$) is possible since $X_k \succ 0$ ($k = 0, \ldots, N-1$). □

8.3.2 Robust Stabilizing Controller Synthesis Using SV-LMIs

Consider the case where the periodic system (8.1) is subject to the polytopic uncertainty of the form

$$M_k := \begin{bmatrix} A_k & B_k & E_k \\ C_k & D_k & F_k \end{bmatrix}, \quad M_k^{[l]} = \begin{bmatrix} A_k^{[l]} & B_k^{[l]} & E_k^{[l]} \\ C_k^{[l]} & D_k^{[l]} & F_k^{[l]} \end{bmatrix} \quad (l = 1, \ldots, L),$$

$$\begin{bmatrix} M_0 \\ \vdots \\ M_{N-1} \end{bmatrix} = \sum_{l=1}^{L} \theta_l \begin{bmatrix} M_0^{[l]} \\ \vdots \\ M_{N-1}^{[l]} \end{bmatrix}, \quad \theta \in \mathbb{E}^L. \tag{8.14}$$

Here, $M_k^{[l]}$ ($k = 0, \ldots, N-1$, $l = 1, \ldots, L$) are known matrices. Again the parameter vector θ is time-invariant and does not depend on the time-instance k. In the sequel we denote by P_θ the periodic system given by (8.4) and (8.14) corresponding to $\theta \in \mathbb{E}^L$. Moreover, $P_{\theta,K}$ stands for the closed-loop system constructed from P_θ and a periodic static state-feedback controller K of the form (8.2).

Our goal in this subsection is to provide LMIs for the synthesis of K such that $P_{\theta,K}$ become stable for all $\theta \in \mathbb{E}^L$. The next result is a direct consequence of Theorem 8.1.

Theorem 8.2 *For the uncertain periodic system described by (8.1) and (8.14), there exists a periodic static state-feedback controller K of the form (8.2) that stabilizes $P_{\theta,K}$ for all $\theta \in \mathbb{E}^L$ if there exist $X_k \succ 0$ and Y_k ($k = 0, \ldots, N-1$) such that*

$$\begin{bmatrix} -X_{k+1} & A_k^{[l]} X_k + E_k^{[l]} Y_k \\ (A_k^{[l]} X_k + E_k^{[l]} Y_k)^T & -X_k \end{bmatrix} \prec 0 \quad (k = 0, \ldots, N-1). \tag{8.15}$$

where $X_N = X_0$. If the system of LMIs (8.15) is feasible, then a robustly stabilizing controller of the form (8.2) is given by (8.12).

It should be noted that the LMI in Theorem 8.2 originates from the one in Theorem 7.2. In view of the discussion just after Theorem 7.2, we see that the approach in Theorem 8.2 is inherently conservative since it relies on a parameter-independent Lyapunov function to ensure robust stability. The next theorem shows that we can reduce the conservatism by using SV-LMI stated in Theorem 7.3.

Theorem 8.3 *For the uncertain periodic system described by (8.1) and (8.14), there exists a periodic static state-feedback controller K of the form (8.2) that stabilizes $P_{\theta,K}$ for all $\theta \in \mathbb{E}^L$ if there exist $X_k^{[l]} \succ 0$ Y_k, and G_k $(k = 0, \ldots, N - 1, l = 1, \ldots, L)$ such that*

$$
\begin{bmatrix} -X_{k+1}^{[l]} & 0 \\ 0 & X_k^{[l]} \end{bmatrix} + \mathrm{He}\left\{ \begin{bmatrix} A_k^{[l]}G_k + E_k^{[l]}Y_k \\ -G_k \end{bmatrix} \begin{bmatrix} 0 & I \end{bmatrix} \right\} \prec 0
$$

$$(k = 0, \ldots, N - 1, l = 1, \ldots, L)$$

(8.16)

where $X_N^{[l]} = X_0^{[l]}$ $(l = 1, \ldots, L)$. If the system of LMIs (8.16) is feasible, then a robustly stabilizing controller K of the form (8.2) is given by

$$K_k = Y_k G_k^{-1}.$$

(8.17)

Moreover, if the LMI 8.15 holds with $X_k = X_k^\star$ and $Y_k = Y_k^\star$ $(k = 0, \ldots, N-1)$, then the LMI (8.16) holds with $X_k^{[l]} = X_k^\star$ $(k = 0, \ldots, N - 1, l = 1, \ldots, L)$, $Y_k = Y_k^\star$ and $G_k = X_k^\star$ $(k = 0, \ldots, N - 1)$.

This result is a direct consequence from Theorem 7.3 and the dual system representation. Similarly to robust stability analysis, we can theoretically ensure that the SV-LMI (8.16) is better (no worth) than the standard LMI (8.15). Qualitatively, such conservatism reduction has been done by employing a Lyapunov function that depends affinely on the uncertain parameter θ.

8.3.3 Difficulties in Controller Synthesis Using Rectangular SVs

For the purpose of robust stability analysis, we introduced SV-LMIs in Theorems 7.4 and 7.5 that outperform the SV-LMI in Theorem 7.3. For robust controller synthesis, we successfully derived Theorem 8.3 by extending Theorem 7.3. However, we cannot extend Theorems 7.4 and 7.5 to robust controller synthesis since these LMIs employ

S-variables of rectangular form. For example, we see from Theorem 7.4 and the results for dual system that there exists a periodic static state-feedback controller K of the form (8.2) that stabilizes $P_{\theta,K}$ for all $\theta \in \mathbb{E}^L$ if there exist $X_k^{[l]} \succ 0 \ Y_k$, and $F_{k,j} \ (k = 0, \dots, N-1, \ l = 1, \dots, L, \ j = 1, 2)$ such that

$$
\begin{bmatrix} -X_{k+1}^{[l]} & 0 \\ 0 & X_k^{[l]} \end{bmatrix} + \mathrm{He} \left\{ \begin{bmatrix} A_k^{[l]} + E_k^{[l]} K_k \\ -I \end{bmatrix} \begin{bmatrix} F_{k,1} & F_{k,2} \end{bmatrix} \right\} \prec 0 \tag{8.18}
$$
$$
(k = 0, \dots, N-1, \ l = 1, \dots, L)
$$

where $X_N^{[l]} = X_0^{[l]} \ (l = 1, \dots, L)$. This system of matrix inequalities hold bilinear terms $K_k F_{k,j} \ (j = 1, 2)$ and they cannot be linearized in a straightforward fashion. Similar difficulty arises when we deal with the LMI in Theorem 7.5. Due to this difficulty, we have disregarded Theorems 7.4 and 7.5 for robust controller synthesis. For the same reason, in the subsequent two sections, we disregard the LMIs with rectangular S-variables for robust H_2 and H_∞ controller synthesis.

One of the possible remedies for enabling controller synthesis is somehow restricting the structure of S-variables so that we can linearize the bilinear terms. Reasonable ways for restricting S-variables can be found, for example, in [2]. It is also possible to construct iterative algorithms by fixing controller variables and S-variables alternatively in the matrix inequalities in Theorems 7.4 and 7.5. However, reasonable choice of initial values for controller variables or S-variables remains to be a critical issue.

To overcome these difficulties and obtain a single-shot LMI condition that outperforms Theorem 8.3, the authors recently proposed a specific memory controller structure and conceived LMI conditions for the memory-type periodic controller synthesis. Details for the memory-type periodic controller synthesis using SV-LMIs are found in [3–5].

8.4 H_2 Controller Synthesis for Linear Periodic Systems

8.4.1 Basic Results

For LPTV system G given by (8.1), consider the synthesis of periodic state-feedback controller K of the form (8.2) such that P_K becomes stable and $\|P_K\|_2 < \gamma$ is satisfied for given $\gamma > 0$. From Theorem 7.7 and the dual system representation, the next theorem follows directly.

Theorem 8.4 *For given $\gamma > 0$ and the LPTV system P represented by (8.1), the following statements are equivalent:*

(i) *There exists a periodic state-feedback controller K of the form (8.2) such that the closed-loop system P_K becomes stable and $\|P_K\|_2 < \gamma$.*

(ii) *There exist $X_k \succ 0$, $Z_K \succ 0$ and K_k ($k = 0, \ldots, N - 1$) such that*

$$-X_{k+1} + (A_k + E_k K_k)X_k(A_k + E_k K_k)^T + B_k B_k^T \prec 0,$$

$$Z_k \succ (C_k + F_k K_k)X_k(C_k + F_k K_k)^T + D_k D_k^T, \tag{8.19}$$

$$\text{trace}\left(\sum_{k=0}^{N-1} Z_k\right) < N\gamma^2 \quad (k = 0, \ldots, N - 1)$$

where $X_N = X_0$.

(iii) *There exist $X_k \succ 0$ $Z_k \succ 0$ and Y_k ($k = 0, \ldots, N - 1$) such that*

$$\begin{bmatrix} -X_{k+1} + B_k B_k^T & A_k X_k + E_k Y_k \\ (A_k X_k + E_k Y_k)^T & -X_k \end{bmatrix} \prec 0,$$

$$\begin{bmatrix} -Z_k + D_k D_k^T & C_k X_k + F_k Y_k \\ (C_k X_k + F_k Y_k)^T & -X_k \end{bmatrix} \prec 0, \tag{8.20}$$

$$\text{trace}\left(\sum_{k=0}^{N-1} Z_k\right) < N\gamma^2 \quad (k = 0, \ldots, N - 1).$$

where $X_N = X_0$. Moreover, if the system of LMIs (8.20) is feasible, then a stabilizing controller K of the form (8.2) is given by (8.12).

8.4.2 Robust H_2 Controller Synthesis Using SV-LMI

Consider the case where the periodic system (8.1) is subject to the polytopic uncertainty of the form (8.14). For given $\gamma > 0$, our goal in this subsection is to provide LMIs for the synthesis of periodic static state-feedback controller K given by (8.2) such that $\|P_{\theta,K}\|_2 < \gamma$ is satisfied for all $\theta \in \mathbb{E}^L$. As in the robust stabilizing controller synthesis, the following two theorems follow.

Theorem 8.5 *For given $\gamma > 0$ and the uncertain periodic system described by (8.1) and (8.14), there exists a periodic static state-feedback controller K of the form (8.2) that achieves $\|P_{\theta,K}\|_2 < \gamma$ ($\forall \theta \in \mathbb{E}^L$) if there exist $X_k \succ 0$ and Y_k ($k = 0, \ldots, N - 1$) such that*

$$\begin{bmatrix} -X_{k+1} + B_k^{[l]} B_k^{[l]T} & A_k^{[l]} X_k + E_k^{[l]} Y_k \\ (A_k^{[l]} X_k + E_k^{[l]} Y_k)^T & -X_k \end{bmatrix} \prec 0,$$

$$\begin{bmatrix} -Z_k + D_k^{[l]} D_k^{[l]T} & C_k^{[l]} X_k + F_k^{[l]} Y_k \\ (C_k^{[l]} X_k + F_k^{[l]} Y_k)^T & -X_k \end{bmatrix} \prec 0, \qquad (8.21)$$

$$\text{trace}\left(\sum_{k=0}^{N-1} Z_k \right) < N\gamma^2 \quad (k = 0, \dots, N-1, \ l = 1, \dots, L)$$

where $X_N^{[l]} = X_0^{[l]}$ $(l = 1, \dots, L)$. If the system of LMIs (8.21) is feasible, then the desired robust H_2 controller of the form (8.2) is given by (8.12).

Theorem 8.6 *For given $\gamma > 0$ and the uncertain periodic system described by (8.1) and (8.14), there exists a periodic static state-feedback controller K of the form (8.2) that achieves $\|P_{\theta,K}\|_2 < \gamma$ $(\forall \theta \in \mathbb{E}^L)$ if there exist $X_k^{[l]} \succ 0$, Y_k and G_k $(k = 0, \dots, N-1, \ l = 1, \dots, L)$ such that*

$$\begin{bmatrix} -X_{k+1}^{[l]} + B_k^{[l]} B_k^{[l]T} & 0 \\ 0 & X_k^{[l]} \end{bmatrix} + \text{He}\left\{ \begin{bmatrix} A_k^{[l]} G_k + E_k^{[l]} Y_k \\ -G_k \end{bmatrix} \begin{bmatrix} 0 & I \end{bmatrix} \right\} \prec 0,$$

$$\begin{bmatrix} -Z_k + D_k^{[l]} D_k^{[l]T} & 0 \\ 0 & X_k^{[l]} \end{bmatrix} + \text{He}\left\{ \begin{bmatrix} C_k^{[l]} G_k + F_k^{[l]} Y_k \\ -G_k \end{bmatrix} \begin{bmatrix} 0 & I \end{bmatrix} \right\} \prec 0, \qquad (8.22)$$

$$\text{trace}\left(\sum_{k=0}^{N-1} Z_k \right) < N\gamma^2 \quad (k = 0, \dots, N-1, \ l = 1, \dots, L)$$

where $X_N^{[l]} = X_0^{[l]}$ $(l = 1, \dots, L)$. If the system of LMIs (8.22) is feasible, then the desired robust H_2 controller of the form (8.2) is given by (8.17). Moreover, if the LMI (8.21) holds with $X_k = X_k^\star$, $Z_k = Z_k^\star$ and $Y_k = Y_k^\star$ $(k = 0, \dots, N-1)$, then the LMI (8.22) holds with $X_k^{[l]} = X_k^\star$ $(k = 0, \dots, N-1, l = 1, \dots, L)$, $Z_k = Z_k^\star$, $Y_k = Y_k^\star$ and $G_k = X_k^\star$ $(k = 0, \dots, N-1)$.

From these theorems, we see that suboptimal robust H_2 controllers can be obtained by solving the following SDPs:

$$\overline{\gamma_2^\star}_{\text{qs}} := \inf_{X_k, Y_k, Z_k} \gamma \quad \text{subject to} \ (8.21), \qquad (8.23)$$

$$\overline{\gamma_2^\star}_{G} := \inf_{X_k^{[l]}, Y_k, Z_k, G_k} \gamma \quad \text{subject to} \ (8.22). \qquad (8.24)$$

Note that $\overline{\gamma_2^\star}_G \leq \overline{\gamma_2^\star}_{\text{qs}}$ is theoretically guaranteed, i.e., the SV-LMI approach (8.22) is better (no worth) than the standard LMI approach (8.21).

8.5 H_∞ Controller Synthesis for Linear Periodic Systems

8.5.1 Basic Results

In this subsection, we give H_∞ counterpart results for the stability and $H2$ case results stated in Theorems 8.1 and 8.4, respectively. The next theorem follows directly from Theorem 7.13 and the dual system representation,

Theorem 8.7 *For given $\gamma > 0$ and LPTV system P represented by (8.1), the following statements are equivalent:*

(i) *There exists periodic state-feedback controller K of the form (8.2) such that the closed-loop system P_K becomes stable and $\| P_K \|_\infty < \gamma$.*
(ii) *There exist $X_k \succ 0$ and K_k $(k = 0, \ldots, N - 1)$ such that*

$$
\begin{bmatrix}
-X_{k+1} + B_k B_k^T & (A_k + E_k K_k) X_k & B_k D_k^T \\
X_k (A_k + E_k K_k)^T & -X_k & X_k (C_k + F_k K_k)^T \\
D_k B_k^T & (C_k + F_k K_k) X_k & D_k D_k^T - \gamma^2 I
\end{bmatrix} \prec 0 \quad (k = 0, \ldots, N - 1)
$$

(8.25)

where $X_N = X_0$.
(iii) *There exist $X_k \succ 0$ and Y_k $(k = 0, \ldots, N - 1)$ such that*

$$
\begin{bmatrix}
-X_{k+1} + B_k B_k^T & A_k X_k + E_k Y_k & B_k D_k^T \\
(A_k X_k + E_k Y_k)^T & -X_k & (C_k X_k + F_k Y_k)^T \\
D_k B_k^T & C_k X_k + F_k Y_k & D_k D_k^T - \gamma^2 I
\end{bmatrix} \prec 0 \quad (k = 0, \ldots, N - 1)
$$

(8.26)

where $X_N = X_0$. Moreover, if the system of LMIs (8.26) is feasible, then a stabilizing controller of the form (8.2) is given by (8.12).

8.5.2 Robust H_∞ Controller Synthesis Using SV-LMI

As in the previous two sections, consider the case where the periodic system (8.1) is subject to the polytopic uncertainty given by (8.14). For given $\gamma > 0$, our goal in this subsection is to provide LMIs for the synthesis of periodic static state-feedback controller K given by (8.2) such that $\| P_{\theta, K} \|_\infty < \gamma$ is satisfied for all $\theta \in \mathbb{E}^L$. The following two theorems follow.

Theorem 8.8 *For given $\gamma > 0$ and the uncertain periodic system described by (8.1) and (8.14), there exists a periodic static state-feedback controller K of the form (8.2) that achieves $\|P_{\theta,K}\|_\infty < \gamma$ ($\forall \theta \in \mathbb{E}^L$) if there exist $X_k \succ 0$ and Y_k ($k = 0, \dots, N - 1$) such that*

$$
\begin{bmatrix}
-X_{k+1} + B_k^{[l]}B_k^{[l]T} & A_k^{[l]}X_k + E_k^{[l]}Y_k & B_k^{[l]}D_k^{[l]T} \\
(A_k^{[l]}X_k + E_k^{[l]}Y_k)^T & -X_k & (C_k^{[l]}X_k + F_k^{[l]}Y_k)^T \\
D_k^{[l]}B_k^{[l]T} & C_k^{[l]}X_k + F_k^{[l]}Y_k & D_k^{[l]}D_k^{[l]T} - \gamma^2 I
\end{bmatrix} \prec 0
$$

$$(k = 0, \dots, N - 1, \; l = 1, \dots, L)$$

(8.27)

where $X_N = X_0$. If the system of LMIs (8.27) is feasible, then the desired robust H_∞ controller of the form (8.2) is given by (8.12).

Theorem 8.9 *For given $\gamma > 0$ and the uncertain periodic system described by (8.1) and (8.14), there exists a periodic static state-feedback controller K of the form (8.2) that achieves $\|P_{\theta,K}\|_\infty < \gamma$ ($\forall \theta \in \mathbb{E}^L$) if there exist $X_k^{[l]} \succ 0$, Y_k and G_K ($k = 0, \dots, N - 1, \; l = 1, \dots, L$) such that*

$$
\begin{bmatrix}
-X_{k+1}^{[l]} + B_k^{[l]}B_k^{[l]T} & 0 & B_k^{[l]}D_k^{[l]T} \\
0 & X_k^{[l]} & 0 \\
D_k^{[l]}B_k^{[l]T} & 0 & D_k^{[l]}D_k^{[l]T} - \gamma^2 I
\end{bmatrix}
$$
$$
+\,\mathrm{He}\left\{
\begin{bmatrix}
A_k^{[l]}G_k + E_k^{[l]}Y_k \\
-G_k \\
C_k^{[l]}G_k + F_k^{[l]}Y_k
\end{bmatrix}
\begin{bmatrix} 0 & I & 0 \end{bmatrix}
\right\} \prec 0 \quad (k = 0, \dots, N - 1, \; l = 1, \dots, L).
$$

(8.28)

where $X_N^{[l]} = X_0^{[l]}$ ($l = 1, \dots, L$). If the system of LMIs (8.28) is feasible, then the desired robust H_∞ controller of the form (8.2) is given by (8.17). Moreover, if the LMI (8.27) holds with $X_k = X_k^\star$ and $Y_k = Y_k^\star$ ($k = 0, \dots, N - 1$), then the LMI (8.28) holds with $X_k^{[l]} = X_k^\star$ ($k = 0, \dots, N - 1, l = 1, \dots, L$), $Y_k = Y_k^\star$ and $G_k = X_k^\star$ ($k = 0, \dots, N - 1$).

From these theorems, we see that suboptimal robust H_2 controllers can be obtained by solving the following SDPs:

$$
\overline{\gamma_{\infty\mathrm{qs}}^\star} := \inf_{X_k, Y_k} \gamma \quad \text{subject to} \quad (8.27),
$$

(8.29)

$$
\overline{\gamma_{\infty G}^\star} := \inf_{X_k^{[l]}, Y_k, G_k} \gamma \quad \text{subject to} \quad (8.28).
$$

(8.30)

Again note that the SV-LMI approach (8.28) is better (no worth) than the standard LMI approach (8.27) and hence $\overline{\gamma^\star_{\infty G}} \le \overline{\gamma^\star_{\infty qs}}$ is theoretically guaranteed.

8.6 Numerical Examples

Let us illustrate the effectiveness of the LMI conditions introduced in the previous sections by numerical examples. In the following numerical computation, we solved SDPs with MATLAB R2011b and SeDuMi [6] and YALMIP [7] on a PC with Intel(R) Core(TM)2 Extreme CPU X9770 3.20GHz.

We consider the periodic system with polytopic uncertainty given by (8.1) and (8.14) where $N = 3$ and $L = 4$ and

$$A_1^{[1]} = \begin{bmatrix} -3 - \overline{\alpha} & 2 \\ -3 & 3 \end{bmatrix}, \quad A_1^{[2]} = \begin{bmatrix} -3 - \underline{\alpha} & 2 \\ -3 & 3 \end{bmatrix}, \quad A_1^{[3]} = A_1^{[1]}, \quad A_1^{[4]} = A_1^{[2]},$$

$$A_2^{[1]} = \begin{bmatrix} -1 - \overline{\alpha} & 2 \\ 0.5 & 0 \end{bmatrix}, \quad A_2^{[2]} = \begin{bmatrix} -1 - \underline{\alpha} & 2 \\ 0.5 & 0 \end{bmatrix}, \quad A_2^{[3]} = A_2^{[1]}, \quad A_2^{[4]} = A_2^{[2]},$$

$$A_3^{[1]} = \begin{bmatrix} 1 - \overline{\alpha} & 2 \\ 2.5 & 3 \end{bmatrix}, \quad A_3^{[2]} = \begin{bmatrix} 1 - \underline{\alpha} & 2 \\ 2.5 & 3 \end{bmatrix}, \quad A_3^{[3]} = A_3^{[1]}, \quad A_3^{[4]} = A_3^{[2]},$$

$$\text{(8.31a)}$$

$$B_1^{[1]} = \begin{bmatrix} 1 \\ \overline{\beta} \end{bmatrix}, \quad B_1^{[3]} = \begin{bmatrix} 1 \\ \underline{\beta} \end{bmatrix}, \quad B_1^{[2]} = B_1^{[1]}, \quad B_1^{[4]} = B_1^{[3]},$$

$$B_2^{[1]} = \begin{bmatrix} 1 \\ -\frac{3\overline{\beta}+2}{10} \end{bmatrix}, \quad B_2^{[3]} = \begin{bmatrix} 1 \\ -\frac{3\underline{\beta}+2}{10} \end{bmatrix}, \quad B_2^{[2]} = B_2^{[1]}, \quad B_2^{[4]} = B_2^{[3]}, \quad \text{(8.31b)}$$

$$B_3^{[1]} = \begin{bmatrix} \frac{\overline{\beta}+1}{2} \\ 1 \end{bmatrix}, \quad B_3^{[3]} = \begin{bmatrix} \frac{\underline{\beta}+1}{2} \\ 1 \end{bmatrix}, \quad B_3^{[2]} = B_3^{[1]}, \quad B_3^{[4]} = B_3^{[3]},$$

$$C_k^{[l]} = \begin{bmatrix} 1 & 0 \end{bmatrix} \quad (k = 1, 2, 3, \ l = 1, \dots, 4), \tag{8.31c}$$

$$D_k^{[l]} = 0 \quad (k = 1, 2, 3, \ l = 1, \dots, 4), \tag{8.31d}$$

$$E_k^{[l]} = B_k^{[l]} \quad (k = 1, 2, 3, \ l = 1, \dots, 4), \tag{8.31e}$$

$$F_k^{[l]} = 0 \quad (k = 1, 2, 3, \ l = 1, \ldots, 4) \tag{8.31f}$$

where $(\underline{\alpha}, \overline{\alpha})$ and $(\underline{\beta}, \overline{\beta})$ are known values that correspond to the range of the variation of the uncertain parameters α and β as in $\underline{\alpha} \le \alpha \le \overline{\alpha}$ and $\underline{\beta} \le \beta \le \overline{\beta}$. This plant is borrowed from [2] with slight modifications.

In the following, we consider robust stabilizing controller synthesis, robust H_2 controller synthesis and robust H_∞ controller synthesis for the plant described by (8.1), (8.14) and (8.31).

8.6.1 Robust Stabilizing Controller Synthesis

Consider the case where $\underline{\beta} = 0.2$ and $\overline{\beta} = 0.8$. By letting $\overline{\alpha} = -\underline{\alpha} := \alpha_{\mathrm{marg}} \ge 0$, our goal in this subsection is to design a robust stabilizing controller K of the form (8.2) that maximizes the stability margin α_{marg}.

To this end, we first carried out a bisection search over α_{marg} by means of the quadratic-stability-based LMI condition (8.15). Then, the bisection search terminated with $\alpha_{\mathrm{marg}} = \alpha_{\mathrm{marg,qs}} = 0.2876$, yielding periodic static state-feedback controller K_{qs} given by

$$K_{\mathrm{qs},0} = \begin{bmatrix} 3.1768 & -2.2357 \end{bmatrix},$$

$$K_{\mathrm{qs},1} = \begin{bmatrix} 0.8834 & -2.4383 \end{bmatrix},$$

$$K_{\mathrm{qs},2} = \begin{bmatrix} -5.1511 & -3.7469 \end{bmatrix}.$$

On the other hand, by bisection search over α_{marg} with SV-LMI (8.16), we obtained $\alpha_{\mathrm{marg}} = \alpha_{\mathrm{marg,G}} = 0.8568$ and periodic static state-feedback controller K_G given by

$$K_{G,0} = \begin{bmatrix} 3.0571 & -2.4104 \end{bmatrix},$$

$$K_{G,1} = \begin{bmatrix} 1.1420 & -1.9360 \end{bmatrix},$$

$$K_{G,2} = \begin{bmatrix} -1.5173 & -2.6735 \end{bmatrix}.$$

It turns out that SV-LMI (8.16) achieves much better stability margin than the standard LMI (8.15).

We show the eigenvalue plots of $P_{\theta,K_{\mathrm{qs}}}$ over $\theta \in \mathbb{E}^3$ for $\alpha_{\mathrm{marg}} = \alpha_{\mathrm{marg,qs}} = 0.2876$ and P_{θ,K_G} over $\theta \in \mathbb{E}^3$ for $\alpha_{\mathrm{marg}} = \alpha_{\mathrm{marg,G}} = 0.8568$ in Figs. 8.1 and 8.2, respectively. From 8.1, we infer that the quadratic-stability-based approach is very conservative, since eigenvalues are located far from stability margin. Figure 8.2 shows that eigenvalues come closer to stability margin and hence the conservatism has been reduced.

Fig. 8.1 Eigenvalue plots of $P_{\theta,K_{qs}}$ over $\theta \in \mathbb{E}^4$ ($\alpha_{marg} = \alpha_{marg,qs} = 0.2876$)

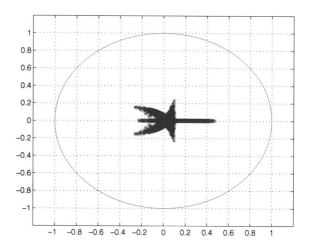

Fig. 8.2 Eigenvalue plots of P_{θ,K_G} over $\theta \in \mathbb{E}^4$ ($\alpha_{marg} = \alpha_{marg,G} = 0.8568$)

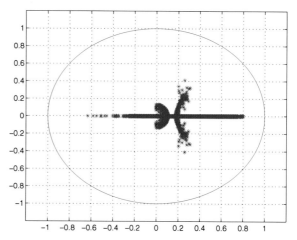

8.6.2 Robust H_2 Controller Synthesis

Consider the case where $\underline{\beta} = 0.2$, $\overline{\beta} = 0.8$ and $\overline{\alpha} = -\underline{\alpha} := \alpha_{marg} = 0.28$. In this subsection, we consider the robust H_2 controller synthesis problem described by

$$\gamma_2^\star := \inf_K \max_{\theta \in \mathbb{E}^4} \| P_{\theta,K} \|_2 .$$

For the upper bound computation of $\gamma_2^\star$, we solved the SDPs (8.23) and (8.24). Then we obtained

Table 8.1 Computation results for robust H_2 controller synthesis

	Upper bound (CPU time [s])	Recomputed upper bounds by Theorem 7.10
SDP (8.23)	$\overline{\gamma_{2\,\mathrm{qs}}^{\star}} = 8.2211$ (0.7300)	2.5717
SDP (8.24)	$\gamma_{2\,G}^{\star} = 1.1434$ (0.9000)	1.1164

$$K_{\mathrm{qs},0} = \begin{bmatrix} 3.1313 & -2.1799 \end{bmatrix},$$

$$K_{\mathrm{qs},1} = \begin{bmatrix} 0.8940 & -2.4241 \end{bmatrix},$$

$$K_{\mathrm{qs},2} = \begin{bmatrix} -5.1177 & -3.7315 \end{bmatrix}.$$

$$K_{G,0} = \begin{bmatrix} 3.0213 & -2.0597 \end{bmatrix},$$

$$K_{G,1} = \begin{bmatrix} 1.0244 & -2.0738 \end{bmatrix},$$

$$K_{G,2} = \begin{bmatrix} -1.3856 & -2.5940 \end{bmatrix}.$$

The resulting upper bounds are summarized in Table 8.1, where we also show the recomputed upper bounds by means of the LMI with rectangular S-variables shown in Theorem 7.10. From this result, we see again that the standard LMI (8.23) is conservative, and the conservatism is drastically reduced by means of the SV-LMI (8.24). This conservatism reduction is gained by introducing additional S-variables, but still the computation times are comparable for both cases in this example.

8.6.3 Robust H_∞ Controller Synthesis

Under the same parameter settings as in the preceding subsection, we next solve the robust H_∞ controller synthesis problem described by

$$\gamma_\infty^{\star} := \inf_K \max_{\theta \in \mathbb{E}^4} \| P_{\theta,K} \|_\infty.$$

For the upper bound computation of $\gamma_\infty^{\star}$, we solved the SDPs (8.29) and (8.30) and obtained

$$K_{\mathrm{qs},0} = \begin{bmatrix} 3.1547 & -2.2087 \end{bmatrix},$$

$$K_{\mathrm{qs},1} = \begin{bmatrix} 0.8895 & -2.4289 \end{bmatrix},$$

$$K_{\mathrm{qs},2} = \begin{bmatrix} -5.1338 & -3.7385 \end{bmatrix}.$$

Table 8.2 Computation results for robust H_∞ controller synthesis

	Upper bound (CPU time [s])	Recomputed upper bounds by Theorem 7.16
SDP (8.29)	$\overline{\gamma^\star_{\infty qs}} = 91.6809$ (0.6900)	5.6726
SDP (8.30)	$\overline{\gamma^\star_{\infty G}} = 1.9055$ (0.7500)	1.7688

$$K_{G,0} = \begin{bmatrix} 3.1450 & -2.1984 \end{bmatrix},$$

$$K_{G,1} = \begin{bmatrix} 1.1231 & -2.0475 \end{bmatrix},$$

$$K_{G,2} = \begin{bmatrix} -1.2757 & -2.5002 \end{bmatrix}.$$

The resulting upper bounds are summarized in Table 8.2, where we also show the recomputed upper bounds by means of the LMI with rectangular S-variables shown in Theorem 7.16. This example clearly shows the effectiveness of the SV-LMI (8.28) over the standard LMI (8.27) for conservatism reduction. Again this conservatism reduction is successfully achieved without much increase in computation time.

References

1. Bittanti S, Colaneri P (2009) Periodic systems: filtering and control. Springer, London
2. Farges C, Peaucelle D, Arzelier D, Daafouz J (2007) Robust H_2 performance analysis and synthesis of linear polytopic discrete-time periodic systems via LMIs. Syst Control Lett 56:159–166
3. Ebihara Y, Peaucelle D, Arzelier D (2011) Periodically time-varying memory state-feedback controller synthesis for discrete-time linear systems. Automatica 47(1):14–25
4. Ebihara Y (2013) Periodically time-varying memory state-feedback for robust H_2 control of uncertain discrete-time linear systems. Asian J Control 15(2):409–419
5. Trégouët JF, Peaucelle D, Arzelier D, Ebihara Y (2013) Periodic memory state-feedback controller: new formulation, analysis and design results. IEEE Trans Autom Control 58(8):1986–2000
6. Sturm JF (1999) Using SeDuMi 1.02, a MATLAB toolbox for optimization over symmetric cones. Optim Meth Softw 11–12:625–653
7. Löfberg J (2004) YALMIP: a toolbox for modeling and optimization in MATLAB. In: Proceedings IEEE Computer Aided Control System Design, pp 284–289, 2004

Index

B

Bilinear matrix inequality (BMI), 168

C

Central polynomial, 171, 172
Characteristic multipliers, 199
Conservatism, 6, 9, 14, 28–32, 47–49, 51, 53,
      57, 59, 107, 108, 113, 123, 124,
      126, 133, 134, 136, 158, 191, 195,
      203, 204, 210, 219, 226, 229, 234,
      241, 243, 244
Coordinate descent-type algorithm, 188

D

Decoupling, 15, 21, 124, 139, 168
Descriptor system, 23, 116
Dilated LMI, 6
Discrete-time periodic system, 3, 7, 199
Discrete-time system lifting, 207, 230
Dual system, 17, 18, 23, 28, 35, 44, 108, 205,
      220, 230–232, 234, 235, 238

E

Elimination lemma, 5, 15, 19, 167, 169–172,
      191
Enhanced LMI, 6
Extended LMI, 6

F

Finsler's lemma, 5, 20, 183
Frequency domain inequality (FDI), 38–41

I

Impulsive mode, 63, 65, 69, 71, 83
Iterative LMI algorithm, viii

K

Kalman-Yakubovich-Popov lemma (KYP
      lemma), 38, 216

L

Linear matrix inequality (LMI), 2, 3, 6,
      12–15, 17, 18, 20, 22–24, 26–
      31, 36, 39, 42, 46, 47, 51, 54,
      57, 58, 108, 111, 113–118, 121–
      126, 130, 132, 133, 139, 141–
      145, 147–153, 157, 166, 168,
      169, 171–173, 175, 177, 187, 188,
      190, 191, 193, 200, 202–206, 208,
      209, 211–213, 216, 217, 219–
      223, 225, 229, 231–241, 243
Linearizing change of variables, 107–109,
      143, 145, 146, 148, 151, 152, 231
Lyapunov function, 4, 12, 14, 17, 18, 21, 46,
      139, 183, 184, 187–189, 203, 234
Lyapunov inequality, 2, 12, 167, 200
Lyapunov variable, 15, 139, 145, 164

M

Monodromy matrix, 199
Multi-objective control design, vii

N

Non-dynamic mode, 63

© Springer-Verlag London 2015
Y. Ebihara et al., *S-Variable Approach to LMI-Based Robust Control*,
Communications and Control Engineering, DOI 10.1007/978-1-4471-6606-1